Physiological Responses of Marine Biota to Pollutants

Academic Press Rapid Manuscript Reproduction

Physiological Responses of Marine Biota to Pollutants

edited by

F. JOHN VERNBERG

Belle W. Baruch Institute for Marine Biology and Coastal Research
University of South Carolina
Columbia, South Carolina

ANTHONY CALABRESE
FREDERICK P. THURBERG

National Marine Fisheries Service
Middle Atlantic Coastal Fisheries Center
Milford Laboratory
Milford, Connecticut

WINONA B. VERNBERG

Belle W. Baruch Institute for Marine Biology and Coastal Research
and School of Public Health
University of South Carolina
Columbia, South Carolina

1977.

ACADEMIC PRESS, INC.   New York   San Francisco   London   1977
A Subsidiary of Harcourt Brace Jovanovich, Publishers

274-287

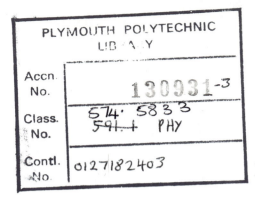
ACADEMIC PRESS, INC.
111 Fifth Avenue, New York, New York 10003

United Kingdom Edition published by
ACADEMIC PRESS, INC. (LONDON) LTD.
24/28 Oval Road, London NW1

Library of Congress Cataloging in Publication Data

Main entry under title:

Physiological responses of marine biota to pollutants.

Proceedings of a symposium held in Connecticut,
Nov. 1975, and sponsored by the Middle Atlantic Coastal
Fisheries Center, National Marine Fisheries Service and
the Belle W. Baruch Institute for Marine Biology and
Coastal Research, University of South Carolina.
Includes index.
1.   Fishes, Effect of water pollution on–Congresses.
2.   Aquatic animals, Effect of water pollution on–
Congresses.   3.   Marine pollution–Environmental aspects
–Congresses.   I.   Vernberg, F. John,   Date
II.   Middle Atlantic Coastal Fisheries Center.
III.   Belle W. Baruch Institute for Marine Biology and
Coastal Research.
SH174.P48   591.5'2636   76-58372
ISBN 0–12–718240–3

PRINTED IN THE UNITED STATES OF AMERICA

Contents

PART IV. FACTOR INTERACTION

PART V. GENERAL

Contributors

J. W. ANDERSON, Department of Biology, Texas A&M University, College Station, Texas 77840

LOWELL H. BAHNER, U. S. Environmental Research Laboratory, Gulf Breeze, Florida 32561

B. L. BAYNE, Institute for Marine Environmental Research, 67 Citadel Road, Plymouth PL1 3DH, England

JOHN R. BEERS, Institute of Marine Resources, University of California, San Diego, California 92093

C. G. BOOKHOUT, Duke University Marine Laboratory, Beaufort, North Carolina 28516

PHYLLIS H. CAHN, Long Island University, C. W. Post College, Marine Science Department, Greenvale, New York 11548

A. CALABRESE, National Marine Fisheries Service, Middle Atlantic Coastal Fisheries Center, Milford Laboratory, Milford, Connecticut 06460

JUDITH McDOWELL CAPUZZO, Woods Hole Oceanographic Institution, Woods Hole, Massachusetts 02543

LINDA COSTON CLEMENTS, National Marine Fisheries Service, Atlantic Estuarine Fisheries Center, Beaufort, North Carolina 28516

DAVID R. COLBY, National Marine Fisheries Service, Atlantic Estuarine Fisheries Center, Beaufort, North Carolina 28516

DAVID L. COPPAGE, United States Environmental Protection Agency, Environmental Research Laboratory, Gulf Breeze, Florida 32561

J. D. COSTLOW, Jr., Duke University Marine Laboratory, Beaufort, North Carolina 28516

M. A. DAWSON, National Marine Fisheries Service, Middle Atlantic Coastal Fisheries Center, Milford Laboratory, Milford, Connecticut 06460

MARILYN DELDONNO, Mount Desert Island Biological Laboratory, Salsbury Cove, Maine 04672

D. B. DIXIT, Department of Biology, Northern Virginia Community College, Sterling, Virginia 22170

DANA DONOVAN, Department of Chemistry, Bowdoin College, Brunswick, Maine 04011

JACK FOEHRENBACH, Long Island University, C. W. Post College, Marine Science Department, Greenvale, New York 11548

R. S. FOSTER, Department of Biology, Texas A&M University, College Station, Texas 77840

C. S. GIAM, Department of Chemistry, Texas A&M University, College Station, Texas 77843

EDWARD S. GILFILLAN, Bigelow Laboratory for Ocean Sciences, West Boothbay Harbor, Maine 04575

EDITH GOULD, National Marine Fisheries Service, Middle Atlantic Coastal Fisheries Center, Milford Laboratory, Milford, Connecticut 06460

D. R. GREEN, Institute of Oceanography, University of British Columbia, Vancouver, British Columbia, Canada

R. A. GREIG, National Marine Fisheries Service, Middle Atlantic Coastal Fisheries Center, Milford Laboratory, Milford, Connecticut 06460

WILLIAM GUGGINO, Long Island University, C. W. Post College, Marine Science Department, Greenvale, New York 11548

M. S. GURAM, Department of Biology, Voorhees College, Denmark, South Carolina 29042

SHERRY HANSON, Bigelow Laboratory for Ocean Sciences, West Boothbay Harbor, Maine 04575

DONALD E. HOSS, National Marine Fisheries Service, Atlantic Estuarine Fisheries Center, Beaufort, North Carolina 28516

WILLIAM B. KINTER, Mount Desert Island Biological Laboratory, Salsbury Cove, Maine 04672

P. KOELLER, Institute of Oceanography, University of British Columbia, Vancouver, British Columbia, Canada

CHRISTOPHER C. KOENIG, Department of Biology, College of Charleston, Charleston, South Carolina 29401

RICHARD F. LEE, Skidaway Institute of Oceanography, Savannah, Georgia 31406

DONALD V. LIGHTNER, University of Arizona Environmental Research Laboratory, Tucson International Airport, Tucson, Arizona 85706

CHARLES W. MAJOR, Zoology Department, University of Maine, Orono, Maine 04473

DANA W. MAYO, Department of Chemistry, Bowdoin College, Brunswick, Maine 04011

DAVID S. MILLER, Mount Desert Island Biological Laboratory, Salsbury Cove, Maine 04672

ROBERT J. MONROE, Department of Experimental Statistics, North Carolina State University, Raleigh, North Carolina 27607

J. M. NEFF, Department of Biology, Texas A&M University, College Station, Texas 77843

DEL WAYNE R. NIMMO, U. S. Environmental Research Laboratory, Gulf Breeze, Florida 32561

DAVID S. PAGE, Department of Chemistry, Bowdoin College, Brunswick, Maine 04011

CARL F. PETERS, John Graham and Company, Environmental Studies Group, 1110 Third Avenue, Seattle, Washington 98101

DONALD J. REISH, Department of Biology, California State University, Long Beach, California 90840

STANLEY D. RICE, Northwest Fisheries Center, Auke Bay Fisheries Laboratory, National Marine Fisheries Service, NOAA, Post Office Box 155, Auke Bay, Alaska 99821

DENNIS J. SABO, Woods Hole Oceanographic Institution, Woods Hole, Massachusetts 02543

JOHN J. SASNER, Jr., University of New Hampshire, Zoology Department, Durham, New Hampshire 03824

A. N. SASTRY, Graduate School of Oceanography, University of Rhode Island, Kingston, Rhode Island 02881

ARDIS SAVORY, Belle W. Baruch Institute for Marine Biology and Coastal Research, University of South Carolina, Columbia, South Carolina 29208

BODIL SCHMIDT-NIELSEN, Mount Desert Island Biological Laboratory, Salsbury Cove, Maine 04672

DON L. R. SEIBERT, Institute of Marine Resources, University of California, San Diego, California 92093

JONATHAN SHELINE, Mount Desert Island Biological Laboratory, Salsbury Cove, Maine 04672

JEFFREY W. SHORT, Northwest Fisheries Center, Auke Bay Fisheries Laboratory, National Marine Fisheries Service, NOAA, Post Office Box 155, Auke Bay, Alaska 99821

JOHN J. STEGEMAN, Woods Hole Oceanographic Institution, Woods Hole, Massachusetts 02543

D. STRAUGHAN, Allan Hancock Foundation, University of Southern California, Los Angeles, California 90007

M. TAKAHASHI, Institute of Oceanography, University of British Columbia, Vancouver, British Columbia, Canada

ROBERT E. THOMAS, Chico State University, Chico, California 95926

WILLIAM H. THOMAS, Institute of Marine Resources, University of California, San Diego, California 92093

F. P. THURBERG, National Marine Fisheries Service, Middle Atlantic Coastal Fisheries Center, Milford Laboratory, Milford, Connecticut 06460

R. K. TUCKER, National Marine Fisheries Service, Middle Atlantic Coastal Fisheries Center, Sandy Hook Laboratory, Highlands, New Jersey 07732

SANDRA L. VARGO, Graduate School of Oceanography, University of Rhode Island, Kingston, Rhode Island 02881

F. JOHN VERNBERG, Belle W. Baruch Institute for Marine Biology and Coastal Research and Department of Biology, University of South Carolina, Columbia, South Carolina 29208

G. S. WARD, Bionomics Marine Laboratory, Pensacola, Florida 32507

J. ERNEST WARINNER, III, Department of Environmental Physiology, Virginia Institute of Marine Science, Gloucester Point, Virginia 23062

DOUGLAS D. WEBER, Northwest Fisheries Center, National Marine Fisheries Service, National Oceanic and Atmospheric Administration, 2725 Montlake Boulevard East, Seattle, Washington 981112

J. WIDDOWS, Institute for Marine Environmental Research, 67 Citadel Road, Plymouth PL1 3DH, England

C. WORRALL, Institute for Marine Environmental Research, 67 Citadel Road, Plymouth PL1 3DH, England

PAUL L. ZUBKOFF, Department of Environmental Physiology, Virginia Institute of Marine Science, Gloucester Point, Virginia 23062

Preface

Since the publication in 1974 of the papers presented at a symposium entitled "Pollution and Physiology of Marine Organisms," a significant scientific surge has taken place in this subject area. As a result of this continuing interest, the editors of this volume felt the need both to have a second symposium and to have the papers published. The symposium was sponsored jointly by the Middle Atlantic Coastal Fisheries Center, National Marine Fisheries Service, and the Belle W. Baruch Institute for Marine Biology and Coastal Research, University of South Carolina.

Pollution of marine waters is continuing, and it will continue into the foreseeable future. In comparison with the 1974 book, the papers included in this volume reflect continuing concern with the influence of petroleum products, heavy metals, pesticides, and PCBs on the physiology of marine organisms. As expected, more information is now available on the functional mechanisms involved in the response to pollutants either acting singly or in concert with other pollutants and/or "normal" environmental factors.

The editors are indebted to the authors for their cooperation and to the staff of the Institute for assistance in many ways, especially Ms. Susan Counts and Ms. Dorothy Knight.

F. JOHN VERNBERG
ANTHONY CALABRESE
FREDERICK P. THURBERG
WINONA B. VERNBERG

Part I.
Pesticides and PCBS

Effects of Malathion on the Development of Crabs

C. G. BOOKHOUT[1] and ROBERT J. MONROE[2]

[1]Duke University Marine Laboratory
Beaufort, North Carolina 28516

[2]Department of Experimental Statistics
North Carolina State University
Raleigh, North Carolina 27607

Malathion, like many other organophosphate insecticides, is used widely to control crop pests, flies and mosquitoes, presumably because it is biodegradable and seldom leaves residues but for a short time. Even though malathion decomposes at high temperatures and with increasing alkalinity (Walker and Stojanovic, 1973), it may be highly toxic to target and nontarget organisms alike. Since it is a non-persistent insecticide, repeated applications may be necessary for control of pests, and cumulative reduction of acetylcholinesterase (AChE) may occur (Coppage and Matthews, 1974). There is also documented evidence from the literature that arthropod pests develop resistance to malathion (Busvine, 1959; Brown and Abedi, 1960; La Brecque and Wilson, 1960; Mengle and Lewallen, 1963; Mount, Seawright, and Pierce, 1974). Hence, increasing concentrations of malathion have to be used to control pests, and the danger to non-target organisms becomes greater.

Few field studies have been conducted on the effect of malathion sprays on estuarine animals. Darsie and Corriden (1959) reported that aerial spraying was toxic to the fish, Fundulus ocellaris, and Conte and Parker (1971) found that aerial application to marsh embayments of Texas killed 14% to 80% of commercial shrimp, Penaeus aztecus and P. setiferus.

3

Tagatz et al. (1974), on the other hand, reported aerial spraying of malathion 95 had no effect on confined animals, such as blue crabs and commercial shrimp.

If aerial spraying can kill commercial shrimp as Conte and Parker (1971) reported, malathion sprays might kill crab larvae. To date, however, there have been no publications on the effects of malathion on the complete larval development of crabs. The objective of this investigation was to determine the effects of malathion on the development of the mud-crab, Rhithropanopeus harrisii Gould, and the commercial blue crab, Callinectes sapidus Rathbun, from the time of hatching to the time the first crab stage is reached.

MATERIALS AND METHODS

The malathion (O,O-dimethyl phosphoro-dithionate of diethyl mercaptosuccinate) used in this investigation was obtained from American Cyanamid Company and had a purity of 96%. Hazleton Laboratories America, Inc., prepared the stock solutions by dissolving a known weight of malathion in pesticide analytical grade acetone and various concentrations were made from this stock solution.

For experiments on the effects of malathion on R. harrisii, one ml of each of the following stock solutions 0.011 parts per thousand (grams/liter), 0.014 ppt, 0.017 ppt, 0.02 ppt and 0.05 ppt malathion was added to 999 ml of 20 o/oo filtered seawater daily to give final concentrations of 0.011 parts per million (milligrams/liter) to 0.05 ppm malathion.

For experiments on the effects of malathion on C. sapidus, one ml of each of the following stock solutions 0.02 ppt, 0.05 ppt, 0.08 ppt and 0.11 ppt malathion was added to 999 ml of 30 o/oo filtered seawater daily to give the final concentrations of 0.02 ppm to 0.11 ppm malathion.

Fresh stock solutions of malathion were shipped to Beaufort, North Carolina from Hazleton Laboratories, Vienna, Virginia before each experiment with larvae of one of the species of crabs. The stock solutions were stored in a cold room at 5°C at Duke University Marine Laboratory (DUML) to prevent deterioration during the course of an experiment.

Acetone control for experiments was prepared by adding 1 ml of full strength pesticide analytical grade acetone to 999 ml of 20 o/oo filtered seawater for R. harrisii larvae or 30 o/oo filtered sea water for C. sapidus larvae to give a final concentration of 1.0 ppt. Previous studies revealed that there was no significant difference in survival of R.

harrisii and C. sapidus larvae reared in seawater and 1.0 ppt
insecticide grade acetone.

Ovigerous R. harrisii, small mud-crabs belonging to the
family Xanthidae, were collected in a small estuary in the
vicinity of Cocoa Beach, Florida. They were shipped by air
freight to Beaufort, North Carolina in January and February,
1975. Each crab was placed in a large finger bowl (19.4 cm
diam) with 20 o/oo seawater and maintained in a constant
temperature culture cabinet at 25°C until hatching. After
hatching, bowls of larvae were also kept in a culture cabinet
under the same conditions of salinity and temperature.
Ovigerous C. sapidus, commercial blue crabs belonging to the
family Portunidae, were collected from the Beaufort Inlet,
North Carolina in the summer of 1974. The method of
developing the eggs has been described previously (Costlow
and Bookhout, 1960; Bookhout and Costlow, 1975). After
hatching, bowls of larvae were maintained in a salinity of
30 o/oo and in a constant temperature culture cabinet at
25°C.

The methods of rearing larvae of R. harrisii and C.
sapidus in acetone control and different concentrations of
malathion were essentially the same. As soon as the eggs
from one ovigerous crab hatched, ten larvae were placed in
each of the finger bowls (8.9 cm diam) to be used in a
series. A series included five bowls of control larvae in
1.0 ppt acetone in filtered seawater and the same number of
bowls of larvae for each concentration of malathion. The
methods of feeding and changing C. sapidus larvae daily
were the same as described previously (Bookhout and Costlow,
1975). R. harrisii were fed only Artemia salina nauplii
hatched from California eggs, otherwise the methods of
handling larvae were the same as described previously
(Bookhout and Costlow, 1975). Each ovigerous crab furnished
enough larvae for one series. There were four replicate
series of R. harrisii larvae designated as Rh I-IV, and four
series of C. sapidus larvae termed Cs I-IV.

In chronic bioassay studies of larval development of
crabs authorities do not often define what they mean by
sublethal and lethal concentrations of an insecticide. In
this study, as in the paper by Bookhout and Costlow (1975),
sublethal concentrations are arbitrarily considered those in
which there is differential survival with increased concentration in relation to survival in the control medium and
those in which more than 10% of the larvae reach the crab
stage; whereas, acutely toxic concentrations are those in
which 10% or less of the larvae reach the first crab stage.

1. Pages 6 and 7 have been inadvertently reversed.

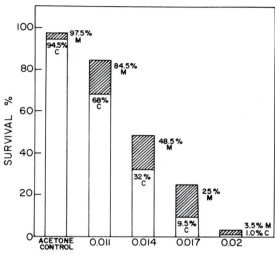

Fig. 1.   Average percent survival of four replicate series of <u>Rhithropanopeus</u> <u>harrisii</u>, Rh I-IV, reared from hatching to megalopa (M) and to first crab (C) in different concentrations of malathion.

(ii) to compare mortalities between stages at the same concentration level. The results are shown in Figure 2. Only the 0.05 ppm concentration of malathion showed significant first stage mortality, while all concentrations showed second stage increases in larval mortality. These zoeae surviving the second stage at 0.014 ppm showed significant increases at Stage IV and the 0.011, 0.014 and 0.017 ppm levels gave increases in megalopa mortality. At 0.02 ppm a significant increase in third stage mortality occurred, and at 0.05 ppm there was no survival after the second stage.

Callinectes sapidus

Survival. The percent survival of each of the four series of larvae which passed through zoeal and megalopa development in each of the four replicates, Cs I-IV, is given in Table 4. The average percent survival of larvae of all replicate series reared in five media is plotted in Figure 3. There was differential survival to the megalopa and to the first crab stage from acetone control to 0.08 ppm malathion.

RESULTS AND DISCUSSION

Rhithropanopeus harrisii

Survival. The percent survival of each of four series
of Rhithropanopeus harrisii larvae, which passed through
zoeal and megalopa development, is given in Table 1, and the
average percent survival of larvae of four replicate series
in five media is plotted in Figure 1. There was a reduction
in survival of larvae with each increase in concentration of
malathion from 0.011 ppm to 0.02 ppm (Fig. 1). Concentrations of 0.011 ppm and 0.014 ppm were sublethal, and
0.017 ppm and 0.02 ppm were acutely toxic. Larvae reared
in 0.05 ppm did not survive beyond the second zoeal stage.
A two-way analysis of variance (ANOVA) with replication
indicated that the differences between the number of larvae
which completed zoeal and megalopa development and the
differences in survival in acetone control and in 0.011 ppm,
0.014 ppm, 0.017 ppm and 0.02 ppm malathion were both highly
significant (0.001 probability).
Duration of Development. Table 1 gives the mean duration of zoeal and megalopa development and the mean time
from hatching to the first crab stage in each of the four
replicate series of larvae reared in five media. An average
duration of zoeal and megalopa development of larvae from all
replicate series is listed in Table 2. It shows that the
duration of zoeal development is shortest in acetone control
and is prolonged with each increase in concentration of
malathion from 0.011 ppm to 0.02 ppm.
Another sublethal effect of malathion was the autotomy
of legs when the larvae were in the megalopa and crab stages.
A two-way ANOVA with replication showed that variation
of duration of R. harrisii larvae reared in acetone control
and in four concentrations of malathion was related to
stages and to treatment at the 0.001 and 0.05 levels of
probability, respectively.
Mortality. Rhithropanopeus harrisii passes through
four zoeal stages and a megalopa stage before molting into a
first crab stage. In an effort to determine if larvae in
one or more stages of development were particularly sensitive
to different concentrations of malathion, a record of deaths
by stage was made for each of the four series of larvae, Rh
I-IV (Table 3).
The cumulative larval mortalities by stages were transformed using the angular transformation, and the analysis of
variance of these furnished an estimate of experimental
error. The necessary standard errors were computed (i) to
compare first stage larval mortalities with the control and

Table 1.
Effect of malathion on percent survival and duration in days through zoeal and megalopa development of <u>Rhithropanopeus harrisii</u> I-IV.

Culture Media Salinity 20 o/oo Temp. 25°C	Initial No. of Larvae Per Series	Mean Duration of Development in Days			% Survival to	
		Zoea	Megalopa	Hatching to First Crab	Megalopa	First Crab
Acetone	Rh I-50	10.92	4.96	15.88	100	100
Control	Rh II-50	11.38	5.28	16.66	100	100
	Rh III-50	10.73	5.47	16.20	98	98
	Rh IV-50	11.02	6.00	16.63	92	80
Malathion	Rh I-50	11.46	4.79	16.30	88	78
0.011 ppm	Rh II-50	11.26	5.20	16.43	84	60
	Rh III-50	11.37	5.00	16.18	82	66
	Rh IV-50	11.24	5.21	16.35	84	68
Malathion	Rh I-50	11.47	5.00	15.96	68	48
0.014 ppm	Rh II-50	10.95	5.13	16.50	44	16
	Rh III-50	11.52	4.88	16.11	44	36
	Rh IV-50	11.63	5.36	16.86	38	28
Malathion	Rh I-50	12.21	5.13	18.00	38	16
0.017 ppm	Rh II-50	10.95	5.13	16.50	30	10
	Rh III-50	12.57	6.00	17.00	14	2
	Rh IV-50	12.56	4.60	17.40	18	10
Malathion	Rh I-50	12.67	5.50	17.00	6	2
0.02 ppm	Rh II-50	12.00	-	-	6	0
	Rh III-50	12.00	-	-	2	0
	Rh IV-50	-	-	-	0	0

Table 2
Average duration in days of zoeal and megalopa development of Rhithropanopeus
harrisii I-IV.

Medium	Duration of Zoeal Development	Duration of Megalopa Development	Time from Hatch to First Crab
Acetone Control	11.01	5.43	16.19
Malathion			
0.011 ppm	11.33	5.05	16.32
0.014 ppm	11.39	5.09	16.36
0.017 ppm	12.07	5.22	17.23
0.02 ppm	12.22	5.50 (one)	17.00 (one)

Table 3
Percent mortality in five developmental stages of <u>Rhithropanopeus</u> <u>harrisii</u> I-IV.

Stage		I	II	III	IV	Megalopa	Total
Acetone Control	Rh I	0	0	0	0	0	0
	Rh II	0	0	0	0	0	0
	Rh III	0	0	0	2	0	2
	Rh IV	2	0	0	6	12	20
Malathion	Rh I	2	8	0	2	10	22
0.011 ppm	Rh II	6	6	4	0	24	40
	Rh III	0	12	2	4	16	34
	Rh IV	2	14	0	0	16	32
Malathion	Rh I	2	22	4	4	20	52
0.014 ppm	Rh II	6	18	4	28	28	84
	Rh III	0	14	12	30	8	64
	Rh IV	0	44	8	10	10	72
Malathion	Rh I	0	58	0	4	22	84
0.017 ppm	Rh II	8	32	12	18	20	90
	Rh III	0	50	14	22	12	98
	Rh IV	2	74	0	6	8	90
Malathion	Rh I	2	82	8	2	2	96
0.02 ppm	Rh II	0	60	26	8	6	100
	Rh III	14	70	12	2	2	100
	Rh IV	0	92	8	0	0	100
Malathion	Rh I	80	20	-	-	-	100
0.05 ppm	Rh II	30	70	-	-	-	100
	Rh III	98	2	-	-	-	100
	Rh IV	80	20	-	-	-	100

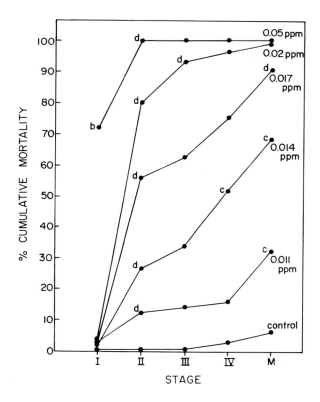

Fig. 2.   Larval mortality of <u>Rhithropanopeus</u> <u>harrisii</u> by
          stages of development and by concentration levels
          of malathion.

          b = significant difference at 0.01 level from the
              control at stage I.

          c or d = significant increase in mortality from
                   previous stage at 0.05 or 0.01 level,
                   respectively.

Survival in 0.11 ppm malathion was somewhat better to the
first crab stage than in 0.08 ppm. Concentrations of 0.02
ppm and 0.05 ppm malathion were sublethal and 0.08 ppm and
0.11 ppm were acutely toxic.
      Since <u>C</u>. <u>sapidus</u> larvae can develop in a range from
0.02 ppm to 0.11 ppm malathion and <u>R</u>. <u>harrisii</u> larvae can
only develop in a range from 0.011 ppm to 0.02 ppm, <u>C</u>.
<u>sapidus</u> larvae are much more resistant than <u>R</u>. <u>harrisii</u>

Table 4
Effect of malathion on percent survival and duration in days through zoeal and
megalopa development of Callinectes sapidus I-IV.

Culture Media Salinity 30 o/oo Temp. 25°C	Initial No. of Larvae Per Series	Mean Duration of Development in Days			# Survival to	
		Zoeae	Megalopa	Hatching to First Crab	Megalopa	First Crab
Acetone	Cs I-50	40.81	7.82	48.41	42	34
Control	Cs II-50	35.45	7.64	43.09	66	66
	Cs III-50	37.17	7.42	44.36	72	66
	Cs IV-50	35.68	8.12	44.00	56	52
Malathion	Cs I-50	41.84	10.38	51.88	38	32
0.02 ppm	Cs II-50	37.68	8.33	45.88	50	48
	Cs III-50	41.43	10.33	52.67	14	12
	Cs IV-50	38.70	7.90	46.60	40	40
Malathion	Cs I-50	44.38	11.57	56.29	16	14
0.05 ppm	Cs II-50	39.30	9.50	48.80	20	20
	Cs III-50	40.25	9.75	50.00	8	8
	Cs IV-50	40.67	8.47	49.13	30	30
Malathion	Cs I-50	48.00	10.50	55.50	6	4
0.08 ppm	Cs II-50	46.33	8.66	55.00	6	6
	Cs III-50	43.00	8.00	49.00	6	4
	Cs IV-50	43.00	7.00	50.00	8	8
Malathion	Cs I-50	41.00	7.80	48.80	10	10
0.11 ppm	Cs II-50	45.25	7.88	53.13	16	16
	Cs III-50	47.25	9.00	56.33	8	6
	Cs IV-50	44.00	7.20	51.20	10	10

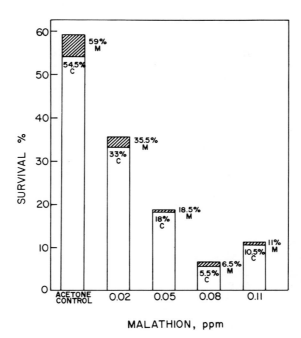

Fig. 3. Average percent survival of four replicate series of
Callinectes sapidus, Cs I-IV, reared from hatching
to megalopa (M) and to first crab (C) in different
concentrations of malathion.

larvae. A concentration of 0.02 ppm malathion was sublethal
to C. sapidus larvae but acutely toxic to R. harrisii larvae,
even though C. sapidus larvae were exposed to daily changes
of 0.02 ppm malathion from 46 to 52 days as compared to an
exposure of 17 days in the case of R. harrisii larvae.
     A two-way ANOVA with replication showed that treatment
accounted for differences in survival of C. sapidus to the
first crab stage (P < 0.001). The stage and interaction of
stage and treatment were not significant. Similar analysis
denoted that differences in survival of R. harrisii larvae
were related to zoeal and to megalopa development (P < 0.001)
and treatment (P < 0.05). This indicates that megalopa of
R. harrisii were sensitive statistically to malathion, but
the megalopa of C. sapidus were not.
     Duration of Development. The mean duration of zoeal and
megalopa development and the time from hatching to the first
crab stage of C. sapidus are given in Table 4 for larvae of

each of the four replicate series, Cs I-IV, which were
reared in five media. The average duration of zoeal and
megalopa development of Cs I-IV is listed in Table 5. The
duration of zoeal development is shortest in acetone control
and is lengthened with each increase in concentration of
malathion from 0.02 ppm to 0.08 ppm. The duration of the
megalopa does not follow this trend, but the total time
from hatching to the first crab stage does (Table 5),
primarily due to the effects of malathion on zoeal
development.

A two-way ANOVA with replication indicated that the
variation of duration of C. sapidus larvae reared in acetone
control and four concentrations of malathion was related to
stages, media and the interaction of the two at the 0.001
level of probability.

The delay in molting rate in C. sapidus and R. harrisii
larvae is considered a sublethal effect of malathion, as is
the tendency to autotomize the legs in megalopa and crab
stages of R. harrisii and, to a lesser extent, in the same
stages of C. sapidus.

Mortality. Callinectes sapidus usually passes through
seven zoeal stages and a megalopa stage before reaching the
first crab stage. A frequent variation from the norm in the
life history of the blue crab is the addition of an eighth
zoeal stage before a molt to a megalopa stage (Costlow and
Bookhout, 1959). To determine if larvae in one or more of
the nine stages of development of C. sapidus were
particularly sensitive to different concentrations of
malathion, a record of deaths by stage was made for larvae
reared from four mother crabs, CS I-IV (Table 6).

The cumulative larval mortalities were treated in the
same way as described for R. harrisii and the results are
shown in Figure 4. Substantial larval mortality of C.
sapidus occurred in the control in zoeal stages II and III
and only the 0.08 ppm and 0.11 ppm concentrations showed
significant first stage mortality. All levels except that
of 0.11 ppm showed significant increases in mortality at
stages II and III and no significant increases thereafter.

The contrast between the two species is evident in
Figure 2 and Figure 4. R. harrisii exhibited early stage
larval mortality only at 0.05 ppm and second stage mortality
at levels from 0.011 ppm to 0.05 ppm. In addition, significant increases in larval mortality occurred at the later
stages at concentrations from 0.011 ppm to 0.017 ppm. C.
sapidus larvae, on the other hand, exhibited significant
increases in mortality through zoeal stages II and III but
no appreciable increases thereafter. It appears that the
R. harrisii larvae are susceptible to malathion at all

Table 5
Average duration in days of zoeal and megalopa development of Callinectes
sapidus I-IV.

Medium	Duration of Zoeal Development	Duration of Megalopa Development	Time from Hatch to First Crab
Acetone Control	36.98	7.72	44.52
Malathion			
0.02 ppm	39.45	8.85	48.16
0.05 ppm	41.05	9.50	50.53
0.08 ppm	44.92	8.27	52.18
0.11 ppm	44.36	7.86	52.10

Table 6
Percent mortality in nine developmental stages of <u>Callinectes</u> <u>sapidus</u> I-IV.

Stage		I	II	III	IV	V	VI	VII	VIII	Megalopa	Total
Acetone	Cs I	2	50	2	0	0	0	4	0	8	66
Control	Cs II	0	4	20	8	0	2	0	0	0	34
	Cs III	8	14	6	0	0	0	0	0	6	34
	Cs IV	2	24	14	0	0	0	4	0	4	48
Malathion	Cs I	4	30	16	4	6	0	2	0	6	68
0.02 ppm	Cs II	8	28	12	2	0	0	0	0	2	52
	Cs III	4	44	22	4	0	4	6	2	2	88
	Cs IV	4	22	30	2	0	0	0	2	0	60
Malathion	Cs I	0	52	28	2	0	0	0	2	2	86
0.05 ppm	Cs II	16	42	20	2	0	0	0	0	0	80
	Cs III	20	46	14	4	0	0	4	4	0	92
	Cs IV	4	22	38	6	0	0	0	0	0	70
Malathion	Cs I	18	72	4	0	0	0	0	0	2	96
0.08 ppm	Cs II	22	40	30	0	2	0	0	0	0	94
	Cs III	16	42	32	4	0	0	0	0	2	96
	Cs IV	14	42	34	0	0	2	0	0	0	92
Malathion	Cs I	44	44	0	0	0	0	0	2	0	90
0.11 ppm	Cs II	46	28	10	0	0	0	0	0	0	84
	Cs III	48	40	2	0	0	0	2	0	2	94
	Cs IV	68	10	8	4	0	0	0	0	0	90

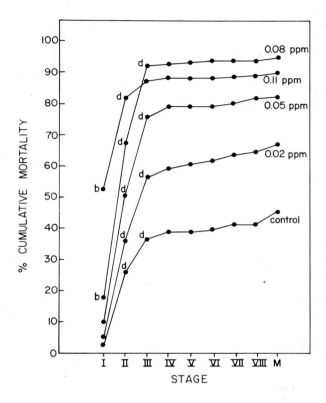

Fig. 4. Larval mortality of <u>Callinectes</u> <u>sapidus</u> by stages
         of development and by concentration levels of
         malathion.

         b = significant difference at 0.01 level from the
             control stage I.

         c or d = significant increase in mortality from
                  previous stage at 0.05 or 0.01 level,
                  respectively.

stages of development and <u>C</u>. <u>sapidus</u> only at the early
stages. Furthermore, 100% <u>R</u>. <u>harrisii</u> mortality was
completed during the second stage at only 0.05 ppm while <u>C</u>.
<u>sapidus</u> mortality was approximately 82% at the same
concentration. Collateral investigations of these same
species using the insecticide methoxychlor (Bookhout <u>et</u> <u>al</u>.,
1976) showed that the species behavior was reversed in the

presence of the chlorinated hydrocarbon from that observed for malathion.

Ecological Implications. It is not known what concentrations of malathion would have sublethal and lethal effects on the larval development of R. harrisii and C. sapidus in the field.

It is possible that crab larvae could be killed by initial spraying of a marsh with malathion and/or by residues of malathion left in the water after spraying, especially if these residues were equal to or above those found to be acutely toxic in this investigation. Tagatz et al. (1974) reported that thermal fogging (420 g/ha) and ULV aerosol spraying (57 g/ha) of malathion 95 on salt marshes left residues as high as 5.2 ppb and 0.49 ppb. Since R. harrisii larvae are affected by 0.011 ppm to 0.02 ppm malathion and the residues Tagatz et al (1974) found in the water would probably not have any direct effect on R. harrisii or C. sapidus larvae. Conte and Parker (1971) reported that aerial application of 256 g/ha to marsh embayments in Texas killed 14% to 80% of commercial shrimp Penaeus aztecus and P. setiferus. They found residues of malathion of 0.8 ppm to 3.2 ppm in water 48 hours after spraying. These concentrations of malathion would undoubtedly kill the larvae of R. harrisii and C. sapidus for they are above the range of concentrations which were found to be acutely toxic to each species.

ACKNOWLEDGMENTS

The research was supported in part by the Environmental Protection Agency, Grant No. R-801128-02-2, through the Gul Breeze Environmental Research Laboratory with Mr. Jack Lowe as Project Officer. We thank Mrs. Karen Johnson, Miss Rebecca Piver, and Mrs. Walter Nelson for their technical assistance.

LITERATURE CITED

Bookhout, C. G. and J. D. Costlow, Jr. 1975. Effects of mirex on the larval development of blue crab. Water Air Soil Poll. 4: 113-126.

_____, _____, and R. Monroe. 1976. Effects of methoxychlor on larval development of mud-crab and blue crab. Water Air Soil Poll. (in press).

Brown, A. W. A. and Z. H. Abedi. 1960. Cross-resistance characteristics of a malathion-tolerant strain developed in Aedes aegypti. Mosquito News 20: 118-124.

Busvine, J. R. 1959. Patterns of insecticide resistance to
    organo-phosphorus compounds in strains of houseflies
    from various sources. Entomologia Exp. Appl. 2: 58-
    67.
Conte, F. S. and J. C. Parker. 1971. Ecological aspects
    of selected Crustacea of two marsh embayments of the
    Texas coast. Texas A. and M. Univ., Sea Grant Publ.
    No. TAMU-SG-71-211. 184 pp.
Coppage, D. L. and E. Matthews. 1974. Short-term effects
    of organophosphate pesticides on cholinesterases of
    estuarine fishes and pink shrimp. Bull. Environ.
    Contam. Toxicol. 11: 483-488.
Costlow, J. D., Jr. and C. G. Bookhout. 1959. The larval
    development of Callinectes sapidus Rathbun reared in
    the laboratory. Biol. Bull. 116: 373-396.
    _____ and _____. 1960. A method for developing
    brachyuran eggs in vitro. Limnol. Oceanogr. 5: 212-
    215.
Darsie, R. F., Jr. and F. E. Corriden. 1959. The toxicity
    of malathion to killifish (Cyprinodontidae) in
    Delaware. J. Econ. Ent. 52: 696-700.
La Brecque, G. C. and H. G. Wilson. 1960. Effect of DDT
    resistance on the development of malathion resistance
    in houseflies. J. Econ. Ent. 53: 320-321.
Mengle, D. C. and L. L. Lewallen. 1963. Metabolism of
    malathion by a resistant and a susceptible strain of
    Culex tarsalis: I. Degradation in vivo and identi-
    fication of organic soluble metabolites. Mosquito
    News 23: 226-233.
Mount, G. A., J. A. Seawright, and N. W. Pierce. 1974.
    Selection response and cross-susceptibility of a
    malathion-resistant strain of Aedes taeniorhynchus
    (Wiedemann) to other adulticides. Mosquito News 34:
    276-277.
Tagatz, M. E., P. W. Borthwick, G. H. Cook, and D. L.
    Coppage. 1974. Effects of ground applications of
    malathion on salt-marsh environments in northwestern
    Florida. Mosquito News 34: 309-315.
Walker, W. W. and B. J. Stojanovic. 1973. Microbial versus
    chemical degradation of malathion in soil. J. Environ.
    Quality 2: 229-232.

Effects of Aroclor® 1016 and Halowax® 1099 on Juvenile Horseshoe Crabs *Limulus polyphemus*

J. M. NEFF[1] and C. S. GIAM[2]

[1]Department of Biology
Texas A&M University
College Station, Texas   77843

[2]Department of Chemistry
Texas A&M University
College Station, Texas   77843

Several industrial halogenated hydrocarbons have been identified as nearly ubiquitous trace contaminants of the marine environment. The polychlorinated biphenyls (PCB's) have aroused concern because of their high toxicity, their persistence, and their tendency to bioaccumulate in marine and freshwater ecosystems. This concern has led the sole United States manufacturer, Monsanto, to restrict sales to uses not likely to result in environmental contamination. In addition, Monsanto has replaced its more highly chlorinated PCB's with a new, presumably less persistent formulation, Aroclor[R] 1016, containing greatly reduced amounts of isomers with 5 or more chlorines per biphenyl. Polychlorinated naphthalenes (PCN's) have physical and chemical properties similar to those of the PCB's and have similar industrial uses. Chlorinated naphthalenes are manufactured in the United States by the Koppers Company under the trade name Halowax[R].

Considerable information has been published concerning the distribution, toxicity, and sublethal biological effects of PCB's in the marine environment (Zitko and Choi, 1971; Nisbet and Sarofim, 1972; Green, 1974; Peakall, 1975). PCN's have recently been identified in environmental samples

21

(Crump-Weisner et al., 1974). However, no information is available concerning the persistence of PCN's in the marine environment or their toxicity and sublethal biological effects to marine animals.

The purpose of the present study was to compare the biological effects of Aroclor 1016 and Halowax 1099 on juvenile horseshoe crabs, Limulus polyphemus. The chronic toxicity of these compounds and their effects on molting and respiration were also investigated.

MATERIALS AND METHODS

Fertilized eggs from a single mating pair of L. polyphemus were collected from Sykes Creek, Brevard County, Florida on January 13, 1975. The eggs had been deposited by the female in about 6 inches of fine sand on the previous day. The sea water salinity at the collection site was 20 o/oo. The eggs were maintained in the laboratory in 22.8 cm glass finger bowls containing 20 o/oo S artificial sea water (Instant Ocean, Aquarium Systems, Inc., Eastlake, Ohio) at a temperature of 20-22°C. The eggs hatched after 28-35 days and the resulting "trilobite larvae" molted to the first tailed stage (T_1) 30-40 days later. At the beginning of the exposure period the horseshoe crabs were either the late T_1 stage or early second tailed stage (T_2).

Two groups (24 animals/group) of horseshoe crabs were exposed to each pollutant concentration. Group A consisted of late T_1-stage crabs (two 11.4 cm finger bowls containing 200 ml 20 o/oo S sea water and 12 animals/bowl at each concentration). Group B consisted of early T_2-stage crabs (three 11.4 cm finger bowls containing 200 ml 200 o/oo S sea water and 8 animals/bowl at each concentration). Each bowl was censused every day. Molts and deaths were recorded, and the animals were fed newly hatched brine shrimp (Artemia salina). The water in each bowl was replaced by a fresh exposure solution every 48 hours. For those groups in which more than 50% of the crabs died in 96 days, median times to mortality (LT50), their 95% confidence intervals and slope functions were computed by the graphic method of Litchfield (1949). The median times from the beginning of the exposure period to each successive molt (ET50), their confidence intervals and slope functions were computed by the same method. The mean durations of intermolt period were then computed as the difference in the ET50 values for successive molts.

At day 50 respiratory rates of those Group A crabs which had molted to the T_2 stage were determined. At day 60 respiratory rates of those Group B crabs which had molted to the T_3 stage were also determined. Individual crabs were placed in either 2.5 ml (T_2) or 5 ml (T_3) all-glass syringes containing 2 or 3 ml respectively of Millipore-filtered 20 o/oo S sea water. Each syringe orifice was sealed with a plastic cap, and the syringe was placed in a 25°C incubator in the dark for 30-60 minutes. Initial and final pO_2 in the syringes were determined by injecting 0.1 ml aliquotes into a Radiometer BMS3 Mark 2 blood gas analyzer thermostatically controlled at 25°C. Crabs were then blotted dry and their weights determined to the nearest 0.5 mg. Oxygen consumption in $ml/O_2/hr$ was determined from the difference in pO_2 between the initial and final reading.

Aroclor 1016 was obtained from the Monsanto Company, St. Louis, Missouri. Halowax 1099 (primarily tri- and tetrachloronaphthalenes, containing approximately 50% chlorine by weight) was obtained from the Koppers Company, Pittsburgh, Pennsylvania. Stock solutions of both compounds were prepared in nanograde acetone. To prepare the exposure solutions, aliquots of the stock solutions were added with a Hamilton microliter syringe to 1 liter of 20 o/oo S sea water in an ehrlenmeyer flask. The flasks were violently shaken to insure thorough dispersion and solution of the compound. The mixtures were used immediately. Calculated (nominal) exposure concentrations of Aroclor 1016 were 0 (control), 10, 20, 40, and 80 µg/l (ppb) and of Halowax 1099 were 0, 20, 40, and 80 µg/l. Initial and 48 hour water samples were taken for analysis near the beginning and near the end of the 96 day exposure period. Analytical results were essentially the same both times. At the end of the 96 day exposure period the surviving crabs in each group at each concentration were pooled, rinsed in nano-grade acetone, weighed and analyzed for Aroclor 1016 or Halowax 1099. Aroclor 1016 residues in water and horseshoe crab tissues were determined by the gas chromatographic method of Giam and Wong (1972). Halowax 1099 in water and tissues was determined spectrophotometrically by a modification of the technique of Neff and Anderson (1975) for naphthalenes. The strong absorbance of Halowax 1099 at a wavelength of 236 nm was used for quantitation.

RESULTS

Analysis of the exposure water (Table 1) revealed that the concentrations of Aroclor 1016 and Halowax 1099 actually present at time = 0 were similar to the calculated (nominal) concentrations. The Aroclor control water contained a small

Table 1
Relationship between the amounts of Aroclor and Halowax
added to the exposure water and the amounts found by gas
chromatography (Aroclor) or UV spectrophotometry (Halowax) at 0 and 48 hours.

Nominal Concentration µg/l (ppb)	Concentration measured µg/l (ppb)	
	0 hours	48 hours
AROCLOR		
0	0.35	0.1
10	12.1	0.85
20	22.4	0.83
40	36.6	3.23
80	71.5	6.76
HALOWAX		
0	<10	<10
20	22	<10
40	40	<10
80	70	38

amount of Aroclor of unknown origin. The UV method for
Halowax is much less sensitive (limits of detection 10-
15 ppb) than the G.C. method for Aroclor (limits of detection
<0.1 ppb), so the presence of small amounts of Halowax in the
control water could not be ascertained. There was a substantial drop in both the Aroclor and Halowax concentrations
in 48 hours. In the Aroclor group 2.96 (control) to 9.4%
(80 ppb) of the amount originally present remained after 48
hours. Halowax could be detected at 48 hours in only the
80 ppb exposure water.

The combined control animals experienced slightly more
than 1% mortality in 96 days, indicating that these animals
are quite hardy under laboratory conditions and that the
experimental design was not itself overly stressful. At all
exposure concentrations of both Aroclor and Halowax Group A
crabs were considerably more sensitive than Group B crabs to
both toxicants (Table 2). No mortalities occurred for nine
days at any exposure concentration. Greater than 50%
mortality was observed only among Group A crabs exposed to
40 and 80 ppb Aroclor and 80 ppb Halowax. The only Group B
crabs experiencing greater than 50% mortality were those
exposed to 80 ppb Aroclor. The median lethal time (LT50) of
Aroclor-exposed Group A crabs was the same at both 40 and
80 ppb. The smaller slope function (S) for the 80 ppb group
indicates a more uniform response to this higher concentration. The Group B crabs at 80 ppb had an LT50 3 times longer

Table 2
Median lethal time (LT50) in days of Group A and B Limulus
polyphemus exposed to 40 and 80 ppb Aroclor 1016 and Halowax 1099. The 95% confidence interval (parentheses) and
slope function (S) are given for each value.

Exposure Concentration (ppb)	LT50, Days	
	Group A	Group B
AROCLOR 40	20.8 (13.4-32.2) S = 2.93	*
80	20.3 (16.9-24.4) S = 1.60	61.0 (47.3-78.7) S = 1.85
HALOWAX 40	*	*
80	27.0 (19.6-37.1) S = 2.18	*

*Less than 50% mortality in 96 days.

than that of the Group A crabs, indicating a much greater
tolerance of these larger crabs to Aroclor. The LT50 of the
Group A crabs exposed to 80 ppb Halowax was nearly seven
days longer than that of similar crabs exposed to the same
concentration of Aroclor. Neither 40 nor 80 ppb Halowax
produced 50% mortality in 96 days among Group B crabs.

Group A crabs were in the late first tailed stage at
the beginning of the exposure period. Molting to the T_2
stage began immediately in the controls and all exposure
groups except the 80 ppb Aroclor and Halowax. The median
time to molt to T_2 (ET50) was not greatly different in the
controls, 10 and 20 ppb Aroclor and 20 ppb Halowax exposure
groups (Table 3). The 40 ppb Halowax group required about 3
days less than the controls to reach ET50. Among those crabs
that molted to the T_2 stage, the median time from the
beginning of the exposure to the next molt to T_3 varied from
48 days for the controls and 20 ppb Halowax group to 41 days
for the 20 ppb Aroclor and 40 ppb Halowax groups. As
indicated by the mean intermolt times, $T_2 \rightarrow T_3$ (column 4 in
Table 3), low concentrations of Aroclor and Halowax seemed
to decrease the intermolt period and thus accelerate growth.
However, at higher concentrations molting was partially
inhibited.

Molting of Group B crabs was also influenced by Aroclor
and Halowax (Table 4). Crabs in this group had molted to the
T_2 stage shortly before the beginning of the exposure period.

Table 3.
Effect of Aroclor 1016 and Halowax 1099 on molting of Group A
Limulus polyphemus. Median time to molt (ET50) is measured from
the beginning of the exposure period. The 95% confidence intervals (parentheses) and slope function (S) are given for each value.

Exposure Concentration (ppb)	ET50, Days from Beginning of Exposure		Mean Intermolt Time, Days
	T_2	T_3	$T_2-- T_3$
0	8.5 (6.7-10.7) S = 1.80	48.0 (43.6-52.8) S = 1.40	39.5
AROCLOR 10	10.5 (8.0-13.8) S = 1.95	43.0 (34.7-53.3) S = 1.71	32.5
20	7.2 (4.0-13.0) S = 4.54	41.8 (36.0-48.5) S = 1.45	34.6
40	*	*	--
80	*	*	--
HALOWAX 20	10.0 (7.5-13.3) S = 2.41	48.8 (42.5-56.0) S = 1.48	38.8
40	5.1 (3.19-8.16) S = 3.16	41.0 (35.2-47.8) S = 1.40	35.9
80	*	*	--

*Less than 50% molting in 96 days.

No molts were observed in these crabs until the tenth day
after the beginning of the exposure period. The ET50 values
for molting to the T_3 stage were similar for the controls,
the 10 and 20 ppb Aroclor and 20 and 40 ppb Halowax groups.
At 40 and 80 ppb Aroclor and 80 ppb Halowax molting was
delayed by 6 to 9 days. However, the ET50 values for the
molt to the T_4 stage were substantially lower than for
controls for all the exposure groups in which more than 50%
molt to T_4 occurred. As a result, there was a drop in the
mean intermolt period, $T_3 \rightarrow T_4$, with increasing concentrations of Aroclor and Halowax. Aroclor at any concentration
had a greater effect in reducing the length of both the
$T_2 \rightarrow T_3$ and the $T_3 \rightarrow T_4$ intermolt periods than did the
same concentration of Halowax.

After 50 days of exposure the respiratory rates of those
Group A crabs which had molted to the T_2 stage were determined. The mean wet weight per individual of these crabs

Table 4.
Effect of Aroclor 1016 and Halowax 1099 on molting of Group B
Limulus polyphemus. Median time to molt (ET50) is measured
from the beginning of the exposure period. The 95% confidence
intervals (parentheses) and slope function (S) are given for
each value.

Exposure Concentration (ppb)	ET50, Days from Beginning of Exposure		Mean Intermolt Time, Days
	T_2	T_3	T_2-- T_3
0	20.0 (18.2-22.0) S = 1.40	82.0 (72.7-92.5) S = 1.44	62.5
AROCLOR 10	19.2 (16.8-22.0) S = 1.40	73.5 (68.2-79.2) S = 1.20	54.3
20	20.7 (18.2-23.6) S = 1.38	67.5 (60.5-75.3) S = 1.30	46.8
40	27.5 (21.3-35.5) S = 1.85	68 (59.1-78.2) S = 1.33	40.5
80	29.5 (26.6-32.7) S = 1.26	*	--
HALOWAX 20	21.8 (19.1-24.9)	69.0 (65.6-72.6) S = 1.10	47.2
40	20.0 (16.8-23.8) S = 1.54	65.2 (61.2-69.4) S = 1.16	45.2
80	26.2 (22.0-30.9) S = 1.52	69.5 (65.3-73.9) S = 1.16	43.3

*Less than 50% molting in 96 days.

was 23 ± 3 mg and was not significantly different in
different exposure groups. Crabs exposed to 10 and 20 ppb
Aroclor or 20 and 40 ppb Halowax had higher mean respiratory
rates than the unexposed controls (Fig. 1). Aroclor had a
more pronounced stimulatory effect on respiration than did
Halowax. At 50 days there were not sufficient numbers of T_2
crabs (due either to molting to T_3 or death) at 40 and 80 ppb
Aroclor and 80 ppb Halowax for determinations of respiration
rate.

The respiratory rates of those Group B crabs in the T_3
stage at day 60 were more variable than those of the Group A
crabs. Mean wet weight per individual of these crabs was
54 ± 14 mg. The mean weight varied in different exposure
groups from 47 mg in 40 ppb Halowax to 68 mg in 80 ppb

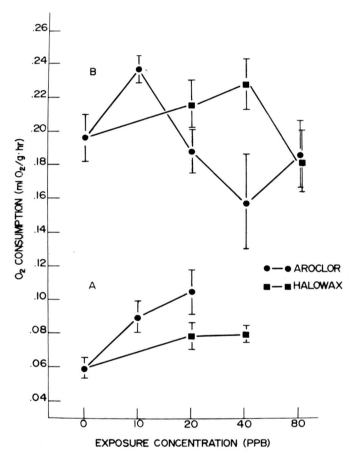

Fig. 1.   Respiratory rates of Group A (A) and
Group B (B) _Limulus polyphemus_ following
50-60 days exposure to Aroclor 1016 or
Halowax 1099. Group A crabs were at the
T_2 molt stage and Group B crabs were at
the T_3 molt stage. Vertical lines
represent standard errors of the means.

Halowax. Unexpectedly, the weight specific respiratory
rates of the heavier T_3 crabs were more than 3 times higher
than those of the T_2 crabs. This is opposite the usual

trend for weight specific metabolic rate to decrease exponentially with increasing mass.

The effects of Aroclor and Halowax on respiratory rates of T_3 crabs were variable. At 10 ppb Aroclor respiration was stimulated. At all higher concentrations it was either unaffected or depressed. Respiratory rate increased with increasing Halowax concentration to 40 ppb and then was slightly below the control rate at 80 ppb.

At the end of the exposure period all the surviving horseshoe crabs were sacrificed and their tissues analyzed for Aroclor or Halowax (Table 5). For the Aroclor-exposed series the pooled Group A and Group B animals at each concentration were analyzed separately. In the Halowax series Group A and Group B animals at each concentration were pooled for analysis. This was necessitated by the much lower sensitivity of the analytical technique for Halowax.

At all Aroclor exposure concentrations for which comparative data are available the Group A crabs accumulated Aroclor to substantially higher concentrations than did the Group B crabs. As a result the bioconcentration factors were much higher for Group A than for Group B crabs.

Halowax concentrations in the tissues of Halowax-exposed crabs were 10-20 times lower than the concentrations of Aroclor in animals exposed to the equivalent concentrations of Aroclor. As a result the bioconcentration factors were low for the Halowax-exposed animals.

DISCUSSION

The concentrations of Aroclor 1016 and Halowax 1099 in the exposure water showed a substantial drop during the 48 hours between water changes. This behavior is typical of sparingly soluble materials like PCB's and PCN's. The solubility of PCB's is inversely proportional to chlorine content, ranging from 1 ppb for hexachlorobiphenyl to 637 ppb for dichlorobiphenyl (Haque and Schmedding, 1975). The solubility of PN's is not known, but it is probably similar to or slightly higher than that of the PCB's. Loss from the aqueous phase may be by adsorption to glass surfaces, volatilization or uptake by the animals. In the present study uptake by the animals can account for a maximum of 6% of the Aroclor lost from the water. As a result of these changes in exposure water PCB and PCN concentrations the crabs were actually exposed to average concentrations considerably lower than the initial values.

Table 5
Concentration of Aroclor and Halowax in the tissues of Group A
(A) and Group B (B) Limulus polyphemus following 96 days exposure. Bioconcentration factors computed as ratio of tissue
concentration to the initial actual exposure concentration.

Initial Exposure Concentration μg/l (ppb)	Tissue Concentration μg/g wet wt. (ppm)		Bioconcentration Factor	
AROCLOR	A	B	A	B
0.35	0.109	0.084	311	240
12.1	2.39	0.496	198	41
22.4	7.70	5.08	344	227
36.6	31.9	11.2	873	305
71.5	*	92.8	---	1298
HALOWAX				
0	--		--	
22	0.51		23	
40	1.00		25	
70	5.20		74	

* No survivors at 96 days.

Juvenile horseshoe crabs were shown in the present
investigation to be extremely hardy and tolerant animals.
The acute toxicity (96 hour LC50) of Aroclor 1016 and
Halowax 1099 for horseshoe crabs could not be estimated,
since no crabs died in 96 hours even at the highest exposure
concentrations used. Similar delayed effects of Aroclor
have been reported by other investigators (Wildish, 1972;
Nimmo et al., 1975). This suggests that Aroclor and Halowax
are cumulative slow acting poisons. Chronic toxicity as
expressed by the median time in days to 50% mortality (TL50)
was only observed at the highest PCB and PCN concentrations
used. Other marine arthropods have been shown to be much
more sensitive to Aroclor 1016. Hansen et al. (1974)
reported 96-hour LC50 values of 10.5 μg/l and 12.5 μg/l for
brown shrimp (Penaeus aztecus) and grass shrimp (Palaemonetes
pugio) respectively.

Horseshoe crabs first exposed at the T_1 stage were
considerably more sensitive to both chlorinated hydrocarbons
than were crabs first exposed at the T_2 stage. This is
particularly interesting since mortality did not begin to
occur among Group A crabs until the ninth day of exposure,
by which time significant numbers of these T_1 crabs had molted
to the T_2 stage. There are several similar reports that

younger life stages of crustaceans are more sensitive to
pollutants than are later stages. Roesijadi et al. (1976)
reported that larval Palaemonetes pugio were substantially
more sensitive to Aroclor 1254 than were the adults. In the
present investigation the greater sensitivity of the Group A
crabs to Aroclor seems to be related to the observation that
they accumulated substantially higher concentrations of
Aroclor in their tissues than did the Group B crabs. The
difference in Aroclor accumulation by the Group A and Group B
crabs may be related to the much larger surface to volume
ratio of the small as compared to the large crabs since uptake is probably a surface area-related phenomenon.

Aroclor 1016 was more toxic than Halowax 1099 to juvenile
horseshoe crabs. The differences in toxicity may be related
to the fact that Halowax was accumulated to a much lesser
extent than was Aroclor at all exposure concentrations. Bioconcentration factors for Halowax ranged from 23-74 while
those for Aroclor were 41-1298. The bioaccumulation factors
for Aroclor, though high, were lower than several published
values. Aroclor bioconcentration factors, usually following
shorter exposure periods than used here, and utilizing flow-
through exposure systems in the range of 5000-25,000 have
been reported for freshwater and marine animals (Sanders and
Chandler, 1972; Nimmo et al., 1974; Hansen et al., 1974).

The lower concentrations of Aroclor and Halowax reduced
both the T_2 --> T_3 and T_3 --> T_4 intermolt periods of the
crabs. Aroclor had a more pronounced effect on intermolt
periods than did Halowax. Mortalities at the higher pollutant
concentrations made estimates of mean intermolt periods
difficult. Arthropod molting is a complicated biochemical
and physiological process involving complex hormonal and
tissue interactions. Any one of a number of the steps
involved may be influenced by pollutant stress. Molting in
Limulus seems to be controlled in part by polyhydroxysteroids
similar to the insect and crustacean ecdysones (Jegla and
Costlow, 1970; Jegla et al., 1972). Whether or not chlorinated hydrocarbons have an influence on steroid metabolism in
horseshoe crabs and other arthropods, as they do in fish
(Freeman and Idler, 1975) and birds (Norwicki and Norman,
1972), is unknown. Roeijadi et al.(1976) showed that Aroclor
1254 at concentrations up to 3.2 µg/l significantly increased
the duration of development to postlarvae of Palaemonetes
pugio. Similarly Laughlin (personal communication) showed
that the mean time to molting to the megalops by Rithropano-
peus harrisi was increased by exposure to Aroclor 1016 and
Halowax 1099. Thus the effects of Aroclor and Halowax on
arthropod molting are variable and the magnitude and direction
of the effect may be different for different larval and post-

larval molt stages. This variability may indicate that a change in molting frequency is a generalized response to pollutant stress.

There was considerable variation in the respiratory rates among individuals of the same molt stage. This respiratory variability seems to be typical of horseshoe crabs (Johansen and Petersen, 1975). Chronic exposure to low concentrations of Aroclor and Halowax increased respiratory rates of both Group A and B crabs above the control values. Higher concentrations generally had little or no effect. If the observed respiratory stimulation is an indication of an increase in overall metabolic rate, then the observed decrease in intermolt periods can be explained as a result of a chlorinated hydrocarbon-induced stimulation of metabolism. Sublethal concentrations of several organochlorine insecticides have been shown to stimulate respiration of freshwater fish (Waiwood and Johansen, 1974) and invertebrates (de la Cruz and Naqvi, 1973) and juvenile blue crabs, Callinectes sapidus, (Leffler, 1975). This stimulatory effect may be due to organochlorine-induced increases in nervous and muscular metabolism.

The ecological significance of reduced intermolt time and thus increased growth rate is obscure. However, a pollutant-induced elevation of metabolic rate might prove harmful to animals experiencing periodic food deprivation or hypoxic stress. Limulus, because of its benthic life style and feeding habits, is periodically exposed to hypoxic conditions (Barthel, 1974; Johansen and Petersen, 1975). Horseshoe crabs are extremely tolerant to hypoxia (Falkowski, 1973, 1974; Mangum et al., 1975; Johansen and Petersen, 1975). PCB's and PCN's by stimulating respiration may reduce this hypoxic tolerance.

The dissolved Aroclor and Halowax concentrations required to cause significant mortality or sublethal effects, although in the low ppb range, are substantially higher than those encountered in all but the most polluted sea water. However, much of the PCB's and PCN's entering the marine environment become adsorbed to organic detritus or inorganic particulates and eventually are deposited in marine sediments. The sedimentary load of these compounds may reach high concentrations. Concentrations of PCB's and PCN's in surface sediments in the range of 100 ppb to 61 ppm have been reported (Nimmo et al., 1971; Hom et al., 1974; Crump-Wiesner et al., 1974). Nimmo et al. (1971, 1974) have shown that sediment adsorbed PCB's are available to marine crustaceans. Thus horseshoe crabs, because of their benthic habits, may encounter potentially harmful concentrations of sediment-adsorbed PCB's and PCN's.

ACKNOWLEDGMENTS

We would like to express our appreciation to Roy Laughlin for collecting the horseshoe crab eggs in Florida and to K. C. Houck for her assistance with the G. C. analyses for Aroclor. This research was supported by Grant # ID075-04890 from the National Science Foundation, International Decade of Oceanic Exploration.

LITERATURE CITED

Barthel, K. W. 1974. Limulus: a living fossil. Horseshoe crabs aid interpretation of an upper Jurassic environment. Die Naturwissen. 61: 428-433.

Crump-Wiesner, H. J., H. R. Feltz, and M. L. Yates. 1974. A study of the distribution of polychlorinated biphenyls in the aquatic environment. Pest. Monit. J. 8: 157-161.

de la Cruz, A. A. and S. M. Naqvi. 1973. Mirex incorporation in the environment: uptake in aquatic organisms and effects on the rates of photosynthesis and respiration. Arch. Environ. Contam. Toxicol. 1: 255-264.

Falkowski, P. G. 1973. The respiratory physiology of hemocyanin in Limulus polyphemus. J. Exp. Zool. 186: 1-6.

_____. 1974. Facultative anaerobiosis in Limulus polyphemus: phosphoenolphruvate carboxykinase and heart activities. Comp. Biochem. Physiol. 48B: 749-759.

Freeman, H. C. and D. R. Idler. 1975. The effect of polychlorinated biphenyl on steroidogenesis and reproduction in the brook trout (Salvelinus fontinalis). Canad. J. Biochem. 53: 666-670.

Giam, C. S. and M. K. Wong. 1972. Problems of background contamination in the analysis of open ocean biota for chlorinated hydrocarbons. J. Chromatog. 72: 283-292.

Green, M. B. 1974. Polychloroaromatics and heteroaromatics of industrial importance. In: Polychloroaromatic Compounds, pp. 403-474, ed. by H. Suschitzky, New York: Plenum Press.

Hansen, D. J., P. R. Parrish, and J. Forester. 1974. Aroclor 1016: Toxicity to and uptake by estuarine animals. Environ. Res. 7: 363-373.

Haque, R. and D. Schmedding. 1975. A method of measuring the water solubility of hydrophobic chemicals:

solubility of five polychlorinated biphenyls. <u>Bull</u>.
<u>Environ</u>. <u>Contam</u>. <u>Toxicol</u>. 14: 13-18.

Hom, W., R. W. Risebrough, A. Soutar, and D. R. Young. 1974.
Deposition of DDE and polychlorinated biphenyls in
dated sediments of the Santa Barbara basin. <u>Sci</u>. 184:
1197-1199.

Jegla, T. C. and J. D. Costlow, Jr. 1970. Induction of
molting in horseshoe crab larvae by polyhydroxy
steroids. <u>Gen</u>. <u>Comp</u>. <u>Endocrinol</u>. 14: 295-302.

_____, _____, and J. Alspaugh. 1972. Effects of
some ecdysones and some synthetic analogs on horseshoe
crab larvae. <u>Gen</u>. <u>Comp</u>. <u>Endocrinol</u>. 19: 159-166.

Johansen, K. and J. A. Petersen. 1975. Respiratory adapta-
tions in <u>Limulus</u> <u>polyphemus</u> (L.). In: <u>Physiological</u>
<u>Ecology</u> <u>of</u> <u>Estuarine</u> <u>Organisms</u>, pp. 129-145, ed. by
F. J. Vernberg, Columbia, S. C.: University of South
Carolina Press.

Leffler, C. W. 1975. Effects of ingested mirex and DDT on
juvenile <u>Callinectes</u> <u>sapidus</u> Rathbun. <u>Environ</u>. <u>Pollut</u>.
8: 283-300.

Litchfield, J. T., Jr. 1949. A method for rapid graphic
solution of time - percent effect curves. <u>J</u>. <u>Pharmac</u>.
<u>Exper</u>. <u>Therap</u>. 97: 399-408.

Mangum, C. P., M. A. Freadman, and K. Johansen. 1975. The
quantitative role of hemocyanin in aerobic respiration
of <u>Limulus</u> <u>polyphemus</u>. <u>J</u>. <u>Exp</u>. <u>Zool</u>. 191: 279-285.

Neff, J. M. and J. W. Anderson. 1975. An ultraviolet
spectrophotometric method for the determination of
naphthalene and alkylnaphthalenes in the tissues of oil-
contaminated marine animals. <u>Bull</u>. <u>Environ</u>. <u>Contam</u>.
<u>Toxicol</u>. 14: 122-128.

Nimmo, D. R., J. Forester, P. T. Heitmuller, and G. H. Cook.
1974. Accumulation of Aroclor 1254 in grass shrimp
(<u>Palaemonetes</u> <u>pugio</u>) in laboratory and field exposures.
<u>Bull</u>. <u>Environ</u>. <u>Contam</u>. <u>Toxicol</u>. 11: 303-308.

_____, D. J. Hansen, J. A. Couch, N. R. Cooley, P. R.
Parrish, and J. I. Lowe. 1975. Toxicity of Aroclor
1254 and its physiological activity in several estuarine
organisms. <u>Arch</u>. <u>Environ</u>. <u>Contam</u>. <u>Toxicol</u>. 3: 22-39.

_____, P. D. Wilson, R. R. Blackman, and A. J. Wilson,
Jr. 1971. Polychlorinated biphenyl absorbed from
sediments by fiddler crabs and pink shrimp. <u>Nature</u>,
<u>Lond</u>. 231: 50-52.

Nisbet, I. C. T. and A. F. Sarofim. 1972. Rates and routes
of transport of PCB's in the environment. <u>Environ</u>.
<u>Hlth</u>. <u>Persp</u>. 1: 21-33.

Norwicki, H. G. and A. W. Norman. 1972. Enhanced hepatic
metabolism of testosterone, 4 androstene-3, 17-dione

and estradiol-17β in chicken pretreated with DDT or
    PCB. Steroids 19: 85-99.
Peakall, D. B. 1975. PCB's and their environmental effects.
    CRC Crit. Revs. Environ. Contr. 5: 469-508.
Roesijadi, G., S. R. Petrocelli, J. W. Anderson, C. S. Giam,
    and G. E. Neff. 1976. Toxicity of polychlorinated
    biphenyls (Aroclor 1254) to adult, juvenile and larval
    stages of the shrimp Palaemonetes pugio. Bull. Environ.
    Contam. Toxicol. 15: 297-304.
Sanders, H. O. and J. H. Chandler. 1972. Biological magni-
    fication of a polychlorinated biphenyl (Aroclor 1254)
    from water by aquatic invertebrates. Bull. Environ.
    Contam. Toxicol. 7: 257-263.
Waiwood, K. G. and P. H. Johansen. 1974. Oxygen consumption
    and activity of the white sucker (Catostomus commersoni),
    in lethal and nonlethal levels of the organichlorine
    insecticide, Methoxychlor. Water Res. 8: 401-406.
Wildish, D. J. 1972. Polychlorinated biphenyls (PCB) in sea
    water and their effect on reproduction of Gammarus
    oceanicus. Bull. Environ. Contam. Toxicol. 7: 182-187.
Zitko, V. and P. M. K. Choi. 1971. PCB and other industrial
    halogenated hydrocarbons in the environment. Fish. Res.
    Bd. Canada, Tech. Rept. #272: 55 pp.

Survival of Larval and Adult Fiddler Crabs Exposed to Aroclor® 1016 and 1254 and Different Temperature-Salinity Combinations

F. JOHN VERNBERG,[1] M. S. GURAM,[2] and ARDIS SAVORY[1]

[1]Belle W. Baruch Institute
for Marine Biology and Coastal Research
and Department of Biology
University of South Carolina
Columbia, South Carolina 29208

[2]Department of Biology
Voorhees College
Denmark, South Carolina 29042

Polychlorinated biphenyls (PCBs) have been introduced into the global ecosystem (Risebrough et al., 1968). Although the toxicity of these substances has been demonstrated in various groups of organisms including rats, birds, fish, and invertebrates (see review of Peakall and Lincer, 1970), relative to some other pollutants, little is known.

Not until 1970 when Duke et al. reported the presence and toxicity of a PCB (Aroclor 1254) in Escambia Bay, Florida, was there published data on the harmful effects of PCBs on estuarine organisms. In a series of papers, Nimmo and associates (1970, 1971a and b, 1974) and Hansen et al. (1971, 1974) further investigated some of the effects of Aroclor 1254 on various estuarine organisms from this region.

The present study was undertaken to determine the relative influence of Aroclor 1254 ,which is no longer in production ,and Aroclor 1016 on the survival of larval and adult fiddler crabs, Uca pugilator, one of the dominant intertidal

---

*® Registered trademark, Montsanto Company, St. Louis,Missouri.

crabs inhabiting most of the estuaries from Massachusetts to Texas. Any adverse influence on this species by a foreign substance would indicate that this substance could adversely affect the marsh-estuarine ecosystem.

MATERIALS AND METHODS

To determine the comparative effects of Aroclor 1254 or 1016 on larvae of Uca pugilator (Bosc), the common sand fiddler crab, the following procedures were used. Ovigerous females were collected in the vicinity of the University of South Carolina Field Laboratory on the edge of the North Inlet Estuary, Georgetown, South Carolina. These animals were maintained in 4 1/2 inch finger bowls in the laboratory at 25°C and at a salinity of 30 °/oo. Each morning the bowls were checked to determine if free-swimming larvae had hatched. Uca pugilator has five larval (zoeal) stages and one megalopa stage. The megalopa then undergoes metamorphosis to the young adult stage. After each hatch, the degree of viability was determined by the larval response to light. A hatch with a low degree of viability, as indicated by a poor photoposi- tive response, was discarded. Viable hatches were pooled and zoeae were randomly selected for experimental purposes. At least three hatches were pooled for each specific experiment. In this manner, a better degree of genetic diversity was achieved. From this pooled sample, twenty zoeae were randomly selected and placed in a 4 1/2 inch finger bowl containing the desired concentration of either a PCB or a control solu- tion. A minimum of 200 zoeae were used per test solution. Each day the zoeae were counted, fed Artemia, and the water changed. The bowls containing zoeae were kept in constant temperature chambers on a 12 hr light-12 hr dark photoperiod. The work reported below was done using zoeae collected between May-September, 1973 and April-September, 1974.

Various concentrations of PCBs (Aroclor 1016 and Arcolor 1254, obtained from Montsanto Chemical Company) were used in different experiments. Since Aroclors are insoluble in sea water, acetone was selected as the carrier. Results of the following tests indicated no effect of acetone on survival.
1. After 5 days exposure to 80 ppm acetone (W/V) and 100 ppm acetone (V/V),the survival level was similar to that of the control group. Additional testing was done routinely on 11 experimental groups with the same general response resulting.
2. In a second test from three mass hatches, three 300 ml Erlenmeyer flasks were set up with each containing 300 zoeae, which were selected at random from the pooled mass hatches. The number of living zoeae were counted every

seventh day for a 21 day period. One flask served as a
control and contained only 30 °/oo filtered sea water, a
second contained 10 ppm acetone (V/V) in 30 °/oo filtered
sea water, and a third contained 1 ppm acetone (V/V) in
30 °/oo filtered sea water. After 21 days, 20-22% survived to the megalops stage in all three flasks. Since
we planned to conduct multiple stress experiments, using
temperature, salinity and a concentration of Aroclor as
variables, a series of experiments were performed at
different temperatures and salinities with 4 ppm acetone
(W/V). These results are summarized in Table 1. Again,

Table 1
The influence of 4 ppm acetone on zoeae of Uca
pugilator exposed to different combinations of
temperature and salinity for 72 hours.

Salinity	Temperature	72 hr Control % Survival	72 hr Acetone % Survival
20 °/oo	20°C	84	89
20 °/oo	25°C	90	93
20 °/oo	30°C	51	74
25 °/oo	20°C	88	85
25 °/oo	25°C	96	84
25 °/oo	30°C	78	75
30 °/oo	20°C	85	93
30 °/oo	25°C	92	96
30 °/oo	30°C	92	91
Average percent survival, all groups combined		84	87

acetone did not appear to adversely influence survival of
these zoeae. These results agree with those of Epifanio
(1971) who reported that 1 ppm acetone did not effect
development of Leptodius floridanus up to the megalopa
stage.
    In addition to studies on larval survival, the influence
of Aroclors on survival of the adult male and female U.
pugilator was examined.

Multiple Factor Interaction

    The influence of stressful but sublethal combinations of
temperature and salinity on the response of zoeae subjected
to sublethal concentrations of Aroclors was investigated during the summers of 1973 and 1974. In general 100 zoeae were
exposed to an Aroclor and 100 zoeae served as controls. Each
group was subjected to the same combinations of temperature
and salinity and an equal concentration of acetone. The zoeae
were kept in a 4 1/2 inch finger bowl (20 individuals per
bowl) in a constant temperature cabinet and exposed to a photoperiod of 12 hrs light and 12 hrs dark. A minimum of three
hatches were pooled and larvae were selected at random and
placed in either experimental or control conditions. Each
day the larvae were counted and mortality noted, fed Artemia,
and the aqueous media changed. Solutions were prepared daily
with sea water of desired salinity and temperature. The
stress combination included five experimental temperatures
(15, 20, 25, 30, and 35°C) and five salinities (15, 20, 25,
30, and 35 °/oo). Two concentrations of Aroclor 1016 and
Aroclor 1254 were used: 0.1 ppb and 1.0 ppb. Data were
analyzed for significant difference between the response of
controls and experimental groups using the standard student's
t-test.

RESULTS

    Figure 1 graphically illustrates the effect of different
concentrations of Aroclor 1254 on the survival of larvae
after 96 hrs exposure. If the survival level of the control
group is considered to be 100%, then the lethal concentration
(LC_{50}) for 96 hrs is approximately 10 ppb.
    Concentrations of 0.1 ppb and 1 ppb had little or no
influence on mortality while a concentration of 5 ppb
increased mortality by about 20% over that of control animals.
A 10 ppb solution of Aroclor 1254 resulted in a 57% reduction
in survival of larvae by 96 hrs, but a 1.0 ppb solution had
no effect on mortality. An increase in concentration to 50
ppb does not drastically increase larval mortality, but in
contrast, 500 ppb is extremely lethal to the larval population.
    In another experiment, Uca larvae were exposed to 5 ppb
Aroclor 1254 for varying periods of time to determine if a
potential toxicant would differentially effect other developmental stages. The results indicate that by day 14 the
survival level was reduced, statistically significant at the
1% level (Fig. 2). This observation would question the use
of the 96 hr level as a valid indicator of biological toxicity
for larvae. At 96 hrs 5 ppb had little effect on survival

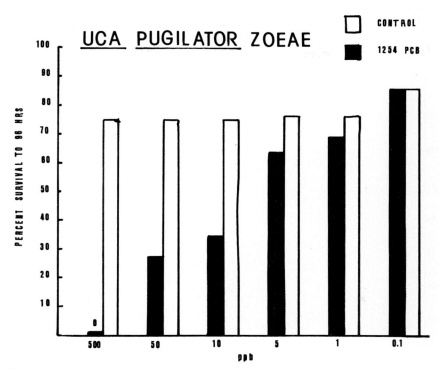

Fig. 1.   Survival of larval <u>Uca pugilator</u> exposed to various
concentrations of Aroclor 1254 for 96 hours.

(Fig. 1), but had a definite effect by day 14. By day 21 the
animals were in the megalopa stage and by day 29 in the first
crab stage. In the control group 41.5% of the larvae reached
the megalopa stage but only 20% of the Aroclor-exposed larvae
reached this late developmental stage. About 18% of the
larvae reached the first crab stage under control conditions,
while only 9% of the experimental group developed to first
crab stage. Therefore, Aroclor 1254 (5 ppb) appeared to
adversely influence development of fiddler crab larvae.
     The results of exposing first stage zoeae to different
concentrations of Aroclor 1016 or Aroclor 1254 for different
periods of time are graphically represented in Figures 3 and
4. The survival values for the experimental groups are
adjusted on the basis that the survival level of the control
group for each experimental group is considered to be 100%,
even though mortality in the control group occurred. Aroclor
1016 in concentrations of 5 ppb, 1 ppb, or 0.1 ppb had little
effect on mortality of zoeae until after an exposure of 120
hrs when 1 ppb resulted in 27% mortality which is not

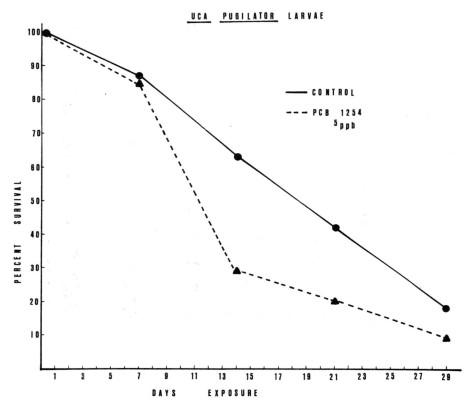

Fig. 2. Survival of larval <u>Uca</u> <u>pugilator</u> exposed to 5 ppb
Aroclor 1254 for various periods of time.

statistically significant (Fig. 3). After 168 hrs of exposure,
the survival level of animals exposed to 5 ppb and 1 ppb was
reduced to 66% and 61%, respectively. A concentration of 0.1
ppb had no visible effect on survival level. Aroclor 1254
also influenced survival rate (Fig. 4). After 96 hrs exposure,
a concentration of 10 ppb resulted in a mortality level of
55% and a similar response was found after 168 hrs exposure.
A concentration of 5 ppb also reduced the survival level at
96 hrs, but to a lesser degree. For example, at 96 hrs the
survivalship was 77%, at 120 hrs 65%, at 144 hrs 81%, and at
168 hrs 60%. This range in survival levels indicates the variability in results that can be expected in experiments dealing
with larvae. However, this range of 60 to 81% does suggest
that 5 ppb Aroclor 1254 has an effect on larval survival.

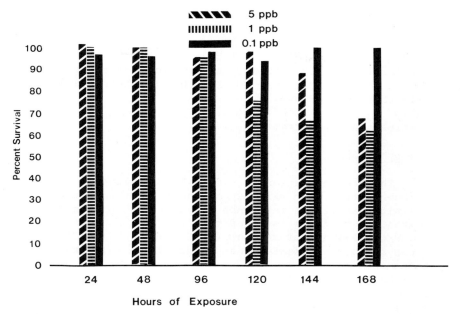

Fig. 3.   Survival of larval Uca pugilator exposed to various
concentrations of Aroclor 1016 for different periods
of time.

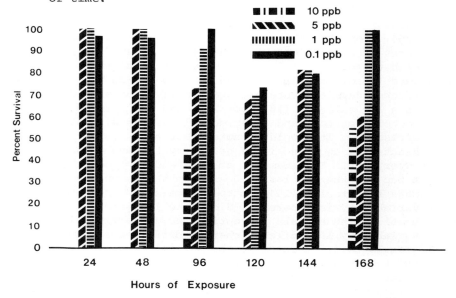

Fig. 4.   Survival of larval Uca pugilator exposed to various
concentrations of Aroclor 1254 for different periods
of time.

In contrast, 1 ppb and 0.1 ppb had a less pronounced effect on survival levels in that after 120 hrs exposure the survival level ranged from 70% to 100%.

A comparison of the relative effects of Aroclor 1254 and Aroclor 1016 would suggest that Aroclor 1254 exerts a toxic effect more rapidly than Aroclor 1016 but both leave similar effects after 168 hrs.

Multiple Factor Stress

0.1 ppb Aroclor

This concentration of either Aroclor 1254 or Aroclor 1016 in combination with various combinations of temperature and salinity had little influence on survivorship of one day old larvae. A tendency is noted that low (15 °/oo) and high (35 °/oo) salinities and low (15°C) and high (35°C) temperatures are more stressful than intermediate salinities and temperatures.

After a three day exposure to 0.1 ppb Aroclor 1254 tended to be more toxic than Aroclor 1016, especially at stressful combinations of temperature and salinity. An unexpected result was observed at 35°C and 35 °/oo in that the organisms exposed to the Aroclors survived better than the control animals. By day 4, the differential effect of these two Aroclors was more pronounced than that after three days of exposure. In general, Aroclor 1016 is less toxic than Aroclor 1254. For the most part the larvae survived best at temperatures of 25 and 30°C when the salinity was 25 °/oo or 30 °/oo. High temperature (35°C) was more stressful than low temperature (15°C) at salinities of 15 °/oo and 35 °/oo. The results after 5 days exposure are graphically represented in Figure 5. At a low salinity (15 °/oo) 15°C and 35°C proved to be lethal to both controls and experimental animals with 35°C being more stressful. It should be noted that there is variability in hatches of eggs and it is necessary to have a control for each experiment. For example, at 15°C and 15 °/oo, the control group for the Aroclor 1016 experiment was much more resistant than the Aroclor 1254 control animals. No reason for this variability is apparent. Since the laboratory technique for handling the animals is routine and uniformly followed, this variability may reflect inherent differences between individuals or differences in the physiological state of the eggs due to previous environmental expsoure. Statistically significant differences (at the 5% level) between the survival levels of controls and Aroclor-exposed were observed at the following experimental combinations: 1) Aroclor 1016, 15 °/oo and 15°C, 20 °/oo and 20°C, 25 °/oo and 30°C, 30 °/oo and

20°C, and 35°/oo and 15°C; 2) Aroclor 1254, 25 °/oo and 20°C,
25 °/oo and 25°C, 25 °/oo and 30°C, 30 °/oo and 20°C, and
35 °/oo and 15°C. Typically these combinations are those
which deviate from the optimal condition of 30 °/oo and 25°C.
Of this ten combination, in seven cases the Aroclor decreased
survival levels but in three combinations the presence of an
Aroclor increased survival. Since we don't know the physiological effect of Aroclors on fiddler crabs, no speculation
on the·reason for this response is given. By day 7 the survival levels of experimental and control groups were equally
reduced and significant differences between the effects of
Aroclor 1254 and 1016 were not uniformly apparent.

In general, 0.1 ppb Aroclor 1016 and Aroclor 1254 did not
appear to have a consistent effect on the survival of larval
Uca pugilator at all temperature-salinity combinations. At
those combinations where an effect was detected, Aroclor 1254
was usually more toxic.

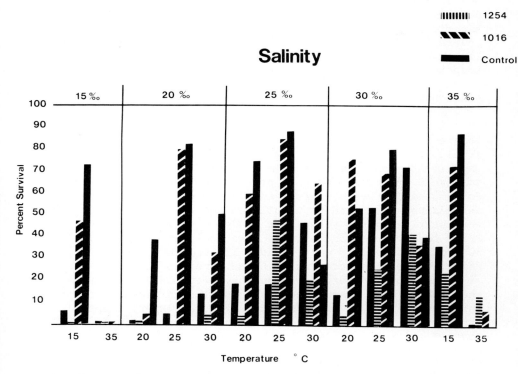

Fig. 5. Survival of larval Uca pugilator subjected to Aroclor
         1016 or 1254 and various salinity-temperature combina-
         tions for five days.

1.0 ppb Aroclor

At combinations of three temperatures (20, 25 and 30°C) and three salinities (20, 25, and 30 ^{o}/oo), a 1.0 ppb solution of Aroclor 1254 negatively influenced survival at 30°C and 20 and 30 o/oo on day 4 and at 20 o/oo and 30°C on day 5 (Fig. 6). By day 7 experimental and control groups were equally influenced. Of the three temperatures 30°C was the most stressful at the three salinities. These experiments were performed during the summer of 1973.

This experiment was repeated during the summer of 1974, but Aroclor 1016 was used. In general the level of survival was not as great as for those experiments of 1973. No definitive explanation is presented at this time. Conceivably these differences reflect yearly fluctuation in response as influenced by yearly differences in environmental conditions. Also, it is possible that the experimental technique varies slightly because of unavoidable differences. Because of the multiplicity of variables (some unknown) it is important to do comparative toxicity studies at the same time using similar experimental organisms. In general, Aroclor 1016 did not adversely effect the survival of larvae after 4 days exposure. If anything they tended to enhance survival slightly, a response not unlike that noted in Figure 5.

Adult Survival

In contrast to the response of larvae, adult male and female fiddler crabs are more resistant to high levels of Aroclor 1016 and 1254 than are the larvae. In separate experiments ten specimens of males and females were able to survive 0.1 ppb, 1.0 ppb, 10 ppb, or 100 ppb of Aroclor 1254 for at least 3 weeks. After the 3-week period of exposure, 5 animals were returned to unpolluted sea water and 5 animals were sacrificed for a separate study on tissue uptake of PCBs. Prior to exposure, the animals were acclimated at 25°C, 30 ^{o}/oo filtered sea water, and a light:dark cycle of 12:12 daily. Exposures were run at these conditions.

Additional observations on the effect of PCBs were made when adults were placed under stressful conditions of temperature and salinity. At 25°C and 5 ^{o}/oo sea water, 5 ppb Aroclor 1016 or 1254 did not increase the mortality level over that of control animals after a 4-week exposure. Nor did two weeks at 35°C, 5 ^{o}/oo and 5 ppb Aroclor 1016 or 1254 cause mortality above that of a control group of animals (35°C and 5 ^{o}/oo). However, at 10°C and 5 ^{o}/oo, animals exposed to 5 ppb Aroclor 1016 reached the 50% mortality level between 21 and 28 days. In contrast Aroclor 1254 had no effect on the

IIIII 1254

▇ Control

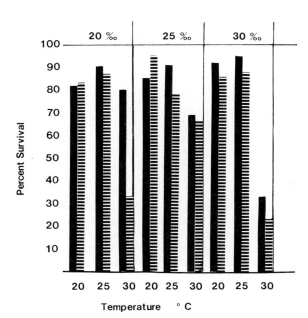

Fig. 6. Survival of larval <u>Uca</u> <u>pugilator</u>
exposed to 1 ppb Aroclor 1254
and different salinity-tempera-
ture combinations for five days.

mortality level. Ten males and 10 females were used in each
control and experimental group. No differences in response
of the two sexes were noted. However, another experiment run
at 7°C and 5 °/oo with or without 5 ppb Aroclor 1016 or 1254
produced more dramatic results. Under these conditions the
50% mortality was now reached after 5-8 days exposure time.
No uniform difference between Aroclor-exposed and unexposed
animals was observed.

When this experiment was repeated with the concentration
of Aroclors increased to 50 ppb and the acclimation time
increased to 5 weeks, Aroclor 1254 was more toxic than Aroclor
1016. The 50% mortality level for males exposed to Aroclor

1254 was 2 days compared to 4-6 days for Aroclor 1016.
Females survived (50% levels) Aroclor 1016 for 7 days but only
4 days in Aroclor 1254. The additional two weeks of acclimation decreased slightly the survival time of the control
animals. For example, the males were relatively uneffected
but the 50% mortality point was reached by females by day 4
when acclimated for five weeks and between day 7 and 8 when
acclimated three weeks. At 35°C and 5 °/oo, 50 ppb Aroclor
1016 or 1254 did not influence survival levels in that little
mortality was observed after exposure for 28 days.

DISCUSSION

Previous studies have demonstrated that Aroclor 1254 has
deleterious effects on certain estuarine organisms. Mosser
et al. (1972) reported species differences in survival of
phytoplankton to PCBs.   Duke et al. (1970) reported that
juvenile shrimp were more sensitive to Aroclor 1254 than
juvenile pinfish. A concentration of 5 ppb killed 72% of a
population sample of juvenile shrimp within 20 days. Although
10 ppb appeared to have no lethal effect on juvenile pinfish,
all of the juvenile shrimp died within 48 hrs and shell
growth of oysters was completely inhibited after 96 hrs exposure. In the present study adult fiddler crabs responded
like the pinfish in that they could survive 100 ppb Aroclor
1254 for at least three weeks when maintained at 25°C and
30 °/oo filtered sea water.
    Not all stages in the life cycle of a species responded
similarly to foreign substances. Nimmo et al. (1971a) found
that juvenile shrimp are more sensitive than adults to Aroclor
1254. For example, 51% of a sample of juvenile Penaeus
duorarum died within 15 days exposure to 0.94 ppb while in
contrast it took 35 days to kill 50% of an adult population at
a stronger concentration (3.5 ppb). As shown in the present
study, larval fiddler crabs are also more sensitive than the
adult form. One example will illustrate this point: the LC_{50}
for 96 hrs is 10 ppb for larvae while adults can withstand
100 ppb for at least three weeks.
    In estuaries, organisms are subjected to numerous widely
varying environmental factors. Hence the influence of a
suspected pollutant may be different at periods of extreme
fluctuation in the normal environmental factors. In general,
multiple factor interaction has been shown to reduce the zone
of compatibility (the biokinetic zone) of a species (Vernberg
and Vernberg, 1970; Vernberg et al., 1974). Nimmo and Bahner
(1974) demonstrated that a sublethal concentration of Aroclor
1254 at a constant salinity of 30 °/oo and 25°C became lethal
to adult shrimp when the salinity was reduced (50% mortality

after 8 hrs of exposure to 10 ^{o}/oo and 7 ^{o}/oo salinity). The present study expands on this general theme and includes temperature and stage of life cycle as variables. Both larvae and adult fiddler crabs demonstrated this same phenomenon with detailed analysis described in the results section. It should be noted that, although the interaction of temperature and salinity did influence the response to Aroclors, not all combinations were effective nor were the results always predictable.

There is little published data on the comparative effects of different Aroclors on organisms. Lichtenstein et al. (1969), working with house flies and fruit flies, found that Aroclor 1221 was more toxic than Aroclor 1268. Hansen et al. (1974) working on oysters, brown shrimp, grass shrimp, and pinfish, concluded that the toxicity, uptake, and retention of Aroclor 1016 are similar to other PCBs. In general, our results would suggest that Aroclor 1016 is less toxic than Aroclor 1254. However, under several combinations of temperature and salinity, Aroclor 1016 was more toxic.

ACKNOWLEDGMENTS

This study was supported by Environmental Protection Agency grant R-802228 to Voorhees College and the University of South Carolina. The dedication to this project by Ms. Sharon T. Maier is gratefully acknowledged.

LITERATURE CITED

Duke, T. W., J. I. Lowe, and A. J. Wilson, Jr. 1970. A poly-
    chlorinated biphenyl (Aroclor 1254) in the water, sedi-
    ment, and biota of Escambia Bay, Florida. Bull. Environ.
    Contam. Toxicol. 5: 171-180.
Epifanio, C. E. 1971. Effects of dieldrin in seawater on the
    development of two species of crab larvae, Leptodius
    floridanus and Panopeus herbstii. Mar. Biol. 11: 356-362.

Hansen, D. J., P. R. Parrish, and J. Forester. 1974. Aroclor
    1016: Toxicity to and uptake by estuarine animals.
    Environ. Res. 7: 363-373.
_____, _____, and J. I. Lowe, A. J. Wilson, Jr., and P. D.
    Wilson. 1971. Chronic toxicity, uptake, and retention
    of Aroclor 1254 in two estuarine fishes. Bull. Environ.
    Contam. Toxicol. 6: 113-119.
Lichtenstein, E. P., K. R. Schultz, T. W. Fuhremann, and
    T. T. Liang. 1969. Biological interaction between plas-
    ticizers and insecticides. J. Econ. Entomol. 62: 761-765.

Mosser, Jerry L., Nicolas S. Fisher, Tzu-Chiu Teng, and Charles F. Wurster. 1972. Polychlorinated biphenyls: toxicity to certain phytoplankters. Science 175: 191-192.

Nimmo, D. R. and L. H. Bahner. 1974. Some physiological consequences of polychlorinated biphenyl- and salinity-stress in Penaeid shrimp. In: Pollution and Physiology of Marine Organisms, pp. 427-443, ed. by F. J. Vernberg and W. B. Vernberg. Academic Press: New York.

_____, A. J. Wilson, Jr. and R. R. Blackman. 1970. Localization of DDT in the body organs of pink and white shrimp. Bull. Environ. Contam. Toxicol. 5: 333-341.

_____, R. R. Blackman, A. J. Wilson, Jr.,and J. Forester. 1971a. Toxicity and distribution of Aroclor 1254 in the pink shrimp Penaeus duorarum. Mar. Biol. 11: 191-197.

_____, _____, _____, and _____. 1971b. Polychlorinated biphenyl absorbed from sediments by fiddler crabs and pink shrimp. Nature, Lond. 231: 50-52.

Peakall, David B. and Jeffrey L. Lincer. 1970. Polychlorinated biphenyls. Another long-life widespread chemical in the environment. BioScience 20: 958-964.

Risebrough, R. W. P. Rieche, D. B. Peakall, S. G. Herman, and M. N. Kirven. 1968. Polychlorinated biphenyls in the global ecosystem. Nature 220: 1098-1102.

Vernberg, F. John and Winona B. Vernberg. 1970. The Animal and the Environment. New York: Holt, Rinehart and Winston. 398 p.

Vernberg, Winona B., Patricia J. DeCoursey, and James O'Hara. 1974. Multiple environmental factor effects on physiology and behavior of the fiddler crab, Uca pugilator. In: Pollution and Physiology of Marine Organisms, pp. 381-425, ed. by F. J. Vernberg and W. B. Vernberg. New York: Academic Press.

PCB Levels in Certain Organs of Some Feral Fish from New York State

PHYLLIS H. CAHN, JACK FOEHRENBACH, AND WILLIAM GUGGINO

Long Island University, C. W. Post College
Marine Science Department
Greenvale, New York 11548

Although PCB's have been used widely since the 1930's (Penning , 1930), it was not until the middle of the 1960's (Jensen, 1966), that they were discovered as ecosystem contaminants. During the past year, there has been much publicity regarding the hazardous concentrations of this pollutant in Hudson River, New York fish, especially in the vicinity of the General Electric plant outfall north of Albany. PCB concentrations of 20 to several hundred ppm were reported in fish (rockbass, smallmouth bass, suckers, and walleye) collected near the plant. These levels are far in excess of the Food and Drug Administration allowable limits of 5 ppm for human consumption (Reid, 1975). In general, it has been reported that the highest PCB residues in fish occur in highly polluted coastal waters, such as Tokyo Bay and Long Island Sound (Hammond, 1972); recent confirmation of this for Long Island Sound is lacking (Foehrenbach, unpublished data).

For reasons yet to be determined, PCB's seem to show a different pattern of movement through marine ecosystems than does DDT (Harvey, 1974), and much of this pattern is still controversial. These differences occur despite similarities in solubility, molecular weight, and other physical constants of both substances (Gustafson, 1970), and even though they show comparable physiological properties, such as their accumulation in tissues high in lipids and in organs such as the liver (Couch, 1975).

51

MATERIALS AND METHODS

The fish studied in this report were selected for several reasons. In the case of the striped bass (Morone saxatilis), PCB's were implicated as dangerous contaminants of the eggs, and possibly responsible for low larval survival (Hansen et al., 1973). Because of the current designation of bluefin tuna (Thunnus thynnus) as a possible endangered species, studies on their biochemistry were initiated. Spiny dogfish (Squalus acanthias) was examined because sharks collected in the Wood's Hole, Massachusetts area contained a high concentration of DDT (Foehrenbach, unpublished data).

The techniques used for tissue preparation, extraction, and gas chromatography were those given in the "Pesticide Analytical Manual for BCF Contracting Agencies". The quantification of the major peaks was done using standards obtained from the Environmental Protection Administration Depository in the Research Triangle Park, North Carolina. There are considerable variations in techniques for PCB quantification (Chau and Sampson, 1975): the method we used involved determination of total peak areas and a comparison of these areas to that of a standard (Zitko, 1971).

When this work was started, no one had reported specific PCB isomers in fish. In the striped bass studies, we examined tissue concentrations of two dichlorobiphenyls, one tetrachlorobiphenyls, three hexachlorobiphenyls, and three heptachlorobiphenyls (Guggino, 1973).

The concentration of AROCLOR 1254 has often been used as a reference standard for PCB's present in tissues; in our studies we compared the concentration of AROCLOR 1254 with that of the total PCB's present and found that the latter values were generally higher. However, the use of a single AROCLOR as a standard may be somewhat misrepresentative (Beezhold and Stout, 1973).

Wherever possible, concentrations of PCB's in gonads, liver, mesenteric fat, viscera, and muscle were determined.

RESULTS

The total PCB concentrations, based on the means of the sum of all nine isomers present, were highest in the mesenteric fat of post-spawning female striped bass (Table 1). The ovaries showed somewhat a reverse pattern in that much of their PCB's were lost following spawning. The liver showed the lowest concentrations of PCB's. Similar trends in the distribution of other organochlorines were found; however the liver seemed to show somewhat higher values of DDE. Because

Table 1

A comparison of concentrations of total PCB's and other organochlorines in mature female striped bass (Morone saxatillis)[a].

	Total PCB's[c] (ppm)		AROCLOR 1254 (ppm)		Other Organochlorines[d] (ppm)	
	Range	Mean	Range	Mean	Range	Mean
PRE-SPAWNING[b]						
ovary	1.23-16.90	4.39	0.60-15.80	3.52	0.16-5.08	2.23
liver	0.10- 2.49	1.38	1.19- 3.32	2.12	0.74-4.56	2.13
fat[e]	2.01-14.90	6.08	2.84- 8.96	5.68	0.55-22.50	7.36
POST-SPAWNING[f,g]						
ovary	0.10- 9.53	1.49	0.10- 8.91	2.29	0.26- 6.67	1.84
liver	0.43- 6.69	1.60	1.08-10.00	3.29	0.51-16.0	3.73
fat	0.10-40.0	10.20	0.10-30.30	11.93	1.43-38.80	12.35

[a]In mg/kg wet weight or ppm.

[b]Hudson River, N. Y., spring 1972; 10 fish, ages 8-13 years.

[c]Including the isomers contained in AROCLOR 1254.

[d]DDE was the major component; lindane and aldrin peaks were also calculated.

[e]This refers to mesenteric fat around the viscera.

[f]Long Island, N. Y., fall 1972; 10 fish, ages 6-12 years.

[g]In one post-spawning, seven-year old fish, the ppm values for fat total PCB's were 276, for AROCLOR 139, and for other organochlorines 133.

of the inhomogeneity of the data, no statistical analyses were
made. There was much variability in organ weights in fish of
similar age and size (Table 2); such variability is common in
feral species. In general, a higher concentration of PCB
isomers containing low percentages of chlorine was noted.

Table 2
Organ weights of female striped bass used for pollutants
analyses (data given in Table 1).

PRE-SPAWNING[a]

Age (yrs.)[b]	Weight (kg)	Total Organ Weight (gms)		
		Liver	Fat[c]	Ovary
8	8.1	21.2	7.7	42.6
9	9.1	16.3	9.3	64.2
9	9.1	29.4	20.0	21.3
9	10.0	17.6	15.3	20.2
12	11.8	66.2	15.3	72.3
12	13.2	65.0	15.4	89.2
12	13.6	59.8	17.8	101.2
12	14.6	59.0	25.2	64.2
13	16.4	69.8	17.5	126.2
13	16.8	79.8	26.2	142.2

POST-SPAWNING[d]

6	2.9	9.4	6.6	3.8
6	4.1	33.0	2.0	21.0
7	4.6	24.3	3.8	4.9
7	5.5	51.6	5.7	9.7
9	7.3	45.0	3.3	41.2
9	10.8	54.6	36.8	18.9
10	8.6	58.3	9.0	28.9
12	13.2	58.8	59.0	38.5
12	13.6	63.3	39.1	38.0
12	14.6	59.5	44.3	18.7

[a]Hudson River, N. Y., spring 1972.
[b]Age determination was based on size data from Mansueti, 1961.
[c]Mesenteric fat around the viscera.
[d]Long Island, N. Y., fall 1972.

Some additional samples of striped bass collected at a
date were found to contain very high concentrations of AROCLOR
1254 in the liver (Table 3). The values however were not too
much higher than the maximal levels for AROCLOR 1254 in the
Fall 1972 post-spawning striped bass (Table 1).

Of the bluefin tuna tissue tested, PCB levels in all
organs sampled, in both sexually immature and mature fish,

were all below the 5 ppm limits for human consumption (Table 4). Muscle of the younger fish had slightly higher levels than the viscera.

Table 3
Concentration of AROCLOR 1254 in livers of some New York area striped bass, summer 1975.

Approx. Age (yrs.)	No. Fish	Collection Site	Liver AROCLOR 1254[a] ppm
3-4	6	Upper N. Y. Bay	11.05
5-9	4	East of Lower, N. Y. Bay (Rowes Shoal)	13.44
5-6	6	Montauk Pt., N. Y.	8.89

[a]Higher concentrations were also found in ovaries and in red muscle in some of the fish sampled.

Table 4
Concentrations of AROCLOR 1254[a] in Atlantic bluefin tuna (Thunnus thynnus) expressed as mg/kg wet weight of tissue (ppm).

Fish Weight(kg)	White Muscle[c]	Liver,Spleen,Pyloric Caecae[d]
ONE YEAR OLD (sexually immature)[b]		
3.41	0.19	0.12
3.41	0.50	0.17
3.53	0.37	0.10
3.53	0.92	0.12
3.55	0.58	0.32
3.97	0.72	0.17
EIGHT TO TEN YEARS OLD (sexually mature females)[e]		
268.00	1.35	1.15
269.00	1.28	1.45

[a]AROCLOR 1254 was used as the standard; only the PCB's present in this AROCLOR are shown.
[b]Caught offshore in vicinity of Babylon, N. Y., July 1975.
[c]Mostly white muscle with some red muscle; 20 gms processed per fish.
[d]Organs pooled; 20 gms processed per fish.
[e]Caught offshore in vicinity of Portland and Boothbay Harbor, Maine.

In the spiny dogfish, the organs analyzed contained relatively low levels of PCB's; somewhat higher concentrations were found in the liver than in the other tissues (Table 5).

Table 5
Concentrations of AROCLOR 1254[a] in spiny dogfish (Squalus acanthias)[b] expressed as mg/kg wet weight of tissue (ppm).

Organ	No. of Determinations	Range	Mean
Liver[c]	4	0.78-1.12	0.92
Spleen[d]	2	0.12	--
Kidney[e]	2	0.12-0.19	0.15
Testes[f]	1	0.27	--
Ovary[g]	1	0.43	--
Muscle[h]	4	0.22-0.27	0.25

[a]AROCLOR 1254 was used as the standard; only the PCB's present in this AROCLOR are shown.
[b]Collected from Atlantic Ocean, three miles south of Fire Island Inlet, N. Y., June 1975. Total length ranged from 53 to 72 cm; 44 females and 16 males, all sexually mature and probably several years old.
[c]Of the 60 fish collected, livers of 15 were pooled per 20 gm sample.
[d]Spleens of 30 fish were pooled per 20 gm sample.
[e]Kidneys of 30 fish were pooled per 20 gm sample.
[f]Testes of 16 males were pooled.
[g]Ovaries of 44 females were pooled.
[h]Muscle per 15 fish was pooled.

DISCUSSION

Although there has been a curtailment in the production of PCB's in both North America and Europe with a concomitant decrease in the concentration present in surface waters (Harvey et al., 1974), the persistence of these substances in organisms and sediments is still very great. No doubt we will have a PCB problem for a very long time. Some of the implications for coastal organisms are summarized in Duke and Dumas (1974).

The literature on PCB's in fish leaves many questions unanswered: Are concentrations greater in pelagic or in benthic species? Is there any food chain magnification of PCB's in fish as occurs with other chlorinated hydrocarbons? Is

the gill route of PCB entry more dangerous to fish than entry
via food? Do older fish concentrate more PCB's than younger
fish? Before discussing these questions, the results of the
present study will be compared with those in the literature.
No attempt will be made to review the PCB literature since
many references can be found in Anderson et al. (1974).

PCB's were considered to be responsible for the poor
reproductive success of Chesapeake Bay striped bass since eggs
contained 2.5 to 8.7 ppm of AROCLOR 1254 (Hansen et al., 1973).
These values are comparable to those we obtained in the
ovaries of sexually mature striped bass. Although these
females were collected in the New York area (pre-spawning fish
were from the Hudson River and the post-spawners were from
the south shore of Long Island), they probably contained mixtures of populations, including some from the Chesapeake area.
In the Atlantic Ocean, there is some mixing of striped bass
following the spawning and migration to the sea of the Hudson
River populations and mixing of those fish from rivers that
empty into the Chesapeake (Hitron, 1975).

The fact that the ovaries of post-spawning females contained a lower concentration of PCB's compared with the pre-
spawning ovaries, does not permit us to conclude that the
reduction occurs because the PCB's are transferred to the eggs.
Experimental work with tracers will be necessary to determine
the fate of ovarian concentrations of PCB's. The origin of
the increased concentrations of PCB's in the mesenteric fat
of post-spawning females cannot be determined without additional studies. After spawning, the fish move down the river
and pass through the heavily polluted waters of New York
Harbor. These polluted pathways, as well as the sojourn in
the polluted Hudson River waters may account for the accumulations noted. Of general interest to this problem is the
comment of Duke and Wilson (1971) that in many pelagic marine
fish, the pre-spawning gonads contain higher pesticide concentrations than the post-spawning gonads.

Prior to this past summer, there was very little evidence
that striped bass contained dangerous levels of PCB's. However, "Stripers Unlimited" (a non-profit service club for
striper fishermen) has shown concern with reports of possibly
high levels of PCB's in striped bass (Pond, personal communication). Other fish in the Hudson River contained high concentrations of PCB's: for example, whole body residues in
goldfish, white perch, and the largemouth bass ranged from
4.8 to 9.5 ppm (Henderson et al., 1971).

There have been sparse reports in the literature on the
concentration of PCB's in tuna: blackfin tuna (Thunnus
atlanticus) muscle contained very low concentrations, e.g.
43 ppb of AROCLOR 1260 was present in Gulf of Mexico and
Caribbean blackfins (Giam et al., 1972). In tissues of the

little tuna (<u>Euthynnus</u> <u>alleteratus</u>), collected from similar
locations, muscle and gonad levels of the same AROCLOR compound were as low as in blackfin; liver concentrations were
considerably higher being 153 ppb (Giam <u>et</u> <u>al</u>., 1972). In
other scombrids, such as in Atlantic mackerel (<u>Scomber</u>
<u>scombrus</u>), whole body residues of PCB's were found to be 0.35
ppm (Zitko, 1972).
      Concentrations of PCB's in elasmobranch liver tissues
reflect the fact that PCB's tend to accumulate in tissues with
high lipid levels (Jensen <u>et</u> <u>al</u>., 1969). For example, levels
of AROCLOR 1254 as high as 218.0 ppm were reported in the
livers of the white shark, <u>Carcharodon</u> <u>carcharias</u> (Zitko,
1972); our studies on spiny dogfish also found higher liver
levels of AROCLOR as compared with muscle and other organs.
      Physiological differences between pelagic and benthic
fish result in differences in sites of lipid concentration:
for example, most of the pelagic bony fish tend to concentrate
their lipids subdermally, and in red muscle tend to have non-
fatty livers during most of their life cycles. Many benthic
teleosts and also the elasmobranchs, store most of their
body lipids in the liver, and have less muscle fat. This may
not be true of some of the sharks with much red muscle, such
as the makos. Elasmobranch lipids are unusual in comparison
with teleosts; many differences have been described including
some major dissimilarities in chemical properties (Love, 1970).
Many of these metabolic variations seem to be reflected in the
different PCB storage patterns. One cannot really say, therefore, whether pelagic or benthic species concentrate greater
quantities; too many other factors are involved.
      As for the problem of food chain magnification, there
seems to be insufficient data for any conclusive opinion.
According to one study, the concentration of PCB's in mixed
marine plankton was about the same as in plankton-feeding fish
(Harvey <u>et</u> <u>al</u>., 1974). Another study on a laboratory population of the mosquito-feeding fish, <u>Gambusia</u>, concluded that
there was some magnification; the data however showed magnification from water to mosquito, but not on the next trophic
level (Metcalf <u>et</u> <u>al</u>., 1975). It is not clear why PCB's do
not show the food chain magnification effects similar to
those found for closely related chlorinated hydrocarbons,
such as DDT. The differences may be related to the greater
metabolic degradability of some of the PCB's compared with
DDE, for example (Metcalf <u>et</u> <u>al</u>., 1975).
      In regard to gill route entry, some investigators believe
that lipid soluble PCB's diffuse from the water into the
exposed gill membranes; for example, they have found that high
concentrations when present in the water seem to be more
toxic than when present in food (Gruger <u>et</u> <u>al</u>., 1975). Data
on gill concentrations of PCB's have not been reported, but

respiratory impairment was demonstrated (Anderson et al., 1974).

Is there an age related differential in regard to the harmful effects of PCB's? It has been found that AROCLOR 1254 was more toxic to fry of the minnow (Cyprinodon variegatus) than it was to juveniles, adults, or to the fertilization success of eggs of that species. Adults, while not killed by high concentrations of the pollutant, showed lethargy and a greater incidence of fin rot disease than non-exposed fish. Concentration factors for the pollutant were found to be almost the same for fry and for adults (Schimmel et al., 1974).

Other physiological effects on fish have been demonstrated including liver degeneration and liver glycogen depletion (Couch, 1975) and pancreatic lesions (Parrish et al., 1974). In instances of chronic exposure, behavioral changes that led to greater susceptibility to predation were found (Hansen et al., 1974).

In conclusion, the results reported in the present paper demonstrated that bottom living teleosts are like the elasmobranchs in regard to the tendency to accumulate PCB's in the liver, a major lipid storage site. Pelagic teleosts instead accumulate more of this pollutant in the muscle where more of their lipid is stored.

ACKNOWLEDGMENT

Thanks are extended to the New York State Department of Environmental Conservation, Stony Brook, N. Y., where Jack Foehrenbach, Senior Chemist in the Department and also an Adjunct Professor at Long Island University, conducted the chemical analyses on all of the samples other than the 1972 striped bass tissue. The latter tissue was analyzed by William Guggino in the course of his Master's thesis research at Long Island University. He is presently a pre-doctoral candidate at the University of North Carolina, Chapel Hill, North Carolina.

The bluefin tuna tissue was collected in conjunction with the extensive biological studies on this species being conducted by Mr. Fred Berry, National Marine Fisheries Service, Miami, Florida, and was kindly made available for PCB analyses.

LITERATURE CITED

Anderson, J. W., M. M. Neff, and S. R. Petrocelli. 1974.
    Sublethal effects of oil, heavy metals, and PCB's on

marine organisms. In: Survival in Toxic Environments, pp. 83-129, ed. by M. A. Q. Khan and J. P. Bederka, Jr. New York: Academic Press.

Beezhold, F. C. and V. F. Stout. 1973. The use and effect of mixed standards on the quantitation of PCB's. Bull. Envir. Contamin. Toxicol. 10: 10-15.

Chau, A. S. Y. and R. C. J. Sampson. 1975. Electron capture gas chromatographic methodology for the quantification of polychlorinated biphenyls: survey and compromise. Environ. Letters 8: 89-102.

Couch, J. A. 1975. Histopathological effects of pesticides and related chemicals on the livers of fishes. In: The Pathology of Fishes, pp. 559-584, ed. by W. Ribelin and G. Migaki. Wisconsin: Univ. Wisconsin Press.

Duke, T. W. and D. P. Dumas. 1974. Implications of pesticide residues in the coastal environment. In: Pollution and Physiology of Marine Organisms, pp. 137-164, ed. by F. J. Vernberg and W. B. Vernberg. New York: Academic Press.

_____ and A. J. Wilson. 1971. Chlorinated hydrocarbons in the livers of fishes from the Northwest Pacific Ocean. Pestic. Monitor. Jour. 5: 228-232.

Giam, C. S., A. R. Hanks, and R. L. Richardson. 1972. DDT, DDE, and polychlorinated biphenyls in biota from the Gulf of Mexico and the Caribbean Sea. Pestic. Monitor. Jour. 6: 139-143.

Gruger, E. H., N. L. Karrick, A. I. Davidson, and T. Hruby. 1975. Accumulation of 3,4,3',4'-tetrachlorobiphenyl and 2,4,5,2'4'5'- and 2,4,6,2',4'6'-hexachlorobiphenyl in juvenile coho salmon. Env. Sci. Tech. 9: 121-127.

Guggino, W. 1973. Identification of polychlorinated biphenyl isomers in Morone saxatilis by gas chromatography. Master's Degree thesis, Long Island University, C. W. Post College, N. Y.

Gustafson, C. G. 1970. PCB-prevalent and persistent. Environ. Sci. Tech. 4: 817-819.

Hammond, A. L. 1972. Polychlorinated biphenyls - chemical pollution. Science 175: 155-156.

Hansen, D. J., S. C. Schimmel, and J. Forester. 1973. AROCLOR 1254 in eggs of sheepshead minnows: effect on fertilization success and survival of embryos and fry. Proceedings of 27th Annual Conf. Southeastern Assn. Game and Fish Commissioners, pp. 420-426.

Hansen, D. J., P. R. Parrish, and J. Forester. 1974. AROCLOR 1016: Toxicity to and uptake by estuarine animals. Envir. Res. 7: 363-373.

Harvey, G. R., H. P. Miklas, V. T. Bowen, and W. G. Steinhauer. 1974. Observations on the distribution of chlorinated hydrocarbons in Atlantic Ocean organisms. Jour. Mar. Res. 32: 103-118.

Henderson, C., A. Inglis, and W. I. Johnson. 1971. Organochlorine insecticide residues in fish - fall 1969. Pest. Monitor. Jour. 5: 1-11.

Hitron, J. W. 1975. Serum transfer in phenotypes in striped bass, Morone saxatilis, from the Hudson River. Ches. Sci. 15: 246-247.

Jensen, S. 1966. Report of a new chemical hazard. New Scientist 32: 612.

_____, A. G. Johnels, M. Olson, and G. Otterlind. 1969. DDT and PCB in marine animals from Swedish waters. Nature 224: no. 5216: 247-250.

Love, R. M. 1970. The Chemical Biology of Fishes, pp. 21-37, 142. New York: Academic Press.

Mansueti, R. J. 1961. Age, growth, and movements of the striped bass, Roccus saxatilus, taken in size selective fishing gear in Maryland. Ches. Sci. 2: 9-36.

Metcalf, R. L., J. R. Sanborn, P. Lu, and D. Nye. 1975. Laboratory model ecosystem studies of the degradation and fate of radio-labeled tri-, tetra-, and pentachlorobiphenyls compared with DDE. Arch. Envir. Contam. Toxic. 3: 151-165.

Parrish, P. R., D. J. Hansen, J. N. Couch, J. M. Patrick, Jr., and G. H. Cook. 1974. Effects of the polychlorinated biphenyl, AROCLOR 1016, on estuarine animals. Assoc. South. Biol. Bull. 21: 74.

Penning, C. H. 1930. Physical characteristics and commercial possibilities of chlorinated diphenyl. Ind. Eng. Chem. 22: 1180-1182.

Reid, O. 1975. Press Release from New York State Commissioner's Office dated 9/8/75 describing the high levels of PCB's in the vicinity of General Electric's Fort Edward, N. Y., and Hudson Falls, N. Y., plants.

Schimmel, S. C., D. J. Hansen, and J. Forester. 1974. Effects of AROCLOR 1254 on laboratory-reared embryos and fry of sheepshead minnows (Cyprinodon variegatus). Trans. Amer. Fish. Soc. 103: 582-586.

Zitko, V. 1971. Polychlorinated biphenyls and organochlorine pesticides in some freshwater and marine fishes. Bull. Environ. Contamin. Toxicol. 6: 464-470.

_____. 1972. Contamination of the Bay of Fundy-Gulf of Maine area with PCB's, polychlorinated terphenyls, chlorinated dibenzodoxins and dibenzofurans. Envir. Health Persp. April, 47-50.

DDT Inhibits Nutrient Absorption and Osmoregulatory Function in *Fundulus heteroclitus*

DAVID S. MILLER and WILLIAM B. KINTER

Mount Desert Island Biological Laboratory
Salsbury Cove, Maine   04672

Fish are particularly sensitive to organochlorine pollutants, such as DDT (Holden, 1973), and recent studies with marine teleosts have suggested that one aspect of DDT toxicity is disruption of osmoregulatory transport in intestine and gill (Kinter et al., 1972; Leadem et al., 1974). A probable site of DDT action is the membrane-bound enzyme, sodium, potassium-activated adenosine triphosphatase (Na, K-ATPase), which appears to mediate two processes in transporting epithelia, intracellular ion regulation and transcellular Na transport. In intestine, uphill absorption of essential nutrients, e.g., amino acids, sugars and water, is Na-dependent and is believed to involve the following mechanisms: 1) organic solutes cross the brush border accompanied by Na (co-transport) and utilize the energy stored in the Na-gradient (low intracellular Na, high extra-cellular Na) to power uphill transport (Schultz and Curran, 1970); 2) water crosses the epithelium in response to osmotic gradients created by transcellular Na transport into the interstitium (Schultz and Curran, 1968). Clearly, DDT inhibition of intestinal Na, K-ATPase could affect both nutrient and osmoregulatory transport. In fact, Janicki and Kinter (1971a) have already reported inhibition of water absorption and Na, K-ATPase activity in eel intestine exposed to DDT in vitro. In the present study we have demonstrated inhibition of intestinal water and amino acid transport and Na, K-ATPase activity in killifish exposed to DDT in vivo.

MATERIALS AND METHODS

Killifish (Fundulus heteroclitus) were captured in an estuary on Mount Desert Island, Maine, during the summer of 1975 and held in running sea water (SW) at 13-15°C. After at least two weeks SW maintenance, fish were assigned to either control or experimental treatments (5-10 fish each), transferred to aluminum pans containing 2 liters of SW and maintained at 15°C without aeration. Experimental fish had p,p'-DDT (99+%, Aldrich Chemical Company) in ethanol (ETOH) added to the SW; the final DDT concentrations were 0.05-5 ppm and the final ETOH concentration was 0.1%. Controls were exposed to 0.1% ETOH (ETOH controls) or SW alone (untreated controls). After 1-24 h of exposure in vivo, blood samples were drawn by heart puncture with lightly heparinized glass capillary tubes, and the plasma spun off and analyzed for Na and K by flame photometry. Fish were decapitated and the proximal one-third of the intestine excised, washed, and processed for enzyme assay or in vitro transport studies. For distribution experiments, fish were exposed to ^{3}H-DDT (New England Nuclear Corporation), and samples of intestinal mucosa, gill, liver, and brain were solubilized and counted using standard liquid scintillation procedures. Preliminary experiments demonstrated that intestinal water and amino acid transport rates and Na, K-ATPase activities for the untreated controls maintained in aluminum pans (above) did not differ from the corresponding values for fish maintained in running SW.

For ATPase assay, intestinal tissue was cut into flat sheets and the mucosa removed by scraping with a glass slide. Mucosal scrapings were homogenized (10-25 mg tissue/ml) in a solution containing 0.25 M sucrose and 5 mM EDTA, and aliquots of homogenate (0.5 ml) were frozen rapidly at dry-ice temperatures, freeze-dried at -20°C for 24 h, and stored at -20°C. For use, freeze-dried homogenates were reconstituted with 10-15 volumes of 0.25 M sucrose, 5 mM EDTA solution. Na, K-ATPase activity was determined as the difference between the inorganic phosphate liberated from ATP in the presence and in the absence of 0.1 mM ouabain (Sigma Chemical Company). Assays were conducted at 15°C. First, 0.2 ml of reconstituted homogenate was added to 0.6 ml of concentrated assay medium, then, the reaction was started by addition of 0.2 ml of 15 mM ATP. The final assay medium (1.0 ml) contained the following concentrations: 120 mM NaCl, 10 mM KCl, 3 mM $MgCl_2$, 1.0 mM EDTA, 3 mM disodium ATP (Sigma Chemical Company), 94 mM Tris (pH 7.5). After 15 min, the reaction was terminated by addition of 4 ml ice cold color reagent, a strongly acidic solution of

1% ammonium molybdate in 1.15 N H_2SO_4 in which 40 mg/ml ferrous sulfate had been dissolved just prior to use (Bonting, 1970). After centrifugation, the absorbance was measured at 700 nm. Protein was determined by the method of Lowry et al. (1951) using bovine albumin as standard. Preliminary experiments had shown that these assay conditions yielded maximal intestinal Na, K-ATPase activities.

Rates of water transport were measured in non-everted sacs of intestinal tissue by the procedure of Janicki and Kinter (1971a). Briefly, sacs of tissue were filled with aerated Forster's marine teleost saline (FMS) containing 135 m$\underline{M}$ NaCl, 2.5 m$\underline{M}$ KCl, 1.5 m$\underline{M}$ $CaCl_2$, 1 m$\underline{M}$ $MgCl_2$, 0.5 m$\underline{M}$ NaH_2PO_4, and 7.5 m$\underline{M}$ $NaHCO_3$ (pH 8.25). The filled sacs were weighed and then incubated in aerated FMS at 15°C. After 1 h of incubation, sacs were removed from the medium, weighed, cut open, blotted and weighed again (tissue weight). The volume of fluid (water) transported was calculated as the weight lost during incubation and the rate of transport is expressed as µl/mg tissue/h.

Amino acid transport rates were determined by measuring accumulation of ^{14}C-cycloleucine (New England Nuclear Corporation) in slices or everted sacs of killifish intestine (procedures modified from those of Miller et al., 1974; Smyth, 1974). Briefly, intestinal tissue (slices or sacs) was incubated at 15°C in oxygenated FMS containing ^{14}C-cycloleucine, unlabelled cycloleucine, glucose (0.3 g/l) and, in some experiments, other unlabelled amino acids or mannitol. After a predetermined incubation period (usually 3 min), tissue was removed from the medium, washed with ice cold FMS, blotted (sacs opened to flat sheets), weighed and solubilized for scintillation counting. In some experiments, Tris-FMS buffer (NaH_2PO_4 and $NaHCO_3$ replaced with 10 m$\underline{M}$ Tris) was used; cycloleucine uptake from this buffer did not differ from that obtained with FMS. In Na-depletion experiments, tissue was preincubated for 10 min at 15°C in Na-free Tris-FMS (NaCl replaced isoosmotically with choline Cl) and then incubated in Na-free Tris-FMS containing cycloleucine. Results from these experiments were compared to those in which tissue was preincubated in Tris-FMS and then incubated in Tris-FMS with cycloleucine; the 10 min preincubation in Tris-FMS did not affect cycluleucine uptake.

RESULTS

Plasma Osmoregulation and Intestinal Osmoregulatory Transport

The ability of killifish to osmoregulate was monitored

by measuring plasma Na and K levels. As shown in Figure 1,
controls maintained a constant plasma Na over the time
course of the experiment; there were no mortalities in either
control group (untreated or ETOH alone). In contrast, plasma
Na levels in experimental fish tended to increase with
increasing level and/or duration of DDT exposure. Because of
the great variability of plasma Na in control as well as
experimental fish, differences at a given time period were
rarely significant ($P<0.05$). Changes in plasma K levels
(not shown) generally paralleled those found for plasma Na,
with 7 h exposure to 1 ppm or 24 h exposure to 0.1 ppm
causing increases from 7 meq/l for controls to 9-10 meq/l
for experimentals. These plasma electrolyte changes suggest
osmoregulatory disruption. Kinter et al. (1972) found
similar evidence of osmoregulatory disruption in eels and
killifish exposed to 0.25-1 ppm DDT for 6-10 h and in killifish exposed to 0.075 ppm for 24 h. Finally, in agreement
with previous studies (Kinter et al., 1972; Crawford, et al.,
1972), DDT was clearly toxic to killifish, causing about 30%
mortality in groups exposed to 1 ppm for 7 h or 0.1 ppm for
24 h.

Since marine teleosts osmoregulate by drinking SW,
absorbing NaCl and water from the intestine and excreting
excess salt at the gills (reviewed in Johnson, 1973a), the
intestine plays an essential role in plasma osmoregulation.
To determine if intestinal fluid transport was inhibited in
killifish exhibiting DDT-induced osmoregulatory disruption,
we measured water absorption in vitro in intestinal sacs
from control and experimental fish. As shown in Figure 2,
sacs from fish exposed to 1 or to 5 ppm DDT (4 h in vivo)
exhibited 44 and 100% inhibition of water absorption,
respectively. The latter inhibitory effect is comparable to
that caused by inclusion of 10^{-4}M ouabain in the incubation
medium (Fig. 2). This ouabain concentration specifically
and totally inhibits Na, K-ATPase activity in homogenates of
killifish intestinal mucosa (see assay procedure).

Na, K-ATPase Activity

Janicki and Kinter (1971a, 1971b) have reported that
DDT in vitro inhibits intestinal Na, K-ATPase in 5 of the 6
species of marine teleost tested and that the eel is the most
sensitive, with 15 ppm causing 50% enzyme inhibition. As
shown in Figure 3, the killifish intestinal enzyme appears
to be even more sensitive to DDT in vitro, since only 4 ppm
caused 50% inhibition. Moreover, when killifish were
exposed to 1 or to 5 ppm DDT (4 h in vivo), intestinal Na,

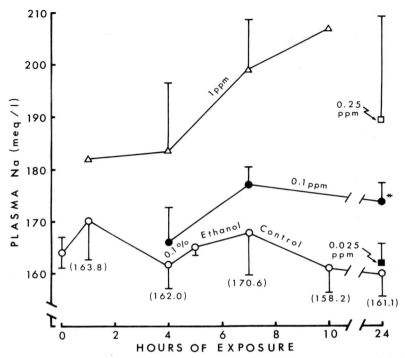

Fig. 1. Effect of DDT exposure on plasma Na levels in SW-adapted killifish. Each point represents the mean value derived from 2 (1 h and 10 h exposure to 1 ppm) or 5-22 fish; variability is given as SE bars. Mean values for groups of 4-5 untreated control fish are given in parentheses; untreated control variability (SE) ranged from 1 to 10 meq/ l.

*Significantly different from controls, P<0.05.

K-ATPase activity was inhibited 29 and 35%, respectively (Fig. 2). With in vivo exposure, mean plasma Na levels were 13-22 meq/l higher than in controls (level for 1 ppm DDT fish in Fig. 1; level for 12 fish exposed to 5 ppm averaged 174 ± 3 meq/l) and intestinal water transport was substantially inhibited (Fig. 2).

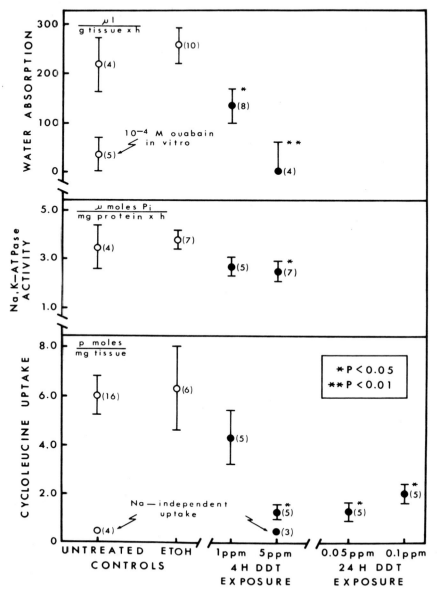

Fig. 2.   Effect of DDT exposure in vivo on water absorption,
Na, K-ATPase activity and amino acid uptake in killifish intestine. Each point represents the mean
value; the variability is given by SE bars and the
number of fish is in parentheses. Amino acid uptake

(3 min incubation) was measured in everted
intestinal sacs as described in the text. Na-
independent uptake was determined as that portion
of the total that was not inhibited by 20 mM L-
leucine.

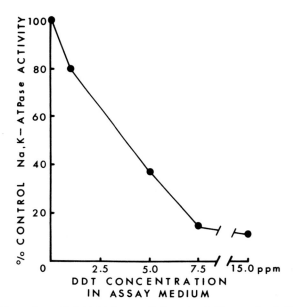

Fig. 3.   Inhibition of Na, K-ATPase in killifish
intestinal mucosa by DDT added in vitro to
the assay medium. Mucosal homogenates
from 2 untreated fish were combined and
each point represents the mean of duplicate
ATPase assays on the pooled homogenate.
DDT was added to the assay medium in N.N.-
dimethylformamide. The final solvent
concentration was 1%; this concentration did
not affect enzyme activity. Control
activity was measured in the presence of
solvent alone.

Amino Acid Transport

Because in vivo exposure to DDT inhibited Na, K-ATPase
activity and Na-coupled water transport in killifish intestine
we were curious to know if another Na-dependent transport

process was similarly affected. The intestinal absorption
of neutral amino acids is Na-dependent in a wide variety of
organisms, including fish (Schultz and Curran, 1970).
Cycloleucine, a non metabolizable synthetic analog of an
essential amino acid, leucine, was chosen as the test
substrate. In initial characterization experiments using
intestinal slices we found that cycloleucine uptake was
saturable and concentrative, i.e., steady state tissue to
medium ratios exceeded unity. Furthermore, uptake was
inhibited by addition of other neutral amino acids to the
incubation medium (competition) or by Na-depletion (Na-
dependence). With 0.01 mM cycloleucine maximal (80%)
inhibition of uptake was observed with 20 mM L-leucine or Na-
depletion. The Na-independent portion of the uptake (20%)
exhibited diffusion kinetics and was not inhibitable by L-
leucine. Since substrate uptake into slices occurs at both
mucosal and serosal surfaces and only mucosal uptake is
physiologically significant, DDT inhibition studies were
carried out using an everted sac preparation in which only
the mucosal (absorptive) surface was exposed to cycloleucine.
Under the same experimental conditions, i.e., 0.01 mM
cycloleucine and 3 min incubation, uptake into sacs was 77%
of that obtained in slices, and, of this, 90% was inhibited
by 20 mM L-leucine and therefore assumed to be Na-dependent
(Fig. 2). In sacs from fish exposed to 1 or 5 ppm DDT for
4 h, cycloleucine uptake was inhibited 32 and 79%
respectively. Significantly, exposure to 5 ppm did not
affect the Na-independent component of uptake; thus, DDT
inhibition of Na-dependent uptake was actually 34 and 86% by
1 and 5 ppm respectively. In addition, 24 h exposure to
either 0.05 or 0.1 ppm DDT caused about 75% inhibition
(Fig. 2), an observation indicating significant impairment
of transport function after long-term exposure to lower DDT
levels.

Tissue DDT Levels

    DDT distribution in tissues of killifish exposed to
^{3}H-DDT are shown in Table 1. Short-term (4 h) exposure to 1
or 5 ppm resulted in levels of about 3-5 ppm in all tissues
sampled; longer exposure (24 h) to 0.05 ppm resulted in
somewhat lower levels in intestinal mucosa, brain, and liver,
but about the same level in gills. Although we did not
measure whole body DDT levels, calculations based on the
data of Crawford et al. (1972) indicate that killifish
exposed to 0.05 ppm for 24 h would have whole body levels of
about 1 ppm. This value is 5-10 times higher than the DDT
residue level reported for killifish collected in the

Table 1
^{3}H-DDT levels in killifish.[*]

DDT Exposure	Intestinal Mucosa	Liver	Gills	Brain
4 Hour				
1 ppm (3)	3.1 ± 0.4	4.4 ± 0.3	-	-
5 ppm (3)	4.9 ± 1.5	5.4 ± 0.6	5.2 ± 0.3	3.7 ± 0.5
24 Hour				
0.05 ppm (4)	1.1 ± 0.2	1.1 ± 0.2	8.0 ± 1.5	1.2 ± 0.1

[*]Values given as mean ± SE; the number of fish is given in parentheses.

relatively unpolluted Mount Desert Island area (Adamson and Guarino, 1972). In comparison, whole body residue levels often exceed 1 ppm in more polluted coastal waters, e.g., the Gulf Coast (Johnson, 1973b). It would be of interest to determine if intestinal function is impaired in fish from highly polluted water.

DISCUSSION

The present results with SW-adapted killifish demonstrate that DDT induced disruption of plasma osmoregulation is accompanied by substantial inhibition of Na-dependent water and amino acid transport in the intestine. These data suggest a cause and effect relationship with regard to impaired intestinal fluid transport and osmoregulatory malfunction. With regard to impaired intestinal amino acid transport, there are at present no definitive data to indicate whether DDT retards growth or protein synthesis in fish. Clearly, nutritional studies should be undertaken to determine the environmental significance of the observed impairment of amino acid transport.

The role of Na,K-ATPase inhibition in the observed impairment of intestinal transport processes by DDT also requires further investigation. In a wide variety of absorbing epithelia, including intestine, the rates of NaCl and water transport are directly proportional (Schultz and Curran, 1968) and inhibition of Na,K-ATPase, e.g., by ouabain, is generally accompanied by a parallel reduction in active Na transport (Bonting, 1970). Thus, if inhibition of fluid transport (water and NaCl) was solely a result of ATPase inhibition, one would expect DDT to affect both fluid transport and enzyme activity to the same extent. The findings that amino acid and water transport were completely inhibited with only 35% ATPase inhibition (Fig. 2) suggest either that DDT may act at membrane sites in addition to the ATPase powered Na pump, or that dissociation of the ATPase-DDT complex occurred during the homogenization and dilution for enzyme assay. At present there is no experimental basis for choosing between these two possibilities.

The results of the present study add to the already substantial body of experimental evidence indicating that organochlorine pollutants, such as DDT, disrupt membrane transport function in sensitive species (the membrane theory of pollutant toxicity has been reviewed by Kinter and Pritchard, 1976). According to the fluid mosaic model of membrane structure (Singer and Nicolson, 1972), membrane phospholipids are aligned in a bilayer with their polar heads at the surface and their fatty acid tails extending into the fluid center; globular proteins are embedded in this phospholipid matrix. To preserve membrane integrity polar regions of the proteins' surface face the aqueous interface while non-polar regions face the hydrophobic interior and interact with the fluid lipid phase of the membrane. Because of hydrophobic interactions between proteins and lipids, changes in the fluidity of the lipid phase (caused, for example, by drugs) may alter protein conformations and, thus, affect transport rates and enzyme activities. DDT is a highly lipophilic molecule which appears to inhibit Na,K-ATPase through the enzyme's associated phospholipid rather than by direct interaction with the protein itself (Sharp et al., 1974). Recently, Schneider (1975) proposed a mechanism for DDT inhibition of Na,K-ATPase in which this pollutant dissolves in the lipid phase of the membrane, decreasing its fluidity, and thus affecting ion-activated allosteric transitions required for enzyme activity.

Schneider also suggested that the well documented neurotoxic effects of DDT could likewise be a result of a DDT-induced decrease in the fluidity of the axon membrane lipid phase. Since DDT disrupts the passive ion-gating mechanism of the axolemma (Matsumura, 1975), the proposed conformational changes would occur in a protein involved in control of membrane permeability rather than in ion pumping.

ACKNOWLEDGMENT

We thank Allyn Seymour, Jr. and David Shoemaker for excellent technical assistance. This investigation was supported by United States Public Health Service Grant ES 00920 and National Science Foundation Grant GB 28139.

LITERATURE CITED

Adamson, R. H. and A. M. Guarino. 1972. Natural levels of DDT-related compounds and polychlorinated biphenyls (PCB's) in various marine species. Bull. Mt. Desert Isl. Biol. Lab., 12: 6-9.

Bonting, S. L. 1970. Sodium-potassium adenosine triphosphatase and cation transport. In: Membranes and Ion Transport, vol. 1, pp. 257-363, ed. by E. E. Bittar. London: Wiley-Interscience.

Crawford, R. B., J. B. Anderson, and A. M. Guarino. 1972. Effects of DDT on Fundulus heteroclitus: survival, uptake and distribution. Bull. Mt. Desert Isl. Biol. Lab., 12: 19-22.

Holden, A. V. 1973. Effects of pesticides on fish. In: Environmental Pollution by Pesticides, pp. 213-253, ed. by C. A. Edwards. London: Plenum Press.

Janicki, R. H. and W. B. Kinter. 1971a. DDT: disrupted osmoregulatory events in the intestine of the eel, Anguilla rostrata adapted to sea water. Science 173: 1146-1148.

_____ and _____. 1971b. DDT inhibits Na^+, K^+, Mg^{2+}-ATPase in the intestinal mucosae and the gills of marine teleosts. Nature 223: 148-149.

Johnson, D. W. 1973a. Endocrine control of hydromineral balance in teleosts. Amer. Zool., 13: 799-818.

_____. 1973b. Pesticide residues in fish. In: Environmental Pollution by Pesticides, pp. 181-212, ed. by C. A. Edwards. London: Plenum Press.

Kinter, W. B., L. S. Merkins, R. H. Janicki, and A. M. Guarino. 1972. Studies on the mechanism of toxicity

of DDT and polychlorinated biphenyls (PCBs):
disruption of osmoregulation in marine fish. Environ.
Health Perspectives, 1(1): 169-173.

_____ and J. B. Pritchard. 1976. Altered permeability
of cell membranes. In: Handbook of Environmental
Physiology. I. Physical and Chemical Agents, in
press. Washington: American Physiological Society.

Leadem, I. P., R. D. Campbell, and D. W. Johnson. 1974.
Csmoregulatory responses to DDT and varying salinities
in Salmo gairdneri - I. Gill Na,K-ATPase. Comp.
Biochem. Physiol., 49A: 197-205.

Lowry, H., N. J. Rosenbrough, A. L. Farr, and R. J. Randall.
1951. Protein measurement with the folin phenol
reagent. J. Biol. Chem. 193: 265-275.

Matsumura, F. 1975. Toxicology of Insecticides. New York:
Plenum Press, pp. 115-123.

Miller, D. S., P. Burrill, and J. Lerner. 1974. Distinct
components of neutral amino acid transport in the
checken small intestine. Comp. Biochem. Physiol., 47A:
767-777.

Schneider, R. P. 1975. Mechanism of inhibition of rat
brain (Na + K) - adenosine triphosphatase by 2,2-bis
(p - chlorophenyl) - 1,1,1 - trichloroethane (DDT).
Biochem. Pharmacol., 24: 939-946.

Schultz, S. G. and P. F. Curran. 1968. Intestinal
absorption of sodium chloride and water. In:
Handbook of Physiology. Alimentary Canal, sect. 6,
vcl. 3, pp. 1245-1275, ed. by C. F. Code, Washington:
American Physiological Society.

_____ and _____. 1970. Coupled transport of
sodium and organic solutes. Physiol. Rev., 50: 637-
718.

Sharp, C. W., D. G. Hunt, S. T. Clements, and W. E. Wilson.
1974. The influence of dichlorodiphenyltrichloroethane,
polychlorinated biphenyls and anionic amphiphilic
compounds on stabilization of sodium - and potassium-
activated adenosine triphosphateses by acidic phospholipids. Mol. Pharmacol., 10: 119-129.

Singer, S. J. and G. L. Nicolson. 1972. The fluid mosaic
model of the structure of cell membranes. Science, 175:
720-731.

Smyth, D. H. 1974. Methods of studying intestinal absorption. In: Intestinal Absorption, pp. 241-283, ed.
by D. H. Smyth. London: Plenum Press.

DDT: Effect on the Lateral Line Nerve of Steelhead Trout

CARL F. PETERS[1] and DOUGLAS D. WEBER[2]

[1]John Graham and Company
Environmental Studies Group
1110 Third Avenue
Seattle, Washington   98101

[2]Northwest Fisheries Center
National Marine Fisheries Service
National Oceanic and Atmospheric Administration
2725 Montlake Boulevard East
Seattle, Washington   98112

The insecticide DDT is known to affect tissues of both the central and peripheral nervous system, although initial symptoms of poisoning are considered to originate from the sensory components (Roeder and Weiant, 1948; Becht, 1958; Narahashi, 1971).   Studies of the lateral line nerve of the clawed toad (Zenopus laevis) have shown that the effects of DDT are observed as the external temperature is lowered and normal, single neural impulses evolve into trains or bursts of repetitive activity (Bercken and Akkermans, 1971; Bercken, Akkermans, and van Langen, 1972).   In brook trout exposed to DDT Anderson (1968) observed an increase in the duration of the burst of action potentials recorded from the lateral line nerve when the mechanoreceptors were stimulated with a low frequency pressure wave.   Bahr and Ball (1971) could not duplicate this prolonged response to a stimulus with fish acclimated to a lower temperature.   Other studies with sublethal concentrations of DDT (5-100 ppb) have indicated that the pesticide influences the temperature selected by salmonids in a thermal gradient (Ogilvie and Anderson, 1965;

Peterson, 1973). Investigations of reflex and motor activity
modification resulting from exposure to sublethal levels of
DDT have also established the importance of the chemical's
action on the central nervous system of fish (Anderson and
Prins, 1970; Hatfield and Johansen, 1972; Weis and Weis,
1974).

In the course of another study concerning function and
morphology of the lateral line of trout (Weber and Schiewe,
1976) we subjected fish to DDT and found a pronounced alteration in lateral line neural activity. Since there is some
conflicting evidence in the literature regarding the effect
of DDT on the lateral line of salmonids, our observations
prompted us to pursue this relationship in more detail. The
objectives of this study were to:
1. document the conditions and characteristics of
   repetitive neural activity,
2. determine the effect of acclimation history on
   neural response to DDT, and
3. determine the effect of DDT on lateral line sensitivity.

MATERIALS AND METHODS

Hatchery reared steelhead trout (<u>Salmo</u> <u>gairdneri</u>)
weighing an average of 100 grams each were acclimated to
either $8^{\circ}C$ or $18^{\circ}C$ for three weeks prior to experimentation.
A stock solution of DDT was prepared by dissolving 1 mg of
99%+ pure p, p'-DDT (2,2-bis (4-chlorophenyl)-1,1,1-trichlo-
roethane) in 1 ml of ethanol. An experiment was initiated by
placing a fish in a 20 l aquarium submerged in a water bath
to maintain the acclimation temperature. An aliquot of the
DDT stock solution was added to the aquarium to obtain concentrations of 0.1 to 1.0 ppm (0.28-2.8M). An airstone was
placed in the aquarium and the fish was held without feeding
for 24 hours.

Following pesticide exposure the fish was first anesthe-
tized with tricaine methanesulfonate (1:15,000) and then
intramuscularly injected with gallamine triethiodide (0.06 mg/
g body weight) to prevent movement during nerve recording.
The fish was then immediately placed on its side in a plexiglass recording chamber and water delivered over its gills at
1 l/min. Water temperature was controlled between $1^{\circ}C$ and
$28^{\circ}C$ with a thermostatic mixing valve and by pumping ice-
chilled water from a reservoir. The water inlet branched
before entering the recording chamber; a small tube supplied
the fish and another supplied 4 l/min to the chamber to
maintain or rapidly change the temperature of the water

surrounding the fish. The water level was adjusted to sub-
merge the fish's body except for the surgical area 3 cm
posterior to the opercular opening. A strip of skin 0.5 cm
by 1.5 cm was then removed directly over the lateral line, and
the trunk nerve was exposed and placed over a spoon shaped,
silver, reference electrode. The portion of the nerve
resting on the reference electrode was then desheathed, and
individual fibers were raised on an uninsulated, electro-
lytically sharpened, stainless steel hook recording electrode.
Neural activity was displayed on an oscilloscope with outputs
to a loudspeaker, storage oscilloscope, and an FM tape
recorder. Continuous recordings and polaroid photography
were used to store data for later analysis.

When a fiber displaying spontaneous activity was located,
the sensory unit length (number of lateral line scales
innervated by the fiber) was determined by gently pressing
the scales and listening for an increase in neural discharge.

A polygraph continuously monitored cardiac function
through volume conduction; the two electrodes were inserted
into a sponge support pad beneath the fish, and the EKG
recording was filtered to include a bandwidth of 10 to 30 Hz.

Following the experiment the fish was wrapped in an air-
tight container and frozen. Whole fish samples were later
prepared using dry ice homogenization (Benville and Tindle,
1970) and were analyzed for total DDT, PCB, and lipid content
(AOAC, 1975).

Repetitive Activity

To detect whether the lateral line nerve had been
affected by DDT the water temperature was lowered slowly
until the normal spontaneous activity, consisting of single
action potentials, evolved into repetitive activity or
"bursts" of spikes. Temperature was then adjusted to
determine, as closely as possible, the point at which this
effect occurred. The fish's internal temperature next to
the lateral line nerve, as measured by an implanted thermis-
ter, approached to within $0.5^{\circ}C$ of the external temperature
when the water temperature was held constant for at least 5
minutes. The experiment was carried out on fish acclimated
to both $8^{\circ}C$ and $18^{\circ}C$ to determine if differences in bursting
threshold were due to acclimation temperature.

Evoked Response

Mechanical stimulation of the lateral line organ was

accomplished by a 1 cm diameter glass sphere attached to the
cone of an 8 inch loudspeaker by a glass rod tripod. After
the group of scales innervated by the fiber was located, the
sphere was positioned midway along the sensory unit and 2 mm
above it. The apparatus was positioned so that the movement
of the pulsating sphere was perpendicular to the plane of the
lateral line.

A pulse generated by an audio oscillator first passed
through an attenuator which allowed approximately 2 db (re
1μbar) variations in sound pressure. Calibration of sound
intensity and frequency range of the stimulus was obscured by
background noise for measurements below -25 db (re 1μbar).
Thus, sound intensity is expressed in relative db units where
0 (relative) equals +10 db (re 1μbar) and -6 (relative)
equals -21 db (re 1μbar) at 80 Hz.

To approximate a stimulus likely to be encountered in
the environment, a square wave form with a pulse length of
0.1 to 0.2 Hz was chosen for the determination of sensitivity
threshold. The "sensitivity threshold" was defined as the
signal attenuation (db, re 1μbar) of the stimulus which
elicited a response from the lateral line unit at least once
in 10 presentations of the stimulus.

RESULTS

Spontaneous Activity

Rate of spontaneous activity is dependent upon temperature and the number of scales innervated by the fiber
(Fig. 1). Fibers from DDT treated fish which exhibited repetitive activity as the temperature was decreased showed
normal spontaneous activity when above the bursting threshold
temperature. Regression coefficients for the spontaneous
activity of treated non-bursting fibers and control fibers
were not significantly different (P<0.01) thus indicating
that DDT does not impair normal nerve activity or temperature
response until repetitive activity occurs.

Repetitive Activity

Fibers from fish exposed to DDT exhibited repetitive
spontaneous activity as the temperature was decreased from
the acclimation temperature (Fig. 2). The bursts of activity
became longer as the temperature was lowered, and the spike
intervals increased toward the end of the bursts. Only
treated fish exhibited this bursting response, and normal

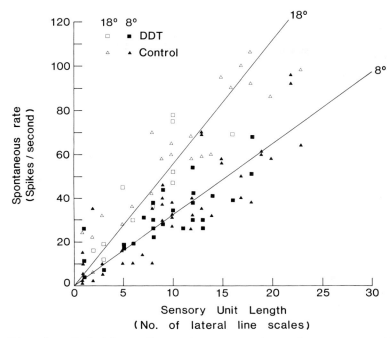

Fig. 1.   Relation of spontaneous rate and sensory unit
length for control and DDT exposed fish at
acclimated temperatures. Lines fitted from
regression equations forced through origin.
$18^{\circ}C$: $Y = 5.57x$, $n = 37$, $r^2 = 0.796$. $8^{\circ}C$:
$Y = 3.25x$, $n = 68$, $r^2 = 0.709$.

spontaneous activity was resumed when the temperature of the
preparation was increased. Repetitive activity was not as
pronounced in fibers from fish exposed to 0.1 ppm DDT as in
those exposed to 1.0 ppm, and frequently, bursts of only 2 or
3 spikes were observed, even at very low temperatures ($1.5^{\circ}C$).
      Fibers which did exhibit repetitive activity are ana-
lyzed in the manner of van den Bercken et al. (1973) in
Figure 3. All fibers are from fish acclimated to $18^{\circ}C$ since
fibers from fish acclimated to $8^{\circ}C$ did not usually exhibit
bursting characteristics. Generally, the effect of DDT was
temperature dependent and occurred when the temperature was
lowered. The number of spikes per burst increased (Fig. 3A),
and the frequency of bursts decreased rapidly with temperature drop (Fig. 3B); both effects were dosage dependent.
During repetitive activity the average number of spikes per
second was temperature independent (Fig. 3C) in contrast to

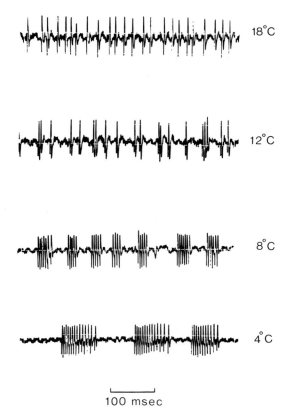

Fig. 2.   Spontaneous activity and temperature effect of DDT.
          Recorded from single lateral line fiber of trout
          acclimated to 18°C and exposed to 1.0 ppm DDT for
          24 hours.

control conditions where a positive correlation (Q_{10} = 1.91)
exists between spontaneous activity and temperature (Weber
and Schiewe, 1976).

    The temperature at which treated fibers began to exhibit
bursts of activity is presented in Table 1. This threshold
temperature for the bursting effect was defined as the point
at which bursts of two or more spikes first occurred as the
temperature was decreased. Fish acclimated to 8°C and
exposed to 0.1 ppm DDT exhibited repetitive activity much
less frequently than those acclimated to 18°C. A high proportion of nerves from 18°C fish were affected, and the
numbers were proportional to exposure concentrations.
Analysis of variance indicated no significant difference in

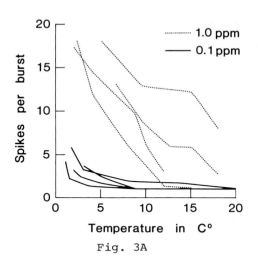

Fig. 3A

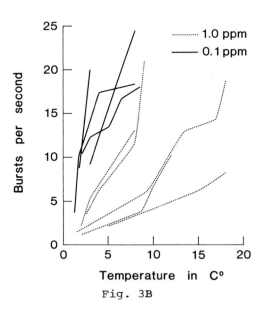

Fig. 3B

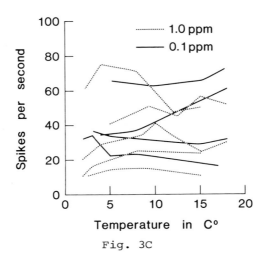

Fig. 3C

Fig. 3.   Effect of temperature on spontaneous
          activity of single lateral line nerve
          fibers after exposure to DDT. Solid
          line - representatives of 22 fibers
          from fish exposed to 0.1 ppm. Dashed
          line - representatives of 9 fibers from
          fish exposed to 1.0 ppm. A: number of
          spikes per burst of repetitive activity;
          B: burst frequency; C: spike
          frequency.

bursting temperature threshold between fibers from 18°C
acclimated fish exposed to 0.1 ppm or 1.0 ppm DDT. Since
the threshold was significantly more variable in fibers
from fish exposed to 0.1 ppm DDT than in fish exposed to
1.0 ppm, the means were compared with an approximate "t"
test (Sokal and Rohlf, 1969) which indicated that the mean
threshold temperatures for the two concentrations were not
significantly different. Data for the groups were pooled to
obtain a mean threshold of 11.0°C $\pm$ 2.2°C (S.E.). This
suggests that increased exposure concentration does not
affect bursting temperature threshold but does increase the
number of fibers affected.

Table 1
Number and Percentage of Bursting Fibers and Bursting
Temperature Threshold of Fibers from Fish Acclimated to
8°C and 18°C.

Acclimation Temperature and Treatment	n Fibers Tested	n Bursting Fibers	% n Bursting	Mean Bursting Threshold Temperature ($^{\circ}$C) ($\pm$ S.E.)
8°C				
Control	15	0	0	–
0.1 ppm DDT	17	1	6	4.5
18°C				
Control	12	0	0	–
0.1 ppm DDT	18	11	61	–
1.0 ppm DDT	9	9	100	11.0 ± 2.2*

*No significant difference in means of 0.1 ppm and 1.0 ppm
treated fish (P<0.05).

DDT Uptake

    The levels of DDT accumulated by fish are known to vary
directly with temperature (Reinert, Stone, and Willford,
1974; Murphy and Murphy, 1971), and this could have resulted
in the different neural responses we observed in fish
exposed to 8°C and 18°C. Results of DDT analysis for 6 fish
are given in Table 2. One fish exposed to 1.0 ppm accumu-
lated approximately twice as much DDT as those exposed to
0.1 ppm at the same temperature; all fibers from this fish
elicited a bursting response as the temperature was
decreased. The slight difference in accumulation between
8°C and 18°C fish exposed to 0.1 ppm does not indicate that
the lack of bursting fibers observed in fish acclimated to
8°C was due to a lower uptake of the pesticide.

Evoked Response
    Examples of evoked response threshold stimulation are

Table 2
Total DDT in Fish Exposed at 8°C and 18°C.

Acclimation Temperature (°C)	Exposure Concentration (ppm)	Total DDT (ppm)*
18	Control	1.2
18	0.1	5.7
18	0.1	6.2
18	1.0	12.1
8	0.1	4.3
8	0.2	6.6

*DDE; o,p'-DDT; p,p'-DDT.

shown in Figures 4 and 5. Both the upward and downward pulse of the stimulus elicited additional single action potentials in control fibers and in those unaffected by DDT treatment (on-off effects). The number of impulses in the response was directly proportional to the stimulus intensity. Fibers exhibiting a bursting response due to DDT exposure responded to the stimulus with either an additional burst of spikes or with a longer burst if the stimulus occurred during an on-going burst. Stimulus threshold was independent of preparation temperature and acclimation history.

Results for the stimulus threshold studies are shown in Figure 6. Analysis of covariance indicated no significant difference between 8°C and 18°C acclimated fish in the control, DDT treated nonbursting, and DDT treated bursting groups. Data for the three groups were pooled and a common regression line was computed. Results indicate that the threshold for stimulation of trout lateral line fibers is inversely related to the number of scale units innervated by the fiber. Treatment with, and the effects of, DDT do not alter the receptor sensitivity of the system.

-13dB

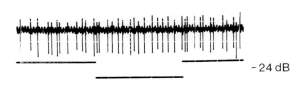

-24 dB

200 msec

Fig. 4. Determination of stimulus threshold in
control or in nonbursting, treated
fibers. Each presentation illustrates
the response of the sensory unit to the
square wave stimulus at the intensity
indicated (db, re lubar).

DISCUSSION

Conditions and Characteristics of Repetitive Neural Activity

The bursting characteristics of lateral line nerve
fibers from DDT treated trout appear to closely resemble
those of lobster and cockroach axons (Narahashi, 1971), and
of lateral line nerves of the clawed toad (Bercken and
Akkermans, 1971; Bercken, 1972). Our biphasic recordings,
however, did not enable detection of the increased negative
after-potential described by Narahashi and Hass (1968) and
Hille (1968). This prolongation of the action potential
reduces the firing threshold and is one of the factors
responsible for repetitive activity. Bercken (1972) recorded
bursting activity in DDT treated toad sciatic nerves and
observed additional action potentials superimposed upon the
prolonged falling phase of the initial action potential.
The interval between successive spikes became greater as the

-10 dB

-10 dB

200 msec

Fig. 5. Determination of stimulus threshold in
fibers from DDT treated fish exhibiting
repetitive activity. TOP: Additional
burst elicited upon stimulation.
BOTTOM: Increased burst length upon
stimulation during repetitive activity.

negative after-potential decreased until a critical interval
was reached and no further spikes occurred within the burst.
A similar spike interval occurred in our studies (Fig. 2)
indicating that a prolonged negative after-potential may also
be induced by DDT in steelhead trout nerves.

Our studies once again illustrate the negative temperature coefficient of the action of DDT. Anderson (1968), after
exposing fish acclimated to 21.5°C under the same conditions
as in our study, reported a prolonged multi-fiber response to
a stimulus with a more pronounced effect at lower temperatures. Bahr and Ball (1971) did not observe this response,
but their trout were acclimated to 13°C and tested at that
temperature. We observed that only 15% of the bursting
fibers from 18°C acclimated fish exhibited bursting at 18°C,
the others beginning repetitive activity as the temperature
was decreased. Our evidence supports the observation of
Bahr and Ball that they may not have perceived a prolonged
response to a stimulus because the temperature was not
lowered.

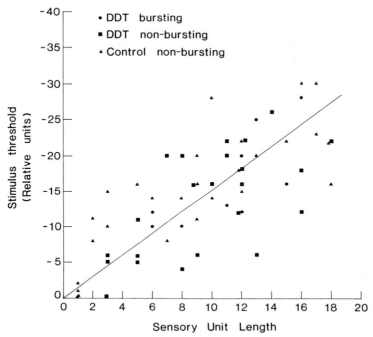

Fig. 6. Evoked response stimulus threshold of lateral line
         sensory units. Line fitted from regression equation
         forced through origin. $Y = 1.48x$, $n = 59$, $r^2 =$
         0.473.

It is well documented that DDT can cause ataxia and convulsions, and that these conditions become more pronounced as
the temperature is reduced. Peterson (1973) observed hyperactivity and convulsions in fish acclimated to $13^{\circ}C$ and
exposed to 0.1 ppm DDT when they were immersed in water colder
than $10^{\circ}C$.

In the cockroach, sensory repetitive discharge alone
apparently does not produce the symptoms of DDT poisoning
since, when the temperature is increased, convulsions and
ataxia disappear while sensory nerves still exhibit bursting
activity (Narahashi, 1971). Thus, even though the temperature threshold of hyperactivity observed by Peterson is
similar to the threshold we observed for lateral line nerve
repetitive activity, it is doubtful that the two are directly
related.

As a basic premise for their DDT studies with trout,
Bahr and Ball (1971) hypothesized that "Although tremors and
convulsions normally reflect events originating in the

central nervous system, it is possible that similar symptoms could arise from post synaptic events initiated by intense afferent activity impinging upon central neurons." It is possible that the aberrant input to the CNS represented by repetitive activity in the lateral line may modify output from the CNS to motor activities. Thus, motor responses thought to receive input from the lateral line sensory system could be effected and may account for the subtle modifications of exploratory behavior and schooling activity observed by Davy and Kleerekoper (1973) and Weis and Weis (1974) in gold-fish following sublethal exposure to DDT.

Acclimation History and Response to DDT

In char (Anderson and Peterson, 1969) and goldfish (Roots and Prosser, 1962) the cold blocking temperature of the simple tail propeller reflex varies with the temperature to which the fish are acclimated. Acclimation temperature can also alter the temperature preferred by trout when placed in a thermal gradient (Javaid and Anderson, 1967). Studies by Ogilvie and Anderson (1965) and Peterson (1973) indicate that exposure to DDT mimics these acclimation temperature responses. Based on this evidence and because the central nervous system is thought to regulate changes in temperature selection (Fisher, 1958), Anderson and Peterson (1969) argued that sublethal amounts of DDT directly affect the CNS and thus alter the thermal acclimation mechanism.

In our studies, the bursting temperature threshold of lateral line fibers from DDT exposed fish varied directly with acclimation temperature, thus providing additional evidence that poikilotherms exhibit meaningful temperature acclimation effects in their peripheral nervous system (Lagerspetz, 1974). Following sublethal exposure to DDT the changed activity in these sensory fibers may contribute to changes in the CNS which then alters the acclimation mechanism.

Effect of DDT on Lateral Line Sensitivity

Previous electrophysiological studies of the trout lateral line nerve using multi-fiber recordings have shown a prolonged burst of neural activity when stimulated by a water drop (Anderson, 1968), and it has been suggested that DDT renders the lateral line hypersensitive (Anderson, 1971). Convulsions and irregular motor activity have also been observed when fish exposed to DDT are subjected to a stimulus such as tapping on the side of the aquarium. In our experiments the sensory units which were affected by the pesticide (i.e., exhibited repetitive activity) had the same stimulus threshold as control fibers. Therefore, exposure to DDT does not cause the lateral line organ itself to become more sensitive to an external stimulus. We believe that uncontrolled reflex reactions to external stimuli result from changes within the central nervous system which may be due to abnormal input from the bursting activity of the sensory system.

LITERATURE CITED

Anderson, J. M. 1968. Effect of sublethal DDT on the lateral line of brook trout, Salvelinus fontinalis. J. Fish. Res. Bd. Can. 25: 2677-2682.
_____. 1971. Assessment of the effects of pollutants on physiology and behavior. Proc. Roy. Soc. Lond. B. 177: 307-320.
_____ and M. R. Peterson. 1969. DDT: Sublethal effects on brook trout nervous system. Science 164: 440-441.
_____ and H. B. Prins. 1970. Effects of sublethal DDT on a simple reflex in brook trout. J. Fish. Res. Bd. Can. 27: 331-334.
Association of Official Analytical Chemists. 1975. Official methods of analysis, 12th ed. Sec. 29.
Bahr, T. G. and R. C. Ball. 1971. Action of DDT on evoked and spontaneous activity from the rainbow trout lateral line nerve. Comp. Biochem. Physiol. 38A: 279-284.
Becht, G. 1958. Influence of DDT and lindane on chorpotonal organs in the cockroach. Nature 181: 777.
Benville, P. E., Jr. and R. C. Tindle. 1970. Dry ice homogenization procedure for fish samples in pesticide residue analysis. Agric. Food Chem. 18: 448-449.
Bercken, J. van den. 1972. The effect of DDT and dieldrin on myelinated nerve fibres. Europ. J. Pharm. 20: 205-214.
_____ and L. M. Akkermans. 1971. Negative temperature coefficient of the action of DDT in a sense organ. Europ. J. Pharm. 16: 241-244.
_____, _____, and R. G. van Langen. 1972. The effect of DDT and dieldrin on skeletal muscle fibres. Europ. J. Pharm. 21: 89-94.
_____, _____, and J. M. van der Zalm. 1973. DDT-like action of allethrin in the sensory nervous system of Xenopus laevis. Europ. J. Pharm. 21: 95-106.

Davy, F. B. and H. Kleerekoper. 1973. Effects of exposure
    to sublethal DDT on the exploratory behavior of goldfish
    (Carassius auratus). Water Resources Res. 9: 900-905.
Fisher, K. C. 1958. An approach to the organ and cellular
    physiology of adaptation to temperature in fish and
    small mammals. In: Physiological Adaptation, pp. 3-49,
    ed. by C. L. Prosser. Washington, D. C.: Amer. Physiol.
    Soc.
Hatfield, C. T. and P. H. Johansen. 1972. Effects of four
    insecticides on the ability of Atlantic salmon parr
    (Salmo salar) to learn and retain a simple conditioned
    response. J. Fish. Res. Bd. Can. 29: 315-321.
Hille, B. 1968. Pharmacological modifications of the sodium
    channels of frog nerve. J. Gen. Physiol. 51: 199-219.
Javaid, M. Y. and J. M. Anderson. 1967. Thermal acclimation
    and temperature selection in Atlantic salmon, Salmo
    salar, and rainbow trout, S. gairdneri. J. Fish. Res.
    Bd. Can. 24: 1507-1519.
Lagerspetz, K. H. 1974. Temperature acclimation and the
    nervous system. Biol. Rev. 49: 477-514.
Murphy, P. C. and J. V. Murphy. 1971. Correlations between
    respiration and direct uptake of DDT in the mosquito
    fish (Gambusia affinis). Bull. Environ. Contam. Toxicol.
    6: 581-588.
Narahashi, T. 1971. Effects of insecticides on excitable
    tissues. In: Advances in Insect Physiology, Vol. 8,
    pp. 1-93. ed. by J. W. L. Beaument, J. E. Treherne, and
    V. B. Wigglesworth. London and New York: Academic
    Press.
_____ and H. G. Haas. 1968. Interaction of DDT with
    the components of lobster nerve membrane conductance. J.
    Gen. Physiol. 51: 177-198.
Ogilvie, D. M. and J. M. Anderson. 1965. Effect of DDT on
    temperature selection by young Atlantic salmon, Salmo
    salar. J. Fish. Res. Bd. Can. 22: 503-512.
Peterson, R. H. 1973. Temperature selection of Atlantic
    salmon (Salmo salar) and brook trout (Salvelinus
    fontinalis) as influenced by various chlorinated hydro-
    carbons. J. Fish. Res. Bd. Can. 30: 1091-1097.
Reinert, R. E., L. J. Stone, and W. A. Willford. 1974.
    Effect of temperature on accumulation of methyl mercuric
    chloride and p, p DDT by rainbow trout (Salmo
    gairdneri). J. Fish. Res. Bd. Can. 31: 1649-1652.
Roeder, K. D. and E. A. Weiant. 1948. The effect of DDT on
    sensory and motor structures of the cockroach leg. J.
    Cell. Comp. Physiol. 32: 175.
Roots, B. I. and C. L. Prosser. 1962. Temperature acclima-

tion and the nervous system in fish. J. Exp. Biol. 39:
617-629.

Sokal, R. R. and F. J. Rohlf. 1969. Biometry - The Principles and Practice of Statistics in Biological Research.
San Francisco, Calif.: W. H. Freeman and Co.

Weber, D. D. and M. H. Schiewe. 1976. Morphology and
function of the lateral line of juvenile steelhead
trout in relation to gas bubble disease. J. Fish Biol.
In press.

Weis, P. and J. S. Weis. 1974. DDT causes changes in activity and schooling behavior in goldfish. Environ. Res.
7: 68-74.

Anticholinesterase Action of Pesticidal Carbamates in the Central Nervous System of Poisoned Fishes

DAVID L. COPPAGE

United States Environmental Protection Agency
Environmental Research Laboratory
Gulf Breeze, Florida 32561

Increased use of carbamate pesticides due to suspended or canceled registration of chlorinated hydrocarbon pesticides generates a need for information on their effects on the aquatic environment. Annual production of carbamate insecticides in the United States was estimated to be more than 3.3×10^7 kg (Lawless et al.,1972). Carbamates may enter surface and estuarine water through industrial effluent, runoff, or direct application (Environmental Protection Agency, 1972; Greve et al.,1972; Caro et al.,1973; Flickinger, 1973; Caro et al.,1974; Coppage and Braidech, 1976). Carbamates are believed to poison animals by inhibition of an enzyme (acetylcholinesterase) that modulates amounts of the neurotransmitter acetylcholine at nerve cell junctions (synapses). Inhibition of the enzyme by carbamates or their metabolites causes accumulation of acetylcholine and disruption of normal neurotransmission (Koelle, 1963; O'Brien, 1967; Karczmar, 1970).

Few studies have been made on acetylcholinesterase inhibition in vivo in fish by carbamate pesticides. Lowe (1967) exposed an estuarine fish, Leiostomus xanthurus (spot), to 100 μg carbaryl/l flowing sea water and found only 17% inhibition of brain acetylcholinesterase after 2.5 months, and normal enzyme activity after 5 months. Exposure to 1000 μg/l for 13 days caused only 32% inhibition. However, the enzyme was not characterized and the method of Hestrin (1949) for acetylcholinesterase assay was used which is less

accurate than kinetic methods such as with the pH stat
(Karczmar, 1970; Augustinsson, 1971; Silver, 1974). Carter
(1971) tested effects of carbamate pesticides on channel catfish, Ictalurus punctatus, by observing brain cholinesterase
inhibition and other signs of poisoning. In static 48-hr
exposures to carbaryl (200-2500 µg/l), carbofuran (200-1000
µg/l), aldicarb (100-2000 µg/l), or methomyl (100-1000 µg/l),
inhibition was proportional to pesticide concentration, and
maximum inhibition was effected in 2 hr with little or no
recovery in 48 hr. Sequential signs of poisoning were:
hyperactivity at 35-75% inhibition, lethargy at 66-80%
inhibition, paralysis of body at 65-86% inhibition, scoliosis
at 62-89% inhibition, loss of equilibrium at 78-92% inhibition, and opercular and mouth paralysis in open position with
some deaths in the 50-95% inhibition range. Replicated
sequential tests in which median effects and acetylcholinesterase inhibition were observed and analyzed statistically
were not reported in the above studies.

More rigorous testing of carbamates is needed to establish whether a causal relationship exists between enzyme
inhibition and functional change in organisms. Also,
sufficient data should be obtained for statistical analysis.
For example, it should be shown by in vitro studies of
enzyme kinetics that (i) acetylcholinesterase is the only
enzyme hydrolyzing the substrate, (ii) pH, temperature,
substrate concentration, and enzyme concentration are close
to optimum, and (iii) enzyme assay is free from error due to
nonenzymatic effects such as substrate hydrolysis, non-linear
reaction rate, and interference with substrate or reaction-
product quantitation (Webb, 1963; Dixon and Webb, 1964). For
purposes of interpreting inhibition in animals sampled from
nature, it is particularly important to show that homologous
inhibitors produce in vivo enzyme inhibition that parallels
functional change, and that functional change occurs at about
the same enzyme inhibition level regardless of the homolog
used (Webb, 1963). In this report, kinetic enzyme methods
and statistical analyses are used to define the relationship
between brain acetylcholinesterase inhibition and near-median
kills in replicate groups of marine fish in the laboratory by
five carbamate pesticides.

METHODS AND MATERIALS

Inhibition of brain acetylcholinesterase was measured
during carbamate poisoning of the marine or estuarine fishes:
pinfish (Lagodon rhomboides; 46-102 mm total length), sheepshead minnows (Cyprinodon variegatus; 35-67 mm), and sailfin

molly (Poecilia latipinna; 55-66 mm). Brain acetylcholinesterase was assayed with a pH stat method previously
developed from kinetic studies of the enzyme (Coppage, 1971).
The assay was carried out by continuous recording of the
"straight line" rate of hydrolysis of acetylcholine at pH 7
and 22°C for 10-20 min. Each sample assayed consisted of
pooled brains taken from 3 fish that survived carbamate
exposure when 40-60% were killed at a convenient time,
percentage inhibition being determined by comparison with
control fish similarly assayed. Dead fish were not used
because data cannot be applied in nature when it is not
known how long fish have been dead and subject to loss of
enzyme activity by autolysis.

     In each test, 3-4 replicates of 10 fish each were
exposed to technical-grade pesticide in 8-liter acrylic
plastic aquaria that received a mixture of flowing sea
water (400 ml/min) and pesticide from a common source.
Pesticides were dissolved in acetone, benzene, or polyethylene glycol and infused into sea water with a syringe
pump or Lambda Pump[R] at solvent concentrations that do not
affect enzyme activity or cause death (Coppage and Matthews,
1975; Coppage et al.,1975). Pesticide concentration in the
water was expressed as µg added per liter. No attempt was
made to identify or quantify pesticides or their transformation products in water or tissue. Temperature range was
18-23°C, and salinity was 17-29 parts per thousand.

     To determine the extent of acetylcholinesterase inhi-
bition resulting from a near-median kill, the survivors of
replicated tests in which 40-60% of populations were killed
by carbaryl (Sevin[R]), carbofuran (Furadan[R]), methomyl
(Lannate[R]), aldicarb (Temik[R]), or Bux[R] were assayed.
Statistical comparisons of acetylcholinesterase activities
of replicate exposed fish were made with replicate unexposed
fish (Student's t-test P <0.005).

RESULTS AND DISCUSSION

     Acetylcholinesterase inhibition in exposed fishes was
similar (77-89%) regardless of compound, species, or period
of exposure (4-48 hr), when a near-median kill occurred
(Table 1). Since these laboratory findings for carbamates
are similar to previous findings for organophosphate inhibitors (Coppage, 1972; Alsen et al.,1973; Coppage and
Matthews, 1974, 1975; Coppage et al.,1975), it becomes more
likely that brain acetylcholinesterase inhibition greater
than 70% is related causally to poisoning in acute exposure
and indicates high probability that some deaths have occurred

Table 1
Inhibition of Acetylcholinesterase in Brains of Fish When 40-60 Percent Were Killed by Carbamates.

Compound	Concentration (μg added/l)	Species	Hours Exposed	Mean Enzyme Activity[a]						Percentage Inhibition[b]
				Control	(N)	SD	Exposed	(N)	SD	
Carbaryl	1333	Pinfish	15	2.00	3	0.23	0.42	3	0.10	79
Carbofuran	100	Pinfish	48	2.01	4	0.33	0.40	3	0.08	80
Methomyl	500	Pinfish	19	2.04	3	0.16	0.35	4	0.12	83
Aldicarb	300	Sheepshead minnow	24	1.64	3	0.24	0.38	3	0.14	77
Bux	100	Sheepshead minnow	4	1.68	3	0.15	0.18	4	0.10	89
	40	Sheepshead minnow	24	1.79	3	0.09	0.33	3	0.03	82
	20	Sailfin molly	46	1.43	3	0.14	0.23	4	0.04	84

[a] Acetylcholinesterase activity was measured with a recording pH stat as μ moles of acetylocholine hydrolyzed/hr/mg brain tissue.

[b] All inhibitions were significant at $p < 0.005$ (Student's t-test).

or will occur in fish populations exposed to anticholinesterase agent. The standard deviation from normal enzyme
activity in a sample size of only 3-4 (3 pooled brains per
sample) was less than 17% in all cases, indicating severely
poisoned fish populations would be readily distinguishable
from unaffected populations.

The chemical mechanism by which carbamates inhibit acetylcholinesterase is believed to be in accordance with the
enzyme-substrate scheme:

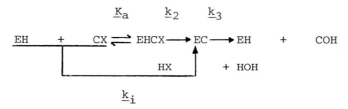

where EH is the enzyme, CX the carbamate, EHCX a complex
controlled by the equilibrium constant K_a, EC the carbamylated active site, HX the leaving group, COH the methyl
carbamic acid, k_2 the carbamylation constant, k_3 the reactivation constant, and k_i the overall bimolecular rate constant
for inhibition ($k_i = k_2/K_a$) (O'Brien et al., 1966; Metcalf,
1971). For example, it has been shown that the "X group"
(p-nitrophenyl group) is released when p-nitrophenyl dime-
thylcarbamate reacts with purified acetylcholinesterase from
a teleost (Electrophorus electricus) in vitro (Bender and
Stoops, 1965; O'Brien et al., 1966).

Carbamates are esters of carbamic acid that may be
considered synthetic analogs of acetylcholine (Fig. 1), and
their complex-formation with acetylcholinesterase is dependent on the similarity of "fit" upon the enzyme surface.
Complex-formation is followed by carbamylation of serine
hydroxyl at the active site. Inhibition results because the
carbamylated enzyme, EC, is several orders of magnitude more
stable than the acetylated enzyme formed in normal acetylcholine hydrolysis (Fig. 2). When the X group is split off
during carbamylation, the pesticide residue would not be
detectable by existing analytical chemical techniques, even
though the enzyme is inhibited and fish are poisoned. Inhibition of fish acetycholinesterase in vitro by carbamate
inhibitor is progressive but levels off before completion,
in contrast to the process for organophosphate inhibitors
(O'Brien et al., 1966), because the decarbamylation rate is
relatively faster than the dephosphorylation rate (Reiner,
1971). However, decarbamylation can be slow; Carter (1971)
reported that inhibition of channel catfish brain

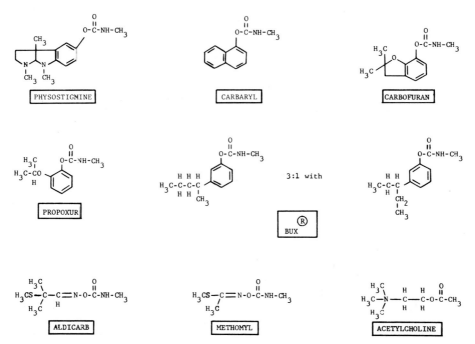

Fig. 1.   Structure of some carbamate anticholinesterase
          agents and the neurotransmitter acetylcholine.

acetylchclinesterase by carbamate pesticides in vivo lasted
for 5.5 - 9 days after exposure was discontinued.

Acetylcholinesterase measurements in fishes have proved
valuable in detecting pollution of fresh and estuarine
waters by some anticholinesterase agents (Williams and Sova,
1966; Holland et al.,1967; Carter, 1971; Coppage and Duke,
1971; Morgan et al.,1973; Coppage and Braidech, 1976). My
findings for carbamate pesticides in this report indicate
that inhibition can be used to determine extent of pollution,
poisoning of fishes, and cause of kills by carbamates in
natural systems. This is of special value for carbamate
pesticides where poisoning results from binding of chemically
undetectable metabolites of carbamates to the enzyme, and
where inadequate data on history of residues and exposure are
available.

I conclude that mechanisms related to injury of fishes
by carbamates are quantifiable in the environment and
should be measured in addition to chemical residues.

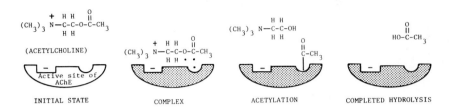

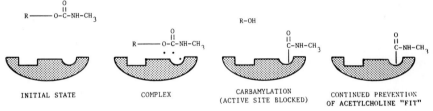

Fig. 2.   Carbamylation of acetylcholinesterase compared to
hydrolysis of normal substrate, acetylcholine, by
acetylation and rapid deacetylation of acetylcholinesterase.

ACKNOWLEDGMENT

I thank Mr. Edward Matthews for assistance in bioassays
and enzyme assays and Mr. Steven Foss for preparing the
figures.

The use of trade names does not constitute endorsement
by the Environmental Protection Agency.

LITERATURE CITED

Alsen, C., A. Herrlinger, and F. K. Ohnesorge. 1973.
Characterization of cholinesterases of the cod (Gadus
callarias) and their in vivo inhibition by paraoxon
and tabun. Arch. Toxicol. 30: 263-275.
Augustinsson, K-B. 1971. Determination of activity of
cholinesterases. In: Analysis of Biogenic Amines and

100   David L. Coppage

Their Related Enzymes, pp. 217-273, ed. by D. Glick.
New York: Wiley.

Bender, M. L. and K. Stoops. 1965. Titration of the active
sites of acetylcholinesterase. J. Amer. Chem. Soc. 87:
1622-1623.

Caro, J. H., H. P. Freeman, D. E. Glotfelty, B. C. Turner,
and W. M. Edwards. 1973. Dissipation of soil-incorporated carbofuran in the field. J. Agr. Food Chem. 21:
1010-1015.

_____, _____, and B. C. Turner. 1974. Persistence
in soil and losses in runoff of soil-incorporated
carbaryl in a small watershed. J. Agr. Food Chem. 22:
860-863.

Carter, F. L. 1971. In vivo studies of brain acetylcholinesterase inhibition by organophosphate and carbamate
insecticides in fish. Ph.D. dissertation. Louisiana
State University, Baton Rouge, Louisiana.

Coppage, D. L. 1971. Characterization of fish brain acetylcholinesterase with an automated pH stat for inhibition
studies. Bull. Environ. Contam. Toxicol. 6: 304-310.

_____ 1972. Organophosphate pesticides: specific
level of brain AChE inhibition related to death in
sheepshead minnows. Trans. Am. Fish. Soc. 101: 534-
536.

_____ and T. Braidech. 1976. River pollution by anticholinesterase agents. Water Res. 10: 19-24.

_____ and T. W. Duke. 1971. Effects of pesticides in
estuaries along the Gulf and Southeast Atlantic Coasts.
In: Proceedings of the 2nd Gulf Coast Conference on
Mosquito Suppression and Wildlife Management, pp. 24-31,
ed. by C. H. Schmidt. Washington, D. C.: National
Mosquito Control - Fish and Wildlife Management Coordi-
nating Committee.

_____ and E. Matthews. 1974. Short-term effects of
organophosphate pesticides on cholinesterases of
estuarine fishes and pink shrimp. Bull. Environ. Contam.
Toxicol. 11: 483-488.

_____ and _____. 1975. Brain-acetylcholinesterase
inhibition in a marine teleost during lethal and sublethal exposures to 1,2-dibromo-2,2-dichloroethyl
dimethyl phosphate (naled) in seawater. Toxicol. Appl.
Pharmacol. 31: 128-133.

_____, _____, G. H. Cook, and J. Knight. 1975.
Brain acetylcholinesterase inhibition in fish as a
diagnosis of environmental poisoning by malathion, O,O-
dimethyl S-(1,2-dicarbethoxyethyl) phosphorodithioate.
Pestic. Biochem. Physiol. 5: 536-542.

Dixon, M. and E. C. Webb. 1964. Enzymes. New York: Academic Press.

Environmental Protection Agency. 1972. The movement and impact of pesticides used in forest management on the aquatic environment and ecosystem. Pesticide Study Series - 7. Washington, D. C.: Office of Pesticide Programs.

Flickinger, E. L. 1973. Annual Progress Report. U. S. Fish and Wildlife Service, Denver Wildlife Research Center, Denver, Colorado.

Greve, P. A., J. Freudenthal, and S. L. Wit. 1972. Potentially hazardous substances in surface waters II. Cholinesterase inhibitors in Dutch surface waters. Sci. Total Environ. 1: 253-265.

Hestrin, S. 1949. The reaction of acetylcholine and carboxylic acid derivatives with hydroxylamine and its analytical application. J. Biol. Chem. 180: 249-261.

Holland, H. T., D. L. Coppage, and P. A. Butler. 1967. Use of fish brain acetylcholinesterase to monitor pollution by organophosphorus pesticides. Bull. Environ. Contam. Toxicol. 2: 156-162.

Karczmar, A. G., ed. 1970. Anticholinesterase Agents. New York: Pergamon Press.

Koelle, G. B., ed. 1963. Cholinesterases and Anticholinesterase Agents. Berlin: Springer-Verlag.

Lawless, E. W., R. Von Rumker, and T. L. Ferguson. 1972. The Pollution Potential of Pesticide Manufacturing. Washington, D. C.: U. S. Environmental Protection Agency.

Lowe, J. I. 1967. Effects of prolonged exposure to Sevin[R] on an estuarine fish, Leiostomus xanthurus Lacepede. Bull. Environ. Contam. Toxicol. 2: 147-155.

Metcalf, R. L. 1971. Structure-activity relationships for insecticidal carbamates. Bull. W. H. O. 44: 43-62.

Morgan, R. P., II, R. F. Fleming, V. J. Rasin, Jr., and D. R. Heinle. 1973. Sublethal effects of Baltimore Harbor water on the white perch, Morone americana, and hogchoker, Trinectes maculatus. Chesapeake Sci. 14: 17-27.

O'Brien, R. D. 1967. Insecticides. New York: Academic Press.

_____, B. D. Hilton, and L. Gilmour. 1966. The reaction of carbamates with cholinesterase. Mol. Pharmacol. 2: 593-605.

Reiner, E. 1971. Spontaneous reactivation of phosphorylated and carbamylated cholinesterases. Bull. W. H. O. 44: 109-112.

Silver, A.   1974.   The Biology of Cholinesterases.   New York:
    American Elsevier.
Webb, J. L.   1963.   Enzyme and Metabolic Inhibitors.   Vol. 1.
    General Principles of Inhibition, New York: Academic
    Press.
Williams, A. K. and R. C. Sova.   1966.   Acetylcholinesterase
    levels in brains of fishes from polluted waters.   Bull.
    Environ. Contam. Toxicol. 1:   198-204.

Part II.
Heavy Metals

Effect of Methylmercury upon Osmoregulation, Cellular Volume, and Ion Regulation in Winter Flounder, *Pseudopleuronectes americanus*

BODIL SCHMIDT-NIELSEN, JONATHAN SHELINE,
DAVID S. MILLER, and MARILYN DELDONNO

Mount Desert Island Biological Laboratory
Salsbury Cove, Maine 04672

Methylmercury occurs as a pollutant in many inland and coastal waters throughout the world (Friburg and Vostal, 1972). A number of studies have shown that methylmercury is concentrated in aquatic organisms through uptake by the gills and through the food chain (Hannerz, 1968; Bache et al., 1971; Hazeltine, 1971; Lockhart et al., 1972; Peakall and Lovett, 1972; Rivers et al., 1972; Fagerstrom and Asell, 1973; Andersen et al., 1974). When the concentration of mercury in the waters is below the acutely lethal limit, fish apparently can accumulate high concentrations of mercury in their tissues. Thus, in Swedish lakes the level of mercury in the muscle of pike averaged 0.5 ppm, and concentrations as high as 17-20 ppm have been recorded (Friburg and Vostal, 1972).

Do concentrations of mercury of these magnitudes in the tissues of fish affect physiological functions? In a previous study (Renfro et al., 1974) we addressed ourselves to the question of how mercury applied acutely might interfere with transepithelial and transmembrane transport of electrolytes. We found that methylmercury and mercuric chloride decreased sodium transport across the gill epithelium of the killifish Fundulus heteroclitus and across the urinary bladder epithelium of the winter flounder Pseudopleuronectes americanus. These mercurials also inhibited the activity of the enzyme Na,K-activated ATPase (Na,K-ATPase), an enzyme which has been implicated in active sodium transport (Schwartz et al., 1975).

From studies of the American eel <u>Anguilla</u> <u>rostrata</u> it
appeared that the mercury in the tissue decreased the ability
of the cells to ion regulate, since a negative correlation was
found between intracellular muscle potassium and mercury concentration of the tissue. The mercury in the muscle ranged
up to 1.5 ppm (on a wet weight basis) and was apparently due
to the contamination in the natural habitat of the eels
(Renfro <u>et</u> <u>al.</u>, 1974).
     In the present study we investigated further the effect
of mercury accumulation upon the overall osmoregulation
(plasma osmolality) in fish and upon intracellular volume and
ion regulation in several tissues. Using the winter flounder
<u>Pseudopleuronectes</u> <u>americanus</u> as an experimental animal, high
intracellular concentrations of mercury (up to 24 ppm in the
gills and 2.5 ppm in the muscles) were obtained by repeated
intramuscular injections of methylmercuric chloride. Neither
overall osmoregulation nor cellular volume and ion regulation
were affected. Measurements of Na,K-ATPase activity in ion
transporting tissues showed no decrease in the enzyme activity
in the mercury treated fish compared to the controls. It
appears that mercury is present in the tissues in a form that
does not inhibit the activity of Na,K-ATPase and the ability
to transport ions.

METHODS

     Winter flounder (<u>Pseudopleuronectes</u> <u>americanus</u>)(body
weight 150-450 g) caught off Mount Desert Island, Maine, were
acclimated in sea water at 10°C in a "living stream" (Frigid
Units, Inc.) for at least one week prior to the experiment.
They were then divided into three groups of four, each group
in separate aerated tanks: 1) control, 2) daily methylmercury
injection, and 3) methylmercury injection every other day.
The tanks were placed in an incubator at 10°C. The injection
fluid contained 4 mg CH_3HgCl/ml in 50% ethanol solution. The
fish were weighed and each injection calculated to correspond
to 1 mg of mercury per kg of wet fish weight (1 ppm Hg).
Controls were injected with a corresponding volume of 50%
ethanol solution every day, and fish given CH_3HgCl on alter-
nate days only (group 3) were injected with 50% ethanol on
the remaining days. All injections were intramuscular. The
water was changed daily for all fish following the injections.
     Half of the fish from each group were killed after four
injections (fish received a total dose of either 0, 2, or 4
ppm of Hg) and half were killed after thirteen injections
(total dose 0, 7, or 13 ppm Hg). Approximately twelve hours
before killing, each fish was injected in the caudal vein

with ^{14}C-labelled polyethylene glycol (PEG, from New England Nuclear), a high molecular weight polymer (M.W.=4000) which does not penetrate the cell membrane and equilibrates between the extracellular space (ECS) and the blood during the twelve hour period (Schmidt-Nielsen et al., 1972).

Blood samples were taken from the caudal vein just prior to killing, and plasma was separated from red cells immediately. Duplicate liver, kidney, gill, intestine and muscle samples (20-500 mg) were taken from each fish. The tissues were weighed and boiled for one minute in 500 µl of distilled water, then left overnight for diffusion. The samples were treated and analyzed as described previously by Schmidt-Nielsen (1976). Samples (20-200 mg) of each type of tissue were also taken for determination of percent water in the tissue. The samples were weighed immediately in preweighed aluminum foil boats and dried in an oven at 100°C for 24-48 hours. Calculations of intracellular water and ion concentrations were carried out as described previously (Schmidt-Nielsen, 1976). Additional tissue was taken for Na,K-ATPase assays. Na,K-ATPase activities were determined in freeze-dried tissue homogenates at 37° by the procedure of Miller and Kinter (1976).

RESULTS

Mercury in the Tissues

Mercury accumulated in the tissues of the flounder is shown in Figures 1 and 2. In muscle, intestine, and liver the increase in mercury concentration in the tissues was roughly proportional to the total dose given (Table 1). In red cells, gills, and kidneys the amounts accumulated were not directly related to the dose given, but also reflected the time elapsed since the last mercury was administered. The fish given 4 ppm received the last injection of mercury less than 24 hours before they were killed while the other groups had not been injected with mercury for 48 hours. Since injected mercury is first taken up in red cells and kidneys, these tissues (Clarkson, 1972) show the higher concentrations in the first hours following injection (Fig. 2).

Effect on Osmo-, Volume- and Ion-regulation

Accumulation of mercury in the gills up to 24 ppm did not decrease the ability of the flounders to hypoosmoregulate. To the contrary, the fish appeared to osmoregulate slightly better since a slight (not statistically significant) decrease in the plasma osmolality was observed with increasing

Table 1
Mercury accumulation in ppm per ppm of
dosage received by the fish (mean $\pm$ S.D.).

Tissue	No. of Determinations	Mean
Muscle	(7)	0.11 $\pm$ 0.02
Intestine	(7)	0.92 $\pm$ 0.12
Liver	(7)	2.00 $\pm$ 0.19

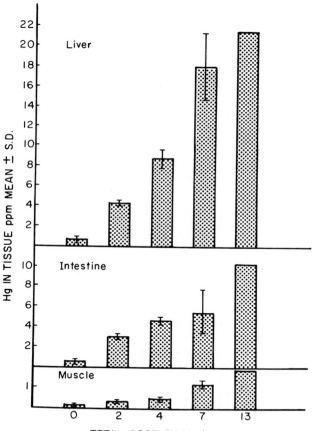

Fig. 1. Mercury in muscle, intestine and liver of the flounder
following the administration of CH_3HgCl i.m. Abscissa:
Total dose administered in parts per million (ppm)
of the wet weight of the whole fish.   Ordinate:
mercury in tissue in ppm on a wet weight basis.   The
values are given as mean of four samples from two
fish $\pm$ S.D.   Only one fish survived the 13 ppm.

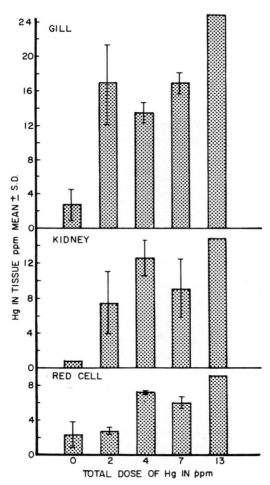

Fig. 2.   Mercury in red cells, gills, and kidney. Legend as
in Figure 1.

concentration of mercury in the gills (Fig. 3). The intracellular potassium concentration was unaffected by the concentration of mercury in all of the tissues analyzed (Figs.3,4).
None of the other intracellular ion concentrations were
affected by the accumulated mercury (Table 2), nor was the
water content of the cells affected (Table 3).

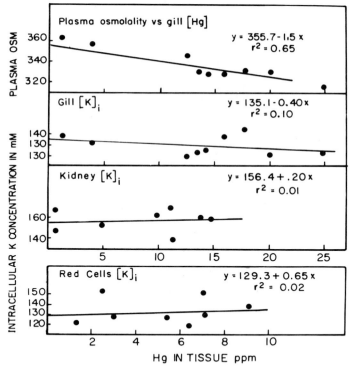

Fig. 3.   The top graph shows the plasma osmolality on the
ordinate, plotted tissue mercury concentration on the
abscissa, one point for each fish. Below the calculated intracellular potassium concentrations (average of 2 samples from each fish) are plotted against
the tissue mercury concentration in ppm on a wet
weight basis. None of the slopes are significantly
different from zero.

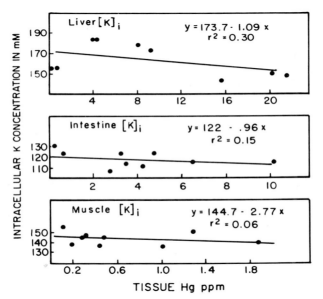

Fig. 4.   Calculated intracellular potassium concentrations
are plotted against tissue mercury concentration on
a wet weight basis.   None of the slopes are significantly different from zero.

Table 2
Calculated intracellular concentrations in mM per liter
intracellular water (mean $\pm$ S.E.).

		K	Cl	Ca	Mg
			Muscle		
Control	(2)	146 ± 9	17.2 ± 4.6	2.0 ± 0.4	7.3 ± 1.2
Hg treated	(7)	142 ± 2	9.8 ± 0.6	1.8 ± 0.3	7.0 ± 0.1
			Intestine		
Control	(2)	127 ± 4	77.5 ± 1.9	6.5 ± 4.6	24.7 ± 11.0
Hg treated	(7)	115 ± 3	59.8 ± 3.5	2.7 ± 0.6	20.9 ± 2.5
			Liver		
Control	(2)	156 ± 0	45.4 ± 1.6	–	7.6 ± 1.5
Hg treated	(7)	166 ± 6.8	32.5 ± 3.3	–	7.6 ± 0.8
			Red Cell		
Control	(2)	121	81.9	2.1	5.6
Hg treated	(7)	134 ± 5	87.8 ± 2.9	2.0 ± 0.6	4.6 ± 0.2
			Gill		
Control	(2)	135 ± 4	116 ± 1	4.3 ± 0.2	8.0 ± 0.8
Hg treated	(7)	128 ± 4	145 ± 16	3.2 ± 1.1	10.4 ± 1.4
			Kidney		
Control	(2)	158 ± 10	49.9 ± 2.1		7.0 ± 1.6
Hg treated	(7)	165 ± 7	43.5 ± 8.2		9.0 ± .41

Table 3
Percent water in tissue and calculated intracellular water expressed as g water per g solute free dry cell solids (mean $\pm$ S.E). Neither in percent water in tissue nor in calculated intracellular water is a statistically significant difference found between control and mercury treated animals.

	Muscle	Liver	Intestine	Red Cell	Gill	Kidney
			% Water in tissue			
Control	0.814±0.0.2	0.772±0.027	0.860±0.004	0.704±0.040	0.838±0.006	0.825±0.004
Hg treated	0.813±0.004	0.786±0.003	0.830±0.004	0.693±0.011	0.838±0.011	0.813±0.012
		Intracellular water g H_2O/g solute free dry cell solids				
Control	4.36 ±0.38	2.85 ±0.44	3.52 ±0.60	1.99 ±0.22	3.78 ±0.42	3.53 ±0.03
Hg treated	4.34 ±0.12	2.66 ±0.07	3.45 ±0.18	2.04 ±0.14	3.12 ±0.33	3.19 ±0.09

ATPase Activity of the Tissues

The Na,K-ATPase activity was determined only in the organs which show transepithelial electrolyte transport (gut, bladder, gill and kidneys). No significant difference in Na,K-ATPase activity was found between mercury treated and control fish in gills and intestine (Table 4). However, in bladder and kidney the Na,K-ATPase activity was higher in mercury treated versus control fish.

Table 4
Na,K-ATPase Activity in Tissues of Flounder (μMPi/mg Protein/ hr) (mean $\pm$ S.E.).

	Gill	Bladder	Intestine	Kidney
Control	17.9	11.9	19.3	13.1
(2)	±1.2	±1.7	±1.4	±1.0
	(8)	(7)	(8)	(9)
Hg treated	15.6	26.3	17.4	20.4
(7)	±1.9	±5.4*	±1.5	±2.2*
	(8)	(7)	(7)	(3)

*$p < 0.05$ compared to controls.

DISCUSSION

Marine teleost fish maintain the osmolality of the body fluids lower than that of the sea water. Water lost from the body by diffusion is replaced by sea water taken up through the alimentary canal; excess sodium and chloride are excreted by the gills. The enzyme Na,K-ATPase appears to be involved in osmoregulatory transepithelial electrolyte transport in the gills, intestine, and urinary bladder, as well as in active sodium-potassium exchange across all cell membranes (Schmidt-Nielsen, 1974). In a wide variety of tissues, this enzyme is sensitive to inhibition by mercurials and other sulfhydryl reagents (Schwartz et al., 1975). In fact, previous studies (Renfro et al., 1974) demonstrated mercurial inhibition of both active sodium transport and Na,K-ATPase in sea-water flounder urinary bladder in vitro, and of sodium trans-

port by freshwater killifish gills in vivo. These observations suggest that inhibition of membrane transport mediated
by Na,K-ATPase is a mechanism of toxicity in acutely exposed
teleosts. However, since apparently viable fish with a high
concentration of mercury in their tissues have been found in
many polluted areas, one might question if mercury, which
accumulates in the tissues of teleost fish during prolonged
low level exposure, continues to have this effect. From the
results reported here, it is obvious that electrolyte transport across the gills was not inhibited, since the fish were
able to maintain the plasma osmolality at the normal level of
350 mOs (650 mOs below the surrounding sea water) even though
tissue mercury levels in the gills approached 24 ppm. Furthermore, gill Na,K-ATPase activity was not inhibited. Several
factors may contribute to the lack of effect of mercury upon
osmoregulation. First, a chronic low level of mercury probably causes protective proteins such as metallothionein to be
synthesized (Piotrowski et al., 1974). Secondly, methylmercury may penetrate the cell membrane rapidly, and thus
membrane effects would be transient. Methylmercury is known
to become highly concentrated in the lysosome-peroxisome
fraction of rat liver cells four days after a single injection
(Norseth and Brendeford, 1971; Fowler et al., 1974). This
intracellular distribution is thus another detoxifying mechanism. Thirdly, if Na,K-ATPase were inhibited during chronic
exposure, it could be replaced by newly synthesized enzyme.
Indeed, recent experiments have demonstrated that flounder
are capable of replacing mercurial-inactivated enzymes by
increasing the rate of enzyme synthesis (Manen et al., in
press).

It is of interest to compare the present data showing no
effect of methylmercury accumulation on intracellular
potassium in the flounder with the previous data in the eel
showing a significantly negative correlation between intracellular potassium and tissue mercury concentration. The
difference in results might be due to species differences.
Another explanation could be that the eel (used in our
previous experiments), having been exposed to mercury in its
natural habitat, had also been exposed to other pollutants,
such as chlorinated hydrocarbons, which appears to have this
effect (unpublished observation).

LITERATURE CITED

Andersen, Arne T. and Ballu B. Neelakantan. 1974. Mercury in
    some marine organisms from the Oslofjord. Norw. J.
    Zool. 22: 231-235.
Bache, C. A., W. H. Gutenmann, and D. J. Lisk. 1971.
    Residues of total mercury and methylmercuric salts in
    lake trout as a function of age. Science 172: 951-952.

Clarkson, T. W. 1972. The pharmacology of mercury compounds. Ann. Rev. Pharmacol. 12: 375-406.

Fagerstrom, R. and B. Asell. 1973. Methylmercury accumulation in an aquatic food chain: A model and some implications for research planning. Ambio. 2: 164-171.

Fowler, Bruce A., Hayes W. Brown, George W. Lucier, and Margaret E. Beard. 1974. Mercury uptake by renal lysosomes of rats ingesting methylmercury hydroxide. Arch. Pathol. 98: 297-301.

Friburg, Lars and Jaroslav Vostal [Eds.]. 1972. Mercury in the Environment. Cleveland, Ohio: Chemical Rubber Co. Press.

Hall, Edward T. 1974. Mercury in commercial canned sea food. J. Assoc. Off. Anal. Chem. 57: 1068-1073.

Hannerz, L. 1968. Experimental investigations on the accumulation of mercury compounds in water organisms. Rep. Inst. Freshwater Res., Sweden, 48: 120-176.

Hazeltine, William. 1971. Mercury in the California environment. Clin. Toxicol. 4: 137-140.

Koirtyohann, S. R., Richard Meers, and L. K. Graham. 1974. Mercury levels in fishes from some Missouri lakes with and without known mercury pollution. Environ. Res. 8: 1-11.

Lockhart, W. L., J. F. Uthe, A. R. Kenney, and P. M. Mehrle. 1972. Methylmercury in northern pike (Esox lucius): Distribution, elimination and some biochemical characteristics of contaminated fish. J. Fish. Res. Bd. Canada 29: 1519-1523.

Manen, Carol-Ann, Bodil Schmidt-Nielsen and Diane H. Russell. 1976. Alterations of polyamine synthesis in liver and kidney of the winter flounder in response to methylmercury. Am. J. Physiol. (in press).

Miller, David S. and William B. Kinter. 1976. DDT inhibits nutrient absorption and osmoregulatory function in Fundulus heteroclitus. This volume.

Norseth, Tor and Margrethe Brendeford. 1971. Intracellular distribution of inorganic and organic mercury in rat liver after exposure to methylmercury salts. Biochem. Pharmacol. 20: 1101-1107.

Peakall, David B. and Raymond J. Lovett. 1972. Mercury: Its occurrence and effects in the ecosystem. Bioscience 22: 20-25.

Piotrowski, Jerzy K., Barbara Trojanowska, and Andrzej Sapota. 1974. Binding of cadmium and mercury by metaliothionein in the kidneys and liver of rats following repeated administration. Arch. Toxicol. 32: 351-360.

Renfro, J. Larry, Bodil Schmidt-Nielsen, David Miller, Dale Benos, and Jonathan Allen. 1974. Methylmercury and inorganic mercury: Uptake, distribution and effect on osmoregulatory mechanism in fishes. In: Pollution and

Physiology of Marine Organisms, pp. 101-122, ed. by
F. John Vernberg and Winona B. Vernberg. New York:
Academic Press.

Rivers, Jerry B., James E. Pearson, and Cynthia D. Shultz.
1972. Total and organic mercury in marine fish. Bull.
Environ. Contamin. Toxicol. 8: 257-266.

Schmidt-Nielsen, Bodil. 1974. Osmoregulation: Effect of
salinity and heavy metals. Fed. Proc. 33: 2137-2146.

_____. 1976. Intracellular concentrations of the salt gland
of the herring gull, Larus argentatus. Am. J. Physiol.
230: 514-521.

_____, J. Larry Renfro, and Dale Benos. 1972. Estimation
of extracellular space and intracellular ion concentration in osmoconformers, hypo- and hyperosmoregulators.
Bulletin, Mt. Desert Island Biol. Lab. 12: 99-104.

Schwartz, A., G. E. Lindenmayer, and J. C. Allen. 1975.
The sodium-potassium adenosinetriphosphatase: Pharmaco-
logic, physiological and biological aspects. Pharmacol.
Rev. 27: 3-134.

Methylmercury-Selenium: Interaction in the Killifish, *Fundulus heteroclitus*

JONATHAN SHELINE and BODIL SCHMIDT-NIELSEN

Mount Desert Island Biological Laboratory
Salsbury Cove, Maine 04672

Selenium is present in the oceans in a concentration of
0.09 ppb, approximately the same concentration as mercury.
Both elements are strongly accumulated by marine mammals.
Furthermore, tuna and swordfish are often found to have high
concentrations of mercury in their tissues. Koeman et al.
(1973, 1975) have shown that in the livers of the marine mammals, such as dolphins, porpoises,and seals, a strong correlation exists between the mercury and the selenium concentration with a molar ratio mercury/selenium of 1:1. This relationship holds true in the wide range of mercury concentrations found in the tissue which ranged from 0.5 to 326 ppm.
Cadmium, arsenic, zinc, and antimony concentrations are not
related to mercury concentration. In herring and mackerel, on
the other hand, selenium levels always exceed mercury levels.
The mercury/selenium ratio is 1:16 on a molar basis (Koeman
et al., 1975). In tuna Ganther et al. (1972) found that
selenium levels also exceed mercury levels. The mercury/
selenium ratio approaches 1:2 on a molar basis with increasing
concentration, but the increments in mercury content between
low and high mercury tuna is in approximately 1:1 molar ratio.
Ganther and Sunde (1974) further showed that tuna in the diet
protected Japanese quail against the toxicity of methylmercury
(the form in which mercury is most commonly found as a pollutant in the environment), while a corn-soya diet did not have
this effect. When selenium was added to the corn-soya diet
in amounts comparable to those in the tuna diet, a similar
decrease in toxicity was observed. Thus, these and several
other studies suggest that selenium protects against the

toxicity of methylmercury. The reverse has also been found
to be true. Mercury protects against the toxicity of selenium (Hill, 1972). The mechanism for this protection is not
fully understood.

Chen et al. (1974) found in the rat that pretreatment
with selenite followed by injections of $^{203}HgCl_2$ markedly
increased the mercury in plasma and testis while significantly
decreasing it in the kidney. They also studied subcellular
binding of mercury, and found that selenium pretreatment
significantly altered mercury distribution in the organs
studied. Mercury content was increased in the crude nuclear,
mitochondrial and microsomal fractions of the liver, and
decreased in the soluble fraction. The mercury in the soluble
fraction of the liver, testis, and kidney was markedly diverted
from a low to a large molecular weight protein which also
contains selenium. They propose that selenium counteracts
mercury toxicity by altering tissue concentrations and divert-
ing tissue mercury to presumably less critical components.

Burk et al. (1974) have presented strong evidence that
the mercury binding protein of $HgCl_2$ injected rats treated
with selenite is a selenoprotein.   They found that simultaneous administration of $^{203}HgCl_2$ and $^{75}SeO_3^{-2}$ greatly increases
the quantities of ^{203}Hg and ^{75}Se in the plasma over that
found when either element is given alone, due to their binding
to a single plasma protein. When doses of the elements are
varied, the molar ratio of selenium to mercury in the protein
remains close to one. Dialysis data showed that the mercury
is attached to the selenium. This supposed selenoprotein may
prevent mercury from reaching target tissues.

The reduction of methylmercury toxicity by selenium may
be due to the catalysis by selenium of the conversion of
methylmercury to a less toxic form (Stillings et al., 1974).
All mercury compounds tested to date undergo some cleavage of
the carbon-mercury bond in animal tissues (Clarkson, 1972).
In particular, rat liver slowly transforms methylmercury to
inorganic mercury and it was found that methylmercury is
excreted much more slowly than inorganic mercury in rats due
to biliary excretion of inorganic mercury and enterohepatic
recirculation of methylmercury (Norseth and Clarkson, 1971).

The present experiments were undertaken to answer some
of the many questions which arise from previous findings.
These questions are:

1. Does selenium cause a change in overall body retention of mercury?

2. Does selenium pretreatment in fish cause a redistribution of mercury among the organs?

3. Does selenium increase the rate with which methylmercury is transformed into inorganic mercury?

METHODS

Animals

Fundulus heteroclitus caught at the mouth of Northeast Creek on Mount Desert Island, Maine, were used in these experiments.[1] Body weights (BW) ranged from 3 to 10 g with an average weight of 5.4 g. The fish were acclimated in running sea water in the laboratory for about 4 weeks prior to the experiment. The temperature of the sea water rose from 8°C to 14°C during this period. The fish were fed twice a week. During the experimental period they were kept in aerated sea water in plastic tanks in an incubator maintained at a constant temperature of 14°C.

Experimental Procedure

The fish were divided into 4 groups according to the type of injections which they were to receive. All injections were administered intramuscularly, using a 50 μl microsyringe. Injections were accurate to 0.2 μl.

Fish were killed by decapitation at varying times following the mercury injection.

Group 1: no selenium, ^{203}Hg labelled methylmercury. An initial sham injection of 2 μl/g BW of a 0.9% saline solution was followed 30 min later by an injection of 1 μl/g BW of ^{203}Hg labelled CH_3HgCl in a 0.00535 molar solution 3.5 μCi/kg BW. This corresponds to a dose of 1.07 mgHg/kg BW (approximately 1 ppm) or 5.35 μmoles/kg BW.

Group 2: selenium pretreated, ^{203}Hg labelled methylmercury. Two μl/g BW of a 0.0025 molar Na_2SeO_3 solution which corresponds to 5 μM/kg BW, was injected and followed 30 min later by an injection of 1 μl/g BW of the ^{203}Hg labelled CH_3HgCl solution described above.

Group 3: no selenium, ^{14}C labelled methylmercury. An initial sham injection of 2 μl/g BW of 0.9% NaCl was followed 30 min later by an injection of 2 μl/g BW of 0.0025 molar ^{14}C labelled CH_3HgCl, which corresponds to 5 μmoles Hg/kg BW. The radioactive dose was 11 μCi/kg BW.

---

[1] Fundulus was chosen for study because it is readily available, the analysis cost per fish is low, and it is euryhaline. Originally a comparison of freshwater and seawater acclimated fish was to be undertaken.

Group 4: selenium pretreated ^{14}C labelled methylmercury.
After an injection of 2 µl/g BW of 0.0025 molar Na_2SeO_3 (as
in Group 2), 30 min later 2 µl/g BW of 0.0025 molar ^{14}C
labelled CH_3HgCl (as in Group 3) was injected.

One experiment tested the effect of selenium on whole
body retention of ^{203}Hg or ^{14}C. There were 32 fish in this
study, 8 in each of the groups mentioned above. Four from
each group were killed five hours after the mercury injection, and the rest 25.5 hrs after the ^{203}Hg or ^{14}C injection.
In a second experiment, the distribution of mercury in
liver, kidney, and gill tissue was measured. This involved
48 fish with 12 fish in each of the groups mentioned above.
Four from each group were killed at 4.5 hrs, at 25.5 hrs, and
at 73 hrs following the mercury injection. After the animals
were killed, samples of liver, kidney, and gill tissue were
taken from each fish. The samples were treated as described
below.
In a third group of 5 fish, the distribution into muscle
was tested. Three fish were treated as Group 1, and 2 fish
were treated as Group 2. These fish were killed after 3 hrs
and distribution of mercury determined in red cells, plasma,
liver, kidney, and muscle.

Sample Treatment

Whole Fish Experiment

Each fish was weighed after it was killed and the whole
fish was put in a Waring blender fitted with a small adapter
cup. Fifteen ml of distilled water was added and the tissue
homogenized to an even consistency. Homogenate (0.45 ml) was
pipetted into a scintillation vial with a 1 ml graduated
disposable pipette. Several samples of 1 ml were weighed to
determine the specific gravity. Two ml of NCS (Nuclear
Chicago Solubilizer) was added to each vial and the tissue was
digested at 45°C in a water bath for 2 hrs. Fifteen ml of
scintillation fluid (PPO, POPOP, toluene) was added and the
sample counted as described below.

Liver, Kidney, Gill, and Muscle Experiments

The tissue samples were taken from each fish, placed in
weighed scintillation vials, and reweighed. All of the kidney
was used, yielding about 10 to 30 mg of tissue; all of the
gills were also taken and the gill arches removed, yielding
15 to 60 mg of gill tissue. About 80 mg of liver tissue was
usually taken, and about 100 mg of muscle. One ml of NCS was
added to each scintillation vial and the tissues were digested

at 45°C for several hours. Next, 0.30 ml of a saturated
toluene solution of benzoyl peroxide (Baker, Reagent Grade)
was added in order to decolorize the digest. The vial was
then returned to the 45°C water bath for an additional 30 min.
Finally, 15 ml of scintillation fluid was added and the
samples counted as described below.

Counting Procedures

All samples were counted in a liquid scintillation counter, Nuclear Chicago Mark 1. Chemical quenching of the β
spectrum emitted by the isotope, which lowers the counting
efficiency, occurs in all samples and is a function of the
volume of tissue added. Colored tissue digests cause color
quenching of the β spectrum, due mainly to the hemoglobin
present in red blood cells. For the whole body retention
experiment both chemical and color quenching effects on counting efficiency were corrected for very simply by using
standards to which an equal volume (0.45 ml) of unlabelled
fish homogenate was added. Thus one uniform counting efficiency value was used since the volume and color composition
of all samples was essentially the same.

For the experiment on fish samples from different organs,
a quench curve was calculated using the external standard
method (Kobayashi and Maudsley, 1974), which relates percent
efficiency to the volume of tissue added, by using a channels
ratio. In this case, the samples were decolorized by adding
benzoyl peroxide to the digested tissues as described above.
This was necessary because different organs contain different
kinds and amounts of color quenching agents.

The concentration of mercury in the tissue was calculated
from the specific activity of the injection solution and from
the count per minute per gram of tissue corrected for quenching and for radioactive decay in the case of ^{203}Hg. All
values are given in μg Hg/g of wet tissue (ppm).

RESULTS

Mercury retained in Whole Body

Selenium pretreatment had no effect upon overall mercury
retention for the first 24 hrs following mercury injection
(Table 1). Mercury retention decreased slightly in both
groups from 5 to 25 hrs, but the difference is insignificant.
Methylmercury labelled by either ^{203}Hg or ^{14}C followed the
identical pattern. The amount retained in the whole body was
the same as the amount injected (within the experimental
error). Thus, there was no appreciable excretion of the

injected methylmercury during the first 25 hrs.

Table 1. Mercury retained in whole body of
          Fundulus following injection (i.m.)
          of 1 ppm in the form of methyl-
          mercury (CH_3HgCl)(mean ± S.D. and
          n = 4).

	Label	ppm Control	Se Pretreated
5 hrs		1.08 ±0.24	0.99 ±0.09
	^{203}Hg		
25 hrs		0.86 ±0.16	0.86 ±0.11
5 hrs		1.04 ±0.12	1.09 ±0.12
	^{14}C		
25 hrs		1.03 ±0.04	0.98 ±0.08

Mercury Distribution Among Organs

    Selenium pretreatment caused a redistribution among the
organs of the fish. Mercury concentration was significantly
reduced in the kidney of selenium pretreated fish compared to
controls (Fig. 1, Table 2). This differnce was evident, both
when the data were expressed as concentration in tissue and
when expressed as the concentration ratio kidney/gill. Already
4.5 hrs after the mercury injection, the mercury concentration
in the kidneys of selenium pretreated fish was 1/2 the concentration in the controls, and remained at this low level
throughout the 73 hrs.
    In the liver the mercury concentration was slightly but
not significantly lower in the selenium pretreated fish compared to the control fish (Table 3, Fig. 2).

Mercury Uptake by Muscles and Blood Cells

    Since mercury is apparently diverted away from the kidney,
it was of interest to see if other organs showed increased
mercury uptake in selenium pretreated fish. In the short-
term experiment of 3 hrs, it was found (Fig. 3) that the

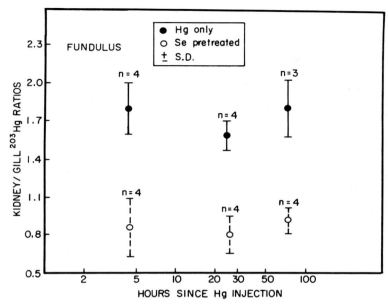

Fig. 1.   Mercury in the kidney. The counts of ^{203}Hg in the
kidneys are divided by the counts in the gills of the
same fish in order to eliminate any small variations
in the amount injected. Throughout the experimental
period the kidney/gill ratio of selenium pretreated
fish was about 1/2 that for control fish.   The difference is highly significant: p < 0.001.

Table 2.   Mercury in kidney of <u>Fundulus</u> calculated from ^{203}Hg
and from ^{14}C following injection of labelled methylmercury (mean ± S.D. and n = 4).

	ppm				Ratio Kidney/gill			
	Hg only		Hg + Se		Hg only		Hg + Se	
	^{203}Hg	^{14}C	^{203}Hg	^{14}C	^{203}Hg	^{14}C	^{203}Hg	^{14}C
5 hrs	0.63	0.47	0.20	0.39	1.78	1.42	0.86	0.87
	±0.07	±0.08	±0.08	±0.18	±0.20	±0.09	±0.23	±0.16
25 hrs	0.91	0.87	0.67	0.58	1.59	1.60	0.80	0.75
	±0.22	±0.21	±0.22	±0.28	±0.12	±0.15	±0.15	±0.11
73 hrs	1.28	1.62	0.67	0.73	1.80	1.76	0.91	0.99
	±0.22	±0.45	±0.24	±0.10	±0.22	±0.55	±0.11	±0.08

Table 3.   Mercury in liver of <u>Fundulus</u> calculated from ^{203}Hg
and from ^{14}C following injection of labelled methylmercury (mean ± S.D. and n = 4).

	ppm				Ratio Kidney/gill			
	Hg only		Hg + Se		Hg only		Hg + Se	
	^{203}Hg	^{14}C	^{203}Hg	^{14}C	^{203}Hg	^{14}C	^{203}Hg	^{14}C
5 hrs	0.52	0.30	0.26	0.38	1.44	0.92	1.12	0.90
	±0.17	±0.06	±0.11	±0.19	±0.35	±0.25	±0.36	±0.28
25 hrs	1.22	0.82	1.12	0.98	2.22	1.48	1.37	1.14
	±0.11	±0.30	±0.30	±0.72	±0.55	±0.19	±0.30	±0.26
73 hrs	1.55	1.98	1.11	1.09	2.23	2.11	1.52	1.46
	±0.06	±0.44	±0.46	±0.34	±0.54	±0.32	±0.39	±0.25

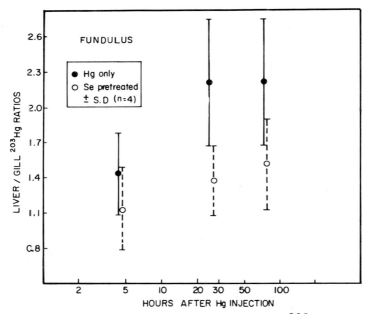

Fig. 2.   Mercury in the liver. The ratio of ^{203}Hg in liver/
gill is shown. The differences are not significant:
p < 0.05 for all three pairs.

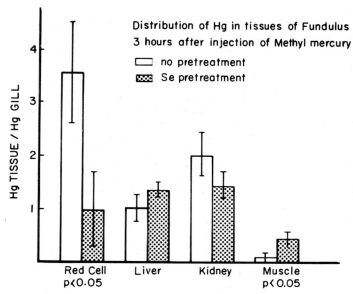

Fig. 3.   Ratios tissue/gill and ± S.D. are shown. Two fish
were selenium pretreated (dotted columns) and
three fish had only methylmercury (white columns).

muscle tissue in the selenium pretreated fish took up considerably more mercury than the muscle of the control animals.
Apparently, mercury is also diverted away from the red cells
in the selenium pretreated fish, since the concentration of
mercury was 3 1/2 times higher in the controls than in the
selenium pretreated fish.

Distribution of ^{203}Hg and ^{14}C methylmercury

From Tables 1, 2 and 3 it is seen that the distribution
of ^{14}C is under all conditions identical to the distribution
of ^{203}Hg within the experimental error. At no time and in no
tissue, whether pretreated with selenium or not, was there a
significant difference between accumulation of ^{203}Hg and ^{14}C.
Had there been a significant breakage of the bond between the
mercury and the methyl group, we should have expected to see
a different distribution of the labelled mercury and carbon.
Particularly, we would expect that with time a difference
would become significant.

DISCUSSION

Retention of Methylmercury

Selenium pretreatment did not change the total body
retention of methylmercury in the Fundulus during the first
24 hrs following the mercury treatment. This finding agrees
with the observation in chicks (Ansari and Britton, 1974)
where it was found that dietary selenium did not influence
^{203}Hg (fed in the form of $HgCl_2$) whole body retention. However, when mercury and selenium were given simultaneously, an
increased excretion of mercury was observed in chickens.
Stillings and Lagally (1974) found that rats fed swordfish
containing methylmercury excreted significantly more and
retained slightly less mercury in the tissue than rats fed
the same amount of methylmercury without the swordfish. Since
the swordfish probably contained selenium, the finding may be
similar to that in the chickens.

It appears then that the various vertebrates react
similarly to selenium with respect to mercury retention.
Selenium pretreatment does not increase the excretion of mercury. Selenium and mercury given simultaneously, however,
result in a slightly increased excretion of mercury.

Distribution in Various Tissues

The results of this study showed that in Fundulus, selenium pretreatment significantly decreased mercury retention in
the kidneys and red cells, and increased mercury retention in
muscle. Similar findings have been made in mammals. Chen et
al. (1974) found in rats that pretreatment with selenium
markedly increased the mercury content in blood plasma and
testis while decreasing it significantly in the kidney. Iwata
et al. (1973) found that sodium selenite in rats accelerated
the accumulation of methylmercury in the grain and significantly decreased retention in various organs.

When selenium is administered alone to weanling rats, it
accumulates to the highest degree in the kidney and to the
lowest degree in brain, muscle, and the eye (Brown and Burk,
1972). Methylmercury and inorganic mercury decreased liver
selenium while increasing muscle and brain selenium (Johnson
and Pond, 1974).

It appears then from the various findings that the
selenium-mercury protein might migrate to other organs after
it has been formed. More studies are needed to verify this.

Cleavage of the C-Hg Bond

In Fundulus we found no difference in concentration of

^{203}Hg and ^{14}C in any of the organs studied (Tables 1, 2, 3). This was true of control as well as of selenium pretreated fish. We feel this finding indicates that selenium caused little or no increased breakage of the C-Hg bond in the methylmercury.

Fang and Fallin (1974) have found a cleavage of phenylmercuric acetate and of ethylmercuric chloride in the kidney and liver of rats, but no measurable cleavage of methylmercuric chloride by kidney and liver.

The effect of dietary selenite on the activity of C-Hg cleavage enzymes in rat liver and kidney was investigated by Fang (1974). The activity of the cleavage enzyme for phenylmercuric acetate was increased by selenite in the liver, but not in the kidney; the activity of ethylmercuric chloride cleavage enzyme was unchanged. There was no measurable cleavage of the methylmercuric chloride either with or without treatment with selenite. Our data in Fundulus are thus in agreement with this finding in rats.

In conclusion, all of the findings in Fundulus are similar to the findings in mammalian tissues. Pretreatment with selenite causes (1) no changes in overall body retention of mercury, (2) a redistribution of mercury among organs, and (3) no measurable increase in C-Hg bond cleavage in methylmercury.

LITERATURE CITED

Ansari, M. S. and W. M. Britton. 1974. Effect of dietary selenium and mercury and ^{203}Hg metabolism in chicks. Poult. Sci. 53: 1134-1137.

Brown, D. G. and R. F. Burk. 1972. Selenium retention in tissues and sperm of rats fed a Torula yeast diet. Fed. Proc. 31: 692.

Burk, R. F., K. A. Foster, P. M. Greenfield, and K. W. Kiker. 1974. Binding of simultaneously administered inorganic selenium and mercury to a rat plasma protein (37894). Proc. Soc. Exp. Biol. Med. 145: 782-785.

Chen, R. W., P. D. Whanger, and S. C. Fang. 1974. Diversion of mercury binding in rat tissues by selenium: A possible mechanism of protection. Pharmacol. Res. Commun. 6: 579.

Clarkson, T. W. 1972. The pharmacology of mercury compounds. Ann. Rev. Pharmacol. 12: 375-406.

Fang, S. C. 1974. Induction of C-Hg cleavage enzymes in rat liver by dietary selenite. Res. Comm. Chem. Pathol. Pharmacol. 9: 579-582.

_____ and E. Fallin. 1974. Uptake and subcellular cleavage of organomercury compounds by rat liver and kidney. Chem-Biol. Interactions 9: 57-64.

Ganther, H. E. and M. L. Sunde. 1974. Effect of tuna fish
    and selenium on the toxicity of methylmercury. J. Food
    Sci. 39: 1-5.
_____, C. Goudie, M. L. Sunde, M. Kopecky, P. Wagner, S. H. Oh,
    and W. G. Hoekstra. 1972. Evidence that selenium in
    tuna decreases mercury toxicity. Fed. Proc. 31: 725.
Hill, C. H. 1972. Interactions of mercury and selenium in
    chicks. Fed. Proc. 31: 692.
Iwata, H., H. Okamoto and Y. Ohsawa. 1973. Effect of selen-
    ium on methylmercury poisoning. Res. Comm. Chem. Pathol.
    Pharmacol. 5: 673-680.
Johnson, S. L. and W. G. Pond. 1974. Inorganic vs. organic
    Hg toxicity in growing rats: Protection by dietary Se
    but not Zn. Nutr. Rep. Int. 9: 135-147.
Kobayashi, Y. and D. V. Maudsley. 1974. Biological Applica-
    tions of Liquid Scintillation Counting. New York:
    Academic Press.
Koeman, J. H., W. H. M. Peeters, and C. H. M. Koudstaal-Hol.
    1973. Mercury-selenium correlations in marine mammals.
    Nature 245: 385-386.
_____, W. S. M. van de Ven, J. J. M. De Goeij, P. S. Tijoe,
    and J. L. van Haaften. 1975. Mercury and selenium in
    marine mammals and birds. Sci. Total Environ. 3: 279-287.
Norseth, T. and T. W. Clarkson. 1971. Intestinal transport
    of Hg-203-labelled methyl mercury chloride--role of
    biotransformation in rats. Arch. Environ. Health 22:
    568-577.
Stillings, B. R. and H. R. Lagally. 1974. Biological avail-
    ability of mercury in swordfish (Xiphias gladius). Nutr.
    Rep. Int. 10: 261-267.
_____, _____, P. Bauersfeld, and J. Soares. 1974. Effect of
    cystine, selenium and fish protein on the toxicity and
    metabolism of methylmercury in rats. Toxicol. Appl.
    Pharmacol. 30: 243-254.

Effects of Cadmium on the Shrimps, *Penaeus duorarum, Palaemonetes pugio* and *Palaemonetes vulgaris*

DEL WAYNE R. NIMMO[1], DONALD V. LIGHTNER[2],
and LOWELL H. BAHNER[1]

[1]U. S. Environmental Research Laboratory
Gulf Breeze, Florida 32561

[2]University of Arizona Environmental
Research Laboratory
Tucson International Airport
Tucson, Arizona 85706

Cadmium has a deleterious effect on most, if not all, marine species tested (Gardner and Yevich, 1969; Gardner and Yevich, 1970; Eisler, Zaroogian, and Hennekey, 1972; Eisler and Gardner, 1973; Newman and MacLean, 1974). Cadmium in the diet can be a major source of cadmium to man and indeed long-term exposures to cadmium in humans has been implicated as contributing to a painful and crippling disease called "itai-itai" by the Japanese (Eisler, 1971). Cadmium occurs in estuarine water and sediments after receipt of industrial wastes (Holmes, Slade, and McLerran, 1974), and this accumulation has resulted in cadmium in sea food (Homes, Slade, and McLerran, 1974).

Histopathological effects of cadmium have been studied in two species of marine finfish (Gardner and Yevich, 1970; Newman and MacLean, 1974); however, no one has reported histopathological effects in any invertebrates. Invertebrates are more sensitive to the metal than are finfish (Eisler, 1971). There are few reports on the rates of accumulation of cadmium in marine species, and the only investigations we are aware of are studies on the flux of cadmium through mussels, shrimp, and euphausids (Fowler and

Benayoun, 1974; and Benayoun, Fowler, and Oregioni, 1974).
This lack of information caused us to undertake the present
study to: (1) determine rates of accumulation and
localization of cadmium in tissues of shrimp (Part I, D. R.
Nimmo); (2) describe the histological effects of cadmium in
shrimp (Part II, D. V. Lightner); and (3) compare the
accumulation of cadmium in shrimp from metal incorporated
in food with that of cadmium administered directly in water
(Part III, L. H. Bahner).

METHODS AND MATERIALS

Experimental Animals

     Juvenile and subadult pink shrimp, Penaeus duorarum,
were used in Parts 1 and 2, and adult grass shrimp,
Palaemonetes vulgaris and P. pugio, were used in Part 3.   All
species were collected by seine and are abundant in grass
beds in Santa Rosa Sound near Pensacola, Florida.   Holding
and acclimation procedures were those of Bahner, Craft, and
Nimmo (1975).

Bioassay Procedures

     Bioassays with shrimp were conducted in intermittently
or continuously flowing seawater (Bahner, Craft, and Nimmo,
1975).   Filtered water (25 $\pm$ 2°C and 20 $\pm$ 2 o/oo salinity)
was delivered to each glass aquarium containing the test
animals.   Intermittent flow-through bioassays with grass
shrimp were conducted using a proportional diluter similar
to that described by Mount and Brungs (1967).
     Concentrations of cadmium (CdCl$_2$.2 $\frac{1}{2}$H$_2$0) mentioned in
this report are calculated concentrations, unless otherwise
indicated.   We found that unless cadmium concentrations
exceeded 500 µg/ℓ, concentrations in the test water were
within 20% of those desired (Table 1.5).
     Procedures for studying the accumulation of Cd from
live food by grass shrimp in the laboratory follow: brine
shrimp, Artemia salina, were hatched in 48 hrs in 2-ℓ
separatory funnels that contained aerated seawater (salinity
20 o/oo).   Appropriate amounts of Cd were added to produce
desired concentrations in the brine shrimp.   The total
hatch in each funnel was collected on a fine nylon screen
(150 µ), rinsed with clean seawater to reduce transfer of
Cd into the feeder water, and then placed in the appropriate
feeder reservoirs that contained clean seawater.   Aeration
of the reservoirs dispersed the shrimp throughout the water.

Table 1.5. 30-day cadmium exposure: Accumulation of Cd in pink shrimp
exposed for 30 days in a flow-through bioassay, temperature
25°C, salinity 20 °/oo.

Concentrations in test water, mg/ℓ		Concentration in muscle, µg/g wet wt. mean ± 2 S.E.M.	Concentration factor [1]
desired	measured		
Control	----	0.4 ± 0.3	----
.075	.079	3.8 ± 0.3	48
.150	.182	10.4 ± 2.4	57
.300	.307	17.0 ± 3.0	55
.500	.586	19.4 ± 6.4	33
.750	.866	30.1 ± 3.4	34
1.000	1.285	30.5 ± 1.4	23

[1]Concentration in muscle divided by measured concentration in water.

Oscillating pumps, activated by a timer, pumped the brine shrimp-seawater mixtures into the test aquaria (Bahner and Nimmo, 1976).

Since we were unable to monitor the amount of food eaten by each grass shrimp, we assumed that, given the opportunity to feed ad libitum, the greater possibility of maximum transfer of Cd from food to carnivore would be achieved.

Preliminary investigations of Cd toxicity to brine shrimp showed that exposure to 100 mg Cd/ℓ allowed an extremely poor hatch in 48 hrs, whereas exposure to 10 mg Cd/ℓ allowed an almost total hatch. Therefore, 10 mg Cd/ℓ was thus chosen as the maximum concentration in which brine shrimp were hatched for the feeding test. A logarithmic series of concentrations, control, 0.32, 1.0, 3.2, and 10 mg Cd/ℓ was chosen. These gave the respective average whole-body bring shrimp residues: 0.75, 27.02, 41.45, 87.50, and 181.87 mg Cd/kg.

We designed specialized aquaria and provided excess water to reduce the likelihood of grass shrimp obtaining Cd from water and from detritus. Seventy-five liters of clean seawater/hr flowed through each 5 ℓ aquarium. This volume allowed the grass shrimp to swim freely and search for live food. Scavenging of detritus that contained Cd was minimized by a false floor of nylon screen, and any uneaten food or detritus were drawn through the screen and carried from the floors of the aquaria every two minutes by an external self-starting siphon.

Since the study required frequent, precise feeding with brine shrimp that contained a range of Cd concentrations, an electronic timer and switch circuit was designed to control the feeding cycle (Bahner and Nimmo, 1976).

Histological Procedures

Juvenile and subadult pink shrimp, Penaeus duorarum, and adult grass shrimp, Palaemonetes vulgaris, were exposed to cadmium in five separate tests. In the first and second, the pink shrimp ranged from 43 to 75 mm total length and in experiments 3 and 4, from 71 to 100 mm. Experiments 1 and 2 lasted 96 hours, and cadmium concentration was 5.0 mg/ℓ Experiments 3 and 4 lasted 21 and 25 days, respectively; concentration was 1.0 mg/ℓ. Grass shrimp were exposed to Cd concentrations between 75 and 300 µg/ℓ for up to 29 days in experiment 5. Cadmium was administered in the water, according to the methods of Bahner, Craft, and Nimmo (1975). In these tests, nominal concentrations varied less than 5% from the measured concentrations.

Some shrimp from the exposure were maintained in normal flowing seawater without cadmium for an additional 14 days to observe any latent effects from the exposure to Cd. Control shrimp were held in similar systems, but were not exposed to cadmium.

Histological material was prepared from shrimp taken at the termination of acute exposure tests (experiments 1 and 2) and from subacute exposure tests (experiments 3, 4, and 5). An occasional moribund animal was observed in the various experiments prior to the termination date and were preserved for histological examination. Also examined histologically were 6 shrimp from the 25-day exposure test (experiment 4) that were maintained in normal flowing seawater without Cd for an additional 14 days prior to being preserved.

Shrimp for histological examination were fixed in Davidson's fixative (Shaw and Battle, 1957) by one of two methods. In one method, the fixative was injected directly into the hepatopancreas and hemocoel prior to immersing the whole shrimp in fixative (experiment 3). The second method gave the best fixation and least artifact: the cuticle over the hepatopancreas and abdominal musculature was opened with scissors to enhance penetration of fixative, then the whole shrimp was immersed into fixative at room temperature. Shrimp were fixed in Davidson's for 24 hours, then stored in 70% ethyl alcohol. Processing for histological examination was accomplished by routine paraffin techniques. Sections were stained with Harris' haematoxylin and eosin or with periodic acid-Schiff.

Two experiments were conducted in static water at the Gulf Breeze Laboratory to determine accumulation of Cd by gills, muscle, hepatopancreas, and exoskeleton of shrimp. These tests were conducted in 30-liter glass aquaria with aerated seawater at a salinity of 20 o/oo and 25°C.

Analytical Methods

Atomic absorption instrumentation and reagents.-- Perkin-Elmer Models 403 and 503 AA spectrophotometers equipped with Models HGA-2000 and HGA-2100 Graphite Furnaces, Deuterium Background Correctors and Model 056 recorders were used for the Cd analyses. Reagents employed were:

Nitric acid (HNO$_3$), concentrated, suitable for mercury
    determination;
Sulfuric acid (H$_2$SO$_4$), concentrated, Baker Ultrex;
Acetic acid (CH$_3$COOH), glacial, reagent grade;
Ammonium-pyrrolidine-dithiocarbamate (APDC), 97% Baker;
Methyl-isobutyl-ketone (MIBK), reagent grade;
Ammonium nitrate 10% (w/v) solution;
Ammonium hydroxide (NH$_4$OH) approximately 10%, prepared

by dissolving of the gas in deionized water; and
Deionized water, prepared by passing tap-water through
2 mixed cation-anion exchange resin columns.

Cd analysis operation.-- Analyses for Cd employed flameless atomic absorption techniques. Operating parameters of
the spectrophotometers were: background correction on; wavelength 228.8 nm; slit 4; UV range; repeat and absorbance
modes; 0.25 A, recorder full scale; recorder response 1,
0.5 nm; chart speed 240 mm/hr. HGA controller was adjusted:
dry 15 sec @ 100, **char** 45 sec @ 300°, and atomize 7 sec
@ 2100. Nitrogen flow-meter was set to 4. Tap water was run
through the Furnace cooling-chamber at approximately 0.5 ℓ/
min.

Cd extraction procedure.-- Samples that contain <100 µg/
ℓ must be concentrated and the Cd separated from major cations
or organic matrices; otherwise, the major cations and organics
in seawater and biological samples interfere with direct
measurement of Cd. The methyl-isobutyl-ketone--ammonium-
pyrrolidine-dithiocarbamate (MIBK-APDC) extraction procedure
(Segar and Gilio, 1973) was modified for this purpose.

Acidified or wet-ashed samples containing too little Cd
for direct analysis were extracted by the following procedure.
One ml glacial acetic acid was pipetted into a 250-ml glass
beaker containing the prepared sample. The pH was adjusted
to 3.0 to 3.5, with NH_4OH. The solution was transferred to a
250-ml separatory funnel and 8 ml of 10% APDC and 5 ml of
glacial acetic acid were added, mixed thoroughly and extracted
twice with 15-ml volumes of MIBK. The organic phases were
combined in a 150-ml beaker and heated at 100°C until a
minimum volume was obtained without scorching the residue.
The hot residue was dissolved in 1 ml H_2SO_4 then ashed by
the addition of 5 ml HNO_3. The solution, heated at 100°C,
was evaporated just to dryness, diluted to 100 ml with
deionized water, and analyzed by atomic absorption spectrometry.

Seawater samples.-- Samples that contained 100 µg Cd/ℓ
were diluted with deionized water to permit direct measurement. A dilution factor of at least 100 eliminated inter-
ferences from the major cations in the sample making
possible direct measurement of minor metals (Segar, 1971).
Samples that contained <100 µg Cd/ℓ and could not be measured
by dilution were analyzed by two methods. (1) The samples
were extracted by the MIBK/APDC procedure for concentration
of Cd and removal of the interfering major cations. Seawater standards containing 500-5000 ng were acidified with
10 ml HNO_3 and diluted to appropriate volume prior to
extraction. (2) The simpler method, which eliminated the

need for extraction, consisted of injecting a 10 µℓ sample, followed by 10 µℓ of the 10% ammonium nitrate directly into the furnace (Table 1.1).

   Tissue samples.-- Tissue residues of shrimp were measured by techniques similar to those used for water samples. Shrimp were weighed to the nearest 0.01 mg, then wet ashed in 10 ml HNO_3 at $100°C$ for 4-8 hrs, or until no flocculent material formed when the ashed sample cooled. The ashed sample was diluted to 100 ml with deionized water. If the Cd concentrations were greater than 10 µg/100 ml, 10 µℓ of the samples were injected directly into the furnace. For samples containing less than 10 µg Cd/100 ml, the entire 100 ml was extracted by the MIBK-APDC method, diluted to the appropriate volume and measured by atomic absorption spectrometry. Recovery of Cd from a fortified solution of bovine albumin is given in Table 1.2 as an indication of the value of the extraction method.

Table 1.1
Recovery of Cadmium from Seawater. Each Datum Represents 6 Replicate Analyses.

Picograms/10 µℓ injected	Recoverable Cd (picograms) ($\bar{X} \pm 2$ S.E.M.)
0	ND[2]
0	ND
25	17.2 ± 2.2
25	16.6 ± 3.6
25	11.1 ± 2.4
25	15.6 ± 2.5
50	46.2 ± 1.9
100	114.0 ± 6.3
100	119.0 ± 3.4
250	221.3 ± 2.8

[1]10.0 µℓ volume of Cd in seawater followed by an equal volume of 10% NH_4NO_3 injected directly into furnace.

[2]ND - not detected (limit of detectability, 20 picograms)

Table 1.2
Recovery of Cadmium from a Fortified Solution of Bovine
Albumin[1].

Spike	Recovery[2]
ng	ng
100	140
200	230
500	480
500	550
500	510
500	510
1000	930
2000	2180

[1]Recovery by the APDC/MIBK extraction from a 10% w/v albumin
solution.

[2]Background concentration of Cd in five samples ranged from 26
to 39 nanograms (average, 33.6).

PART I. ACCUMULATION OF CADMIUM FROM WATER BY PENAEUS
DUORARUM AND LOCALIZATION IN VARIOUS TISSUES

RESULTS

Toxicology

    The toxicity of cadmium to pink shrimp has been reported
elsewhere, the 96-hr LC50 being 3.5 mg/ℓ and the 30-day LC50
720 µg/ℓ (Bahner and Nimmo, 1975; Nimmo and Bahner, in press).

Effect of Anions on Cadmium Toxicity and Accumulation

    Since the possibility existed that the cadmium anion may
have contributed to the toxic response and affected
accumulation of cadmium in tissues, we conducted four parallel
experiments to test this possibility (Table 1.3). Results
indicated no significant differences in toxicity of cadmium

Table 1.3
Cadmium Toxicity to and Accumulation by Pink Shrimp from
Various Anionic Solutions.*

Test Solution	Measured conc. (mg/ℓ)	Percentage Killed	Whole Body concentration (µg/g wet wt. $\bar{X} \pm$ 2 S.E.M.)
Control	0	0	0.3 ± 0.2
Cadmium acetate	1.6	75	6.8 ± 1.4
Cadmium chloride	1.8	65	8.5 ± 4.4
Cadmium sulfate	1.6	55	10.5 ± 3.0
Cadmium nitrate	1.7	75	10.8 ± 3.2

*Test conditions: 7-day flowing water bioassay conducted at
20 o/oo and 25°C; 20 animals per solution.

as the acetate, chloride, sulfate, or nitrate, nor was there
any appreciable difference in accumulation. In all
successive tests we tested cadmium as the chloride, which is
the anion most widely reported in the literature.

Localization in Tissues

     Analyses of shrimp exposed to cadmium for 96 hrs showed
(1) tissue concentration was directly proportional to that
administered in water; (2) Cd concentrations in the four
tissues analyzed differed by three orders of magnitude among
tissues; the ranking is hepatopancreas > exoskeleton >
muscle > serum (Fig. 1.1).
     After 96-hrs exposure, half of the shrimp were trans-
ferred to cadmium-free flowing seawater and held for 7 days.
After depuration cadmium in serum and exoskeleton became
significantly lower (Table 1.4), hepatopancreas remained
unchanged, whereas that in the muscle increased. In all
control tissues cadmium decreased significantly during the
tests.

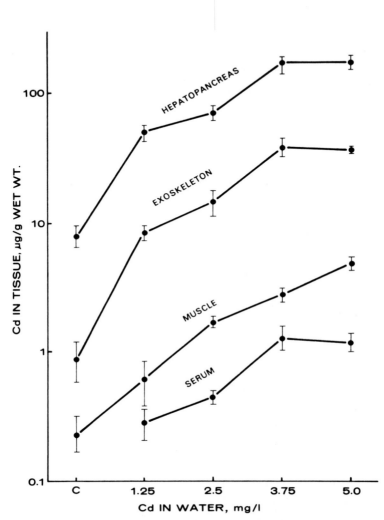

Fig. 1.1 <u>Penaeus</u> <u>duorarum</u>: uptake of cadmium from 20 o/oo seawater at 25°C for 96 hrs. Each datum = $\bar{X} \pm 1$ S.E.M.

Table 1.4.   Cd concentrations in tissues after 96-hr exposure to various Cd concentrations in sea water and after 7 days of depuration in Cd-free sea water.   (μg/g wet wt.; $\overline{X}$ with standard error in parentheses)

CD IN SEAWATER (mg/ℓ)	HEPATOPANCREAS		EXOSKELETON		MUSCLE		SERUM	
	After exposure	After depuration	After exposure	After depuration	After exposure	After depuration	After exposure	After depuration
0	7.9 (1.6)	4.1 (0.59)	0.86 (0.30)	0.37 (0.07)	0.24 (0.07)	.07* (.02)	ND**	ND
1.25	49.5 (6.8)	93.8 (23.9)	8.4 (1.1)	5.4 (0.8)	0.61 (0.23)	1.24* (0.10)	0.28 (0.07)	ND*
2.5	69.5 (9.1)	79.7 (12.7)	14.5 (3.4)	10.2 (1.1)	1.69 (0.17)	3.30* (0.61)	0.44 (0.05)	0.10* (0.06)
3.75	169.0 (33.3)	175.6 (29.4)	37.3 (5.7)	24.4 (6.8)	2.73 (0.35)	4.33* (0.43)	1.29 (0.27)	0.21* (0.05)
5.0	165.9 (19.7)	169.1 (27.5)	35.0 (3.5)	24.5 (5.1)	4.69 (0.54)	9.67* (1.09)	1.15 (0.21)	0.20* (0.03)

*Significant change (α = .05, analysis of variance) after 7 day depuration period.

**ND = non detectable

Accumulation in Muscle

Pink shrimp exposed to a range of 75 to 1000 µg Cd/ℓ accumulated quantities in muscle directly proportional to the concentration in the water (Table 1.5). Concentration factors ranged from 23 to 57.

Pink shrimp were exposed to a range of 2 to 30 µg Cd/ℓ in 2 experiments to determine the minimum concentration of the metal in water which would cause an accumulation in shrimp. At these concentrations, shrimp accumulated Cd in direct proportion to that in the water (Figs. 1.2 and 1.3); however, the concentration factors were slightly greater than those of tests at higher concentrations. When exposed to water that contained 2.0 µg/ℓ, Cd in muscle ranged from 250 ng/g within 50 days and when exposed to 30 µg/ℓ, 1200 ng/ g in 28 days. In the first test, concentrations of Cd in muscle of control animals remained unchanged (Fig. 1.2), while in the second test, they decreased with time (Fig. 1.3).

DISCUSSION

Cadmium can be concentrated by shrimp from water that contains 30 µg Cd/ℓ or less (Figs. 1.2 and 1.3), and concentrations of Cd continue to increase in shrimp after 50-days exposure. These data support those of Fowler and Benayoun (1974), who found that the benthic shrimp, Lysmata reticaudata, had not reached equilibrium after 2 months exposure. Our data suggest that if Cd were present in the environment in the form administered in the tests, then concentrations in shrimp would continue to increase for at least one to two months.

Cadmium concentrations measured in some estuaries approach concentrations toxic to shrimp or are sufficient to produce significant accumulation in shrimp (Holmes, Slade, and McLerran, 1974). In Corpus Christi Bay, Texas, water contained as much as 78 µg Cd/ℓ and in sediments as much as 120 mg Cd/kg (dry weight). The concentration toxic to pink shrimp in a 30-day bioassay was 720 µg Cd/ℓ. If shrimp were exposed to the concentration found in Texas, it is conceivable that it could be directly toxic, accumulated, and passed through the food web.

Depuration of Cd from control shrimp held in the laboratory in our experiments was more rapid than the loss from shrimp held in the field (Fowler and Benayoun, 1974). Our data show that most laboratory-held animals in control aquaria lost Cd. The single exception (an apparent state of

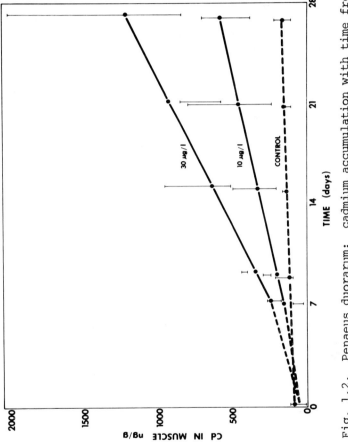

Fig. 1.2.   Penaeus duorarum: cadmium accumulation with time from water containing the metal. Bars indicate $\bar{X} \pm 2$ S.E.M.

143

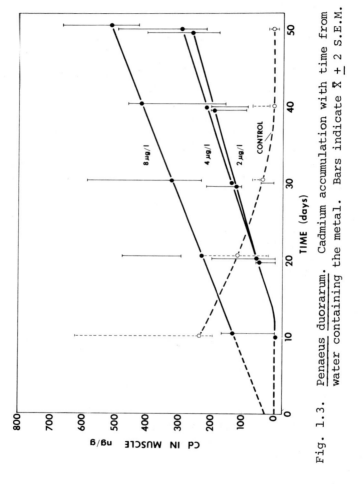

Fig. 1.3. Penaeus duorarum. Cadmium accumulation with time from water containing the metal. Bars indicate $\bar{X} \pm 2$ S.E.M.

equilibrium, Fig. 1.2) could have been due to low concentrations of Cd in a commercial food used to maintain these shrimp; although an analysis of Cd in the food proved <0.2 µg/g. In all other tests we fed shrimp muscle of fish, also <0.2 µg/g as to detectable Cd. Another explanation is that shrimp used in this experiment had been maintained in the laboratory for sufficient time to reach a low equilibrium level of Cd in the tissue.

The rate of cadmium loss in control Penaeus duorarum in the laboratory was much greater than that reported in the euphausiid, Meganyctiphanes norvigica (Benayoun, Fowler, and Oregioni, 1974). For the euphausiid, cadmium excretion was 97% after 96 days; whereas for pink shrimp, we measured a 50% loss in 7-10 days. Three possible reasons for this discrepancy exist: (1) The metabolic rates of these two crustaceans differed with regard to Cd; (2) In the test with euphausiids, the Cd was administered as ^{109}Cd in food. In our studies, we measured the "natural" Cd flux from control shrimp. The difference could be due to what was termed as "compartments" of the metal (Benayoun, Fowler, and Oregioni, 1974). Loss of Cd from these compartments could correspond to what Bryan (1971) has proposed as loss via the urine, gut, and excretion across the body surface and gills. (3) In our studies of Cd flux, we did not take into account the concommitant rate of growth during the time of depuration since our measurements were made on the wet-weight basis.

Cadmium is selectively localized in the tissues of decapod crustaceans (pink shrimp, present study; shrimp, Lysmata seticaudata, Fowler and Benayoun, 1974; lobster, Homarus americanus, Eisler, Zaroogian, and Hennekey, 1972). Cadmium was localized in the tissue of L. seticaudata in the following order: viscera > exoskeleton > muscle > eyes. A similar order was observed in the pink shrimp; hepatopancreas > exoskeleton > muscle > serum. Interestingly, a similar pattern exists in oysters exposed to ^{203}Hg; the metal being concentrated in tissues in the following order: gill > digestive system > mantle > gonad > muscle (Cunningham and Tripp, 1975a). We have found no report of apparent mobilization of Cd in the tissues of an arthropod, but in this study when shrimp were transferred to Cd-free water, Cd content in muscle increased, whereas Cd content of other tissues (e.g. hepatopancreas and exoskeleton) remained unchanged or decreased. During depuration oysters which had ingested ^{203}Hg-labeled algal cells lost the metal from gill and digestive tissue, but concentrations remained constant in the mantle and actually increased in gonad and muscle (Cunningham and Tripp, 1975b).

PART 2.   HISTOLOGICAL EFFECTS OF ACUTE AND SUB-ACUTE EXPOSURE
TO CADMIUM IN THE PINK SHRIMP

RESULTS

Histology

Branchia of pink shrimp exposed to cadmium in acute and
subacute tests consistently developed blackened foci or
melanized lamellae (= filaments) (Fig. 2.1). Shrimp exposed
to Cd in acute exposures (Experiments 1 and 2) developed
"black gills" 5 to 7 days after initiation of exposure. From
Experiment 1, two Cd-exposed and two unexposed control shrimp
were removed for histological examination. Blackened lesions
were grossly visible in the gills of both Cd-exposed shrimp
but were absent in the control shrimp. Similarly, in
Experiment 2, the six Cd-exposed shrimp examined histologically had blackened lesions in their gills, but those of the
control shrimp did not.
        The occurrence of black gills following subacute
exposure to Cd was less consistent. In Experiment 3, five of
eight Cd-exposed shrimp had blackened gills, and none of the
4 controls had blackened gills; whereas in Experiment 4, only
two of seven Cd-exposed shrimp examined histologically had
blackened lesions that were grossly visible.
        None of the three Cd-exposed shrimp held in normal
flowing seawater for 14 days following exposure to 1.0 mg
Cd/ℓ for 25 days, then sampled for histology, had black gills.
        Gills.-- Cross sections prepared through the gill region
of shrimp exposed to lethal concentrations of cadmium
consistently contained sections of damaged gill processes.
The least severe type of lesion observed was congestion and
blackening of branchial lamellae, accompanied by necrosis and
sloughing of individual lamellae (Fig. 2.2). This type of
lesion was observed commonly, particularly near the distal
ends of otherwise normal gill processes. Usually, multiple
lesions of this sort were observed on a single gill process.
Often, the distal portion of a gill process was congested
with hemocytes, blackened and necrotic (Fig. 2.3). In some
shrimp, whole gill processes were affected, while the
adjacent gill process appeared normal in the section studied
(Fig. 2.5).
        Congestion of gill lamellae and processes with large
numbers of hemocytes resulted in distention of the lamellae
and gross enlargement of gill processes, often to nearly
twice the relative size of unaffected gill lamellae and
processes (Fig. 2.4). Often the distortion was so pronounced

Fig. 2.1.   Normal pink shrimp, Penaeus duorarum, (bottom) and
pink shrimp with blackened branchial lesions (top)
that resulted from exposure to near lethal
concentrations of cadmium.

that the affected gill process bore little resemblance to
normal gills (Figs. 2.3, 2.4, and 2.5).

   The type of hemocyte involved in the hemocyte accumulations in the gills was not determined. However, large
numbers of hemocytes accumulated in certain gill lamellae and
gill processes, resulting in blockage of hemolymph channels
in gills where hemocyte accumulations were heavy. We presume
that such gills were nonfunctional.

   Sloughing of gill lamellae, portions of gill process,
and whole gill processes was observed. In every case the
process or lamella sloughed was heavily congested with large
numbers of necrotic and blackened hemocytes. The process
began once a blackened band had been formed on the proximal
side of the hemocyte accumulation (Figs. 2.2, 2.3, 2.4, 2.5,
and 2.6). Often a series of bands were formed as necrosis of
the gill process proceeded proximally (Fig. 2.5). The epidermal covering of the gill process migrated inward from
opposite sides of the gill process along the innermost
melanized band until the advancing epidermal segments met and
the epidermis became continuous proximal to the portion of
the gill process to be sloughed (Fig. 2.6). In these areas,
epidermal cells were cuboidal, rather than the usual
squamous epithelium found in normal gills, and were hyper-

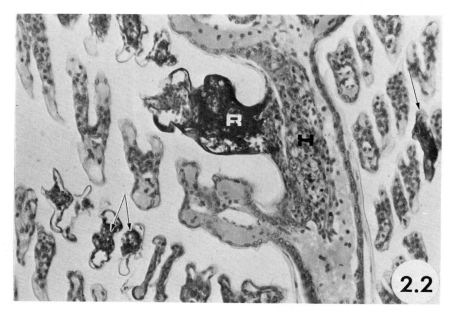

Fig. 2.2.   Gill process from a pink shrimp exposed to 5 ppm cadmium for 96 hrs. A secondary rachis (R) of a gill process is congested with hemocytes, blackened, and necrotic. Proximal to the origin of the secondary rachis in the hemolymph channel of the main gill process is a heavy accumulation of hemocytes (H) that have not become blackened. A few lamellae that are congested with hemocytes and blackened are also shown (arrows). Hematoxylin and eosin. X250.

chromatic as well as hypertrophic. Mitotic figures were occasionally observed along the leading edge of the advancing epidermis (Fig. 2.7). Following the advancing edge of the epidermis, a thin hyaline cuticle was secreted by the epidermis. Once the cuticle was in place, the overlying material began to slough (Fig. 2.6).

Shrimp held in normal flowing cadmium-free seawater for 14 days following exposure to 1 mg Cd/ℓ for 24 days appeared grossly normal. The blackened portions of the branchial processes had been sloughed by the 14th day, and histologically, the gills appeared like those of control

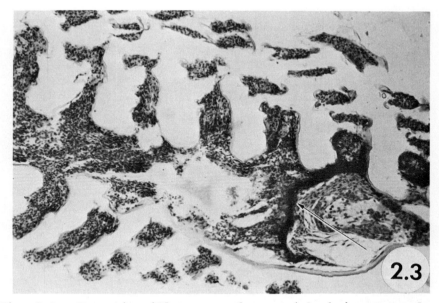

Fig. 2.3.   Necrotic gill process from a pink shrimp exposed
to 5 ppm cadmium for 96 hrs. The hemolymph
channels and supportive tissue elements in the
apical portion of the gill process have been
replaced by massive accumulations of hemocytes.
A dark blackened band (arrow) marks the boundary
between necrotic and viable tissue. Hematoxylin
and eosin. X160.

shrimp, except for occasional accumulations of debris
between the lamellae and an occasional blackened "scab" or
remnant of a lamella that had not yet been sloughed (Figs. 2.8
and 2.9).

        Fusarium.-- Of the shrimp examined histologically in
this study, the gills and adjacent tissues of 3 individuals
were infected by a Fusarium sp. Infections in 2 of the 3
were not advanced, but in the third shrimp, the infection was
advanced, the shrimp dying after 11 days of exposure to 1 mg
Cd/ℓ. The 2 other shrimp survived throughout the 25-day
exposure period. Control shrimp infected with Fusarium sp.
were not observed.

        Establishment of a Fusarium infection in these shrimp
was presumably related to the presence of necrotic gill
tissue. Mycotic infections of gills and adjacent tissues of

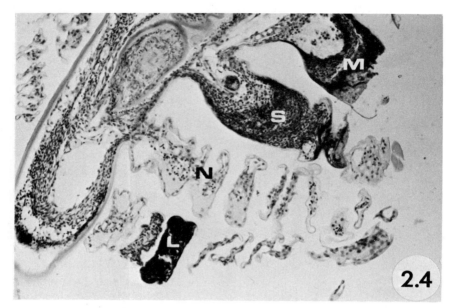

Fig. 2.4.   Section of a gill process from a pink shrimp that
was moribund after exposure to 1 ppm cadmium for
11 days.   Enlargement of a gill lamella (L) and
secondary gill processes (S) due to hemocytic
congestion is demonstrated by comparison to an
adjacent normal secondary process and lamellae (N).
The distal portion of one of the secondary gill
processes has been sloughed leaving a blackened
stump (M).   Hematoxylin and eosin.   X160.

several species of marine crustacea due to Fusarium molds
have been reported (Egusa and Udea, 1972; Johnson, 1974;
Lightner, 1975; Lightner and Fontaine, 1975; Lightner,
Fontaine, and Hanks, 1975).   Fusarium molds are common
marine fungi capable of opportunistic invasion in dead or
dying tissue of marine crustacea.

The gills of shrimp infected with the Fusarium sp.
contained large amounts of hyphae within the gill processes
and gill lamellae (Fig. 2.10).   Lamellae or gill processes
with heavy hemocyte accumulations were assumed to be the site
of initial infection by the mold (Fig. 2.11).   From the
initial site of infection hyphae spread into adjacent gills,
gill cover, and body wall.   The spread of hyphae into

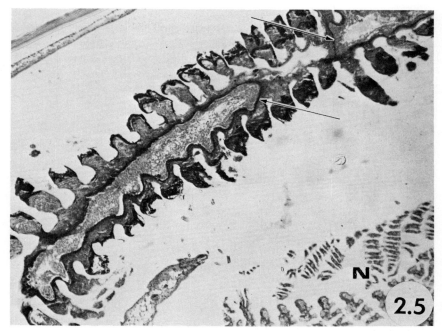

Fig. 2.5.   Low magnification photomicrograph of a single
             heavily congested gill process. A series of
             blackened bands (arrows) indicate the progression
             of hemocytic activity as necrosis of the gill
             process proceeded proximally. A portion of a
             normal gill process (N) is also present in the
             section. Hematoxylin and eosin. X45.

tissues of the body wall was accompanied by an extremely
heavy cellular inflammatory response (Figs. 2.10 and 2.12).
Hyphae in the subcuticular and muscle tissue proximal to the
gills were heavily encapsulated by multiple layers of hemocytes. Frequently, these encapsulations were blackened
(Fig. 2.12). Hemocytes accumulated in large numbers in the
vicinity of the expanding mycelium and replaced normal
tissues with a granulatomotous mass of hemocytes, necrotic
tissue debris, and hyphae (Fig. 2.12).
    Gills of shrimp that died after 11 days exposure to
cadmium were destroyed by the Fusarium mold, but showed only
a light inflammatory response. Gill processes and lamellae
of this shrimp were nearly filled with hyphae and conidia.
Only the cuticular covering of the most severely infected

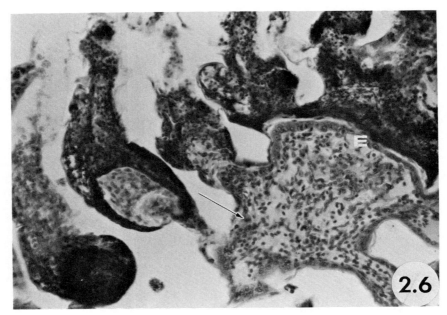

Fig. 2.6. Gill process from a pink shrimp exposed to 5 ppm
           cadmium for 96 hrs. The apical portion of the
           main rachis and the apical portions of each
           secondary rachis shown is being sloughed. The
           epidermal covering of the gill process has
           migrated inward and joined (E) and thereby
           isolated the portion of the process being sloughed.
           Hemocyte activity and accumulation is heavy, even
           proximal to the newly formed epidermis. Arrow
           indicates a mitotic figure on the leading edge of
           the migrating epidermal epithelium as it grows
           under a secondary gill process that is being
           sloughed. Hematoxylin and eosin. X250.

portions of the gills remained intact, leaving only
eosinophilic debris of what were supportive and epithelial
cells (Figs. 2.13 and 2.14).
    Micro- and macronidia were formed within the gill
processes and gill lamellae and on the surface of mycotic
lesions on the body wall. The fungus was classified as a
Fusarium sp. on the basis of morphology of the macroconidia
observed within and on the surface of lesions (Fig. 2.14).

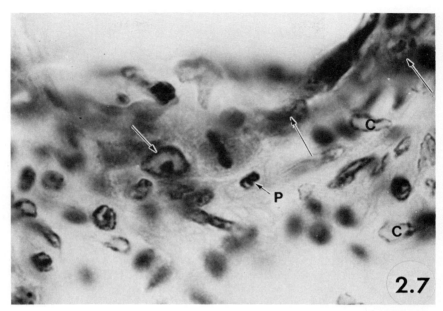

Fig. 2.7.   Higher magnification of the epidermal cell under-
            going mitosis that was shown in Figure 2.6.
            Nuclei of other epithelial cells of the new
            epidermis being formed are indicated by arrows.
            Most or all of the remaining cells are hemocytes,
            many of which show signs of necrosis, such as
            nuclear pyknosis (P) and chromatic margination
            (C). Hematoxylin and eosin. X1600.

Canoe or boat-shaped macroconidia are characteristic of
members of the genus Fusarium (Toussoun and Nelson, 1968).
    Occasionally a band of hematopoietic tissue was obtained
within the cross sections of the cephalothorax (Figs. 2.15
and 2.16). This tissue occurs as a band or nodule of baso-
philic tissue near the base of the gills in the body wall and
in the basal segments of the walking legs. Usually, these
hematopoietic nodules are in contact with a hemolymph vessel
that supplies the gill process. Mitotic figures were
numerous in this tissue in shrimp that had Fusarium infection.
We assume that this is due to the large number of hemocytes
involved in the inflammatory response to the mold, since
mitotic figures were seldom observed in this tissue in control
shrimp.

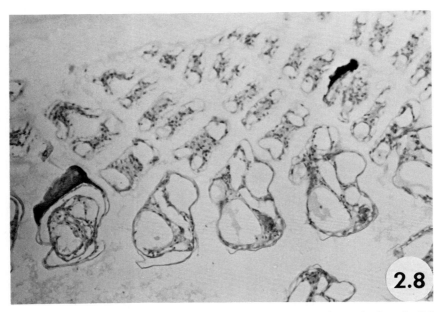

Fig. 2.8.   Section of gill lamellae from a pink shrimp held
            in normal flowing seawater for 14 days after a
            24-day period of exposure to 1 ppm cadmium. The
            lamellae appear normal except for two "scab-like"
            structures which represent sites of heavy hemo-
            cyte activity within the lamellae that have not
            yet sloughed. Hematoxylin and eosin. X160.

        Cuticular lesions.-- Occasional blackened cuticular
lesions were observed on appendages and general body surface
of shrimp exposed to cadmium in both acute and subacute
experiments. Comparable lesions were not observed in
unexposed (control) groups. Cuticular lesions were blackened
(melanized), were soft compared to adjacent unaffected areas,
and were somewhat friable. Histological sections through
cuticular lesions showed a perforation or dissolution of the
exocuticular and endocuticular layers of the exoskeleton
(Fig. 2.17). In some cuticular lesions, the endocuticular
layer of exoskeleton had been completely dissolved, while
leaving the relatively thin epicuticle essentially intact
(Fig. 2.17). In other lesions, the epicuticle was also
breached or eroded away from the underlying endocuticle. A

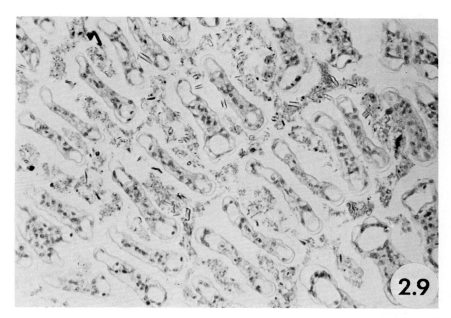

Fig. 2.9. Section of gill lamellae from the same shrimp as
in Figure 2.8. Excessive amounts of food and
other debris are present between the lamellae.
Hematoxylin and eosin. X200.

brownish scab-like structure was typically present within such lesions. This scab-like structure approximately
filled the space left by dissolution of the endocuticle, and
was composed of masses of hemocytes. Hemocyte activity in
underlying subcuticular tissues and muscle was noted but not
pronounced.
    The epidermis underlying some cuticular lesions was
necrotic or absent. In such lesions, hemocyte activity was
more pronounced than in lesions in which the epidermis
remained intact. Apparently, hemocyte accumulations served
as both a wound plug and as a base for ingrowth of epidermis
from the area surrounding the lesion until the epidermis
again became entire. Newly formed epidermis was composed of
a single layer of hypertrophic, hyperchromatic epithelial
cells that formed a new cuticle between the apical surface of
the epidermis and the blackened hemocytic "scab" that filled
the perforation in the old cuticle (Fig. 2.17).

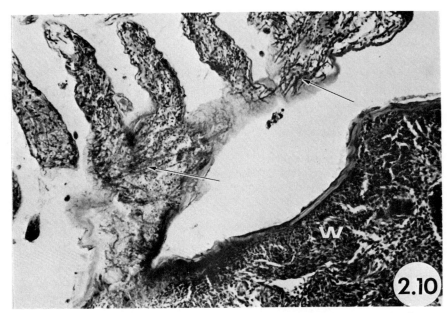

Fig. 2.10.   A gill process at its point of origin from the
             body wall.   Hyphae (arrows) of a <u>Fusarium</u> sp. are
             present in large amounts within the gill process.
             The section is from a pink shrimp that was
             moribund after 11 days of exposure to 1 ppm
             cadmium.   An intense cellular inflammatory
             response to the mold has occurred in the body
             wall (W).   Periodic acid-Schiff.   X160.

Midgut lesions.-- Inflammation of the anterior portion
of the midgut was observed in 2 of 8 shrimp exposed to
cadmium (Experiment 3), then examined histologically.
However, the condition was also observed in one control
shrimp from the same experiment.   The condition was not seen
in shrimp from the other experiments.

The affected portion of the midgut extended posteriorly
from its junction with the hepatopancreatic ducts and ventral
to the heart.   The lumen of the midgut at this location was
congested with masses of irregularly arranged hemocytes that
were becoming necrotic near the apical surface of the mass
and were being sloughed into the midgut lumen (Fig. 2.18).
A lining epithelium of the midgut was not demonstrable in
the affected portion of the midgut.   A blackened band was
present in the hemocyte accumulation in one of 3 shrimp.

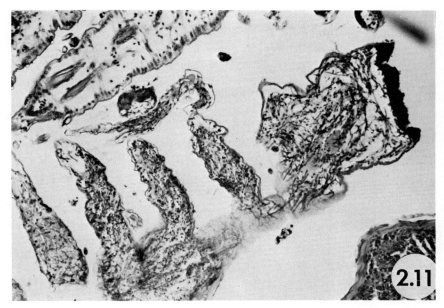

Fig. 2.11.   Distal portion of the gill process shown in
             Figure 2.10, showing a melanized hemocytic
             "scab". The apical portion of the gill process
             has been sloughed. This "scab" may represent the
             site of infection by the Fusarium mold. Periodic
             acid-Schiff. X160.

Accumulation Studies

     Cadmium was accumulated in the gills, hepatopancreas,
muscle, and exoskeleton to concentrations significantly
greater than was found in the same organs or tissues of
untreated control shrimp (Fig. 2.19; Table 2.1). Of these
tissues, the gills and hepatopancreas were the organs in
which highest concentrations of cadmium were found.

DISCUSSION

     Two hypotheses can be made to explain the histological
lesions observed in shrimp exposed to cadmium. The first
hypothesis is that soluble cadmium ions have a toxic effect
on the epidermal tissues of the gills and general body surface
of shrimp. Damage to these tissues would result in a "route

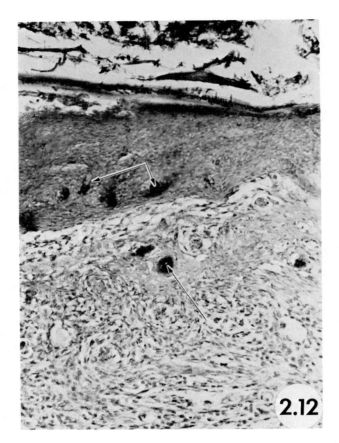

Fig. 2.12.   Fusarium lesion in the ventrolateral portion of
the body wall adjacent to the origin of a gill
process. The lesion is composed of masses of
hemocytes, hyphae, and tissue debris that have
replaced the muscle and subcuticular tissues at
this location. Hyphae (arrows) within the
lesion are characteristically surrounded by
multiple layers of encapsulating hemocytes.
Some of these encapsulations are melanized near
their centers. Hematoxylin and eosin. X280.

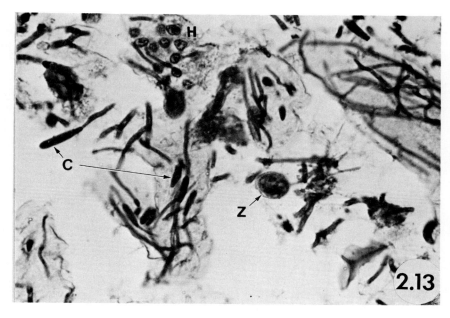

Fig. 2.13. Necrotic gill lamellae that contain hyphae and
conidia (C) of the Fusarium sp. A few hemocytes
(H) are present in one lamella and a Zoothamnium
trophont (Z) is present on the surface of another.
Periodic acid-Schiff. X640.

of entry" for chitinoclastic bacteria, which would in turn
result in "shell disease" lesions that are histologically
indistinguishable from the cuticular lesions observed in this
study. Chitinoclastic bacteria are a normal part of the
bacterial flora of shrimp (Hood and Meyers, 1974) and hence
are opportunistic pathogens of shrimp. This hypothesis,
however, does not explain the lesions observed in the gills.
    The second hypothesis is that the gills serve as a
route of elimination of cadmium. Accordingly cadmium would
be collected from the general body circulation by hemocytes,
transported to the gills, accumulated within the gill
lamellae and gill processes, and finally eliminated by
sloughing off affected portions of the gills. Evidence for
this hypothesis is: 1) gill lesions were the only lesions
consistently present in shrimp exposed to cadmium, whereas
lesions of the cuticle elsewhere on the general body surface
were not consistently present; 2) large numbers of hemocytes
were accumulated in the gills and eliminated by sloughing of

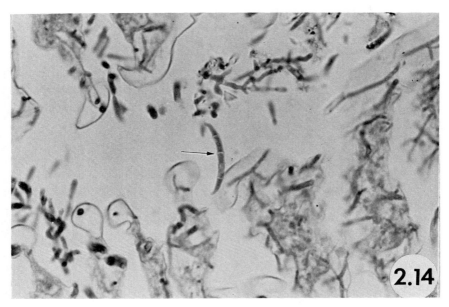

Fig. 2.14.   Necrotic gill lamellae from the same pink shrimp
             as in Figure 2.13. A macroconidium (arrow), the
             characteristic spore of the genus <u>Fusarium</u>, is
             clearly shown. Periodic acid-Schiff. X640.

the affected portions of the gill processes; and 3) cadmium
was concentrated within the gills.
       Accumulation of cadmium in the hepatopancreas, as well
as the gills, implies that both organs may serve as sites of
accumulation and elimination of cadmium (Fig. 2.19;
Table 2.1). The gills and hepatopancreas were shown to be
involved with the hemocytes in accumulation and elimination
of carmine particles that were injected into shrimp
(Fontaine and Lightner, 1974). Furthermore, in the lobster
(<u>Homarus</u> <u>americanus</u>), cadmium was reported to be
accumulated to a higher concentration in the gills (78% by
wet weight and 82% by ash weight greater than in the control
lobsters) than in muscle (25% and 17%), exoskeleton (49% and
7%), and the viscera, the last presumably including the
hepatopancreas, (9% and 11%) (Eisler, Zaroogian, and
Hennekey, 1972). These findings support the hypothesis that
gills serve with the circulating hemocytes to accumulate and
eliminate cadmium.

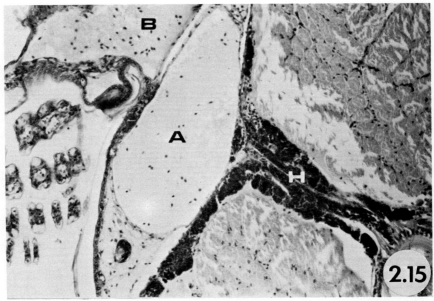

Fig. 2.15.   Section through the body wall at the point of
             origin of a gill process. A nodule of hemato-
             poietic (H) tissue is present in intimate
             association with a branchial artery (A).
             Opposite is the origin of a secondary artery
             (B) that supplies the gill process. Periodic
             acid-Schiff. X160.

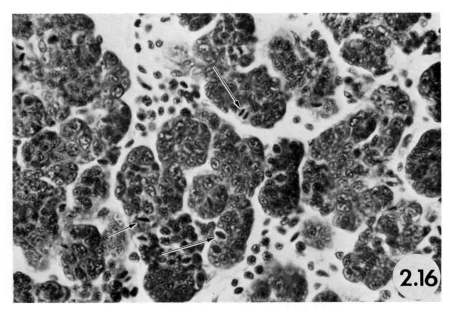

Fig. 2.16.   High magnification of hematopoietic cells
             shown in Figure 2.15.   Numerous mitotic figures
             (arrows) are present.   Periodic acid-Schiff.
             X640.

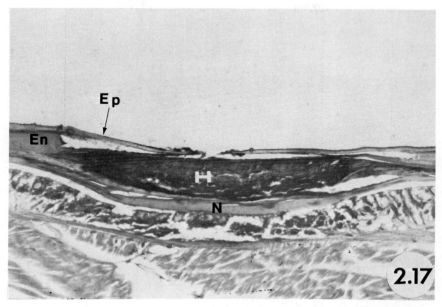

Fig. 2.17.   A cuticular lesion on the sixth abdominal
             segment of a pink shrimp exposed to 1 ppm
             cadmium for 21 days. The endocuticular (En)
             layer of the exoskeleton has been dissolved
             leaving the overlaying epicuticle (Ep)
             unsupported. A melanized accumulation of
             hemocytes, a "scab" (H), has formed a wound
             plug within the lesion. A new cuticle (N) has
             been formed basal to the scab. Hematoxylin
             and eosin. X280.

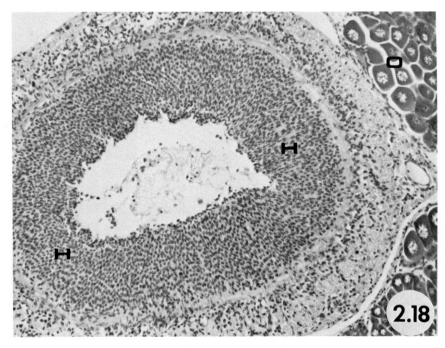

Fig. 2.18.   Section of the anterior portion of the midgut
near its junction with the hepatopancreatic
ducts.   A lining epithelium is absent.   The gut
lumen is nearly filled with a mass of hemocytes
(H).   Ovarian (O) lobes are present dorsal to
the midgut.   Hematoxylin and eosin.   X180.

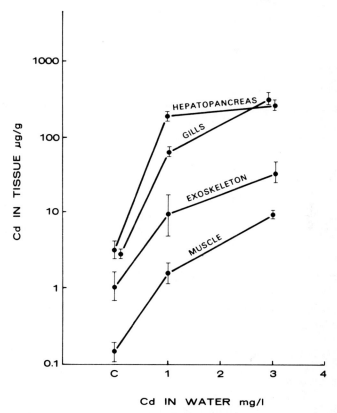

Fig. 2.19.   *Penaeus* *duorarum*.   Accumulation of cadmium in
various organs in comparison to that in the
gills.   Bars indicate $\bar{X} \pm 2$ S.E.M.

Table 2.1.
Accumulation of Cadmium in Gill Tissue of Pink Shrimp
Exposed to Various Concentrations of Cadmium in Flowing
Seawater for 7 Days and for 14 Days.

Test Conc. mg/ℓ		Tissue Conc. µg/g, wet wt.		Conc. Factor at 14 days
Nominal	Measured	7 days	14 days	
Control		1.7	2.6	-
.250	.224	13.6	51.6	230
.500	.515	17.4	60.3	121
.750	.736	33.5	149.8	200
1.000	1.010	74.7	176.5	177

PART 3.   TOXICITY AND ACCUMULATION OF CADMIUM FROM WATER
AND LIVE FOOD BY PALAEMONETES SPP.

RESULTS AND DISCUSSION

Toxicity in Seawater

    LC50's of Cd in seawater for P. vulgaris exposed for
various lengths of time were determined by probit analysis
of mortality data. The LC50 $\pm$ 95% CL at 96 hrs was 0.76 $\pm$
0.09 mg Cd/ℓ (Fig. 3.1), at 12 days was 0.18 $\pm$ 0.03 mg Cd/ℓ
(Fig. 3.2), and at 29 days was 0.12 $\pm$ 0.02 mg Cd/ℓ (Fig. 3.3).
The lowest lethal concentration tested was 50 µg Cd/ℓ.

Accumulation from Water

    In each experiment whole-body residues of P. vulgaris
and P. pugio increased as Cd concentration in the seawater
increased (Figs. 3.4 and 3.5). Cd was concentrated by grass
shrimp from seawater concentrations as low as 7.9 µg/ℓ.
Residues in all samples of Cd-exposed shrimp were
significantly higher ($\alpha = 0.05$) than residues in control
shrimp, indicating that sublethal Cd concentrations were

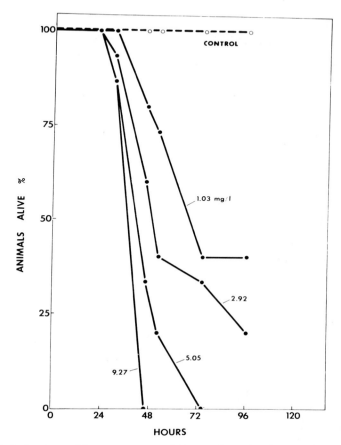

Fig. 3.1.   Mortality curves of P. vulgaris in acute
            (96-hr) bioassay at various measured
            concentrations of Cd in water.   96-hr LC50 –
            0.76 mg Cd/ℓ.

bioaccumulated during 35 days of exposure.   All animals were
dead in the 470 and 227 μg Cd/ℓ exposures after 5 and 10
days, respectively.   An apparent maximum limit of 40 mg Cd/kg
in whole-body residues was achieved in the 227 μg/ℓ exposure
(Fig. 3.5).

Accumulation from Food

     Cadmium residues in P. pugio fed for 14 days with Cd-
dosed brine shrimp increased with time (Table 3.1).

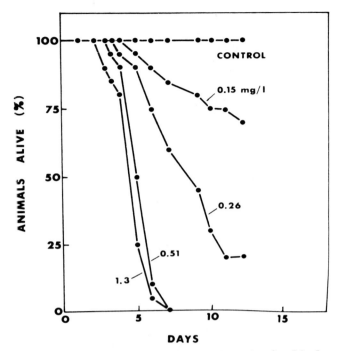

Fig. 3.2.   Mortality curves of P. vulgaris in 12-day
            bioassay at various measured concentrations
            of Cd in water. 12-day LC50 = 0.18 mg Cd/ℓ.

Although concentrations of Cd in brine shrimp remained
consistent throughout the test, Cd residues in grass shrimp
continually increased. The Cd concentration in grass shrimp
came from ingested food and not from Cd from water because
Cd concentrations in water from control and experimental
aquaria were not different.
    Cadmium concentrations in P. pugio in a second feeding
test increased rapidly for up to 7 days, and were stable for
the remainder of the 28-day accumulation study (Fig. 3.6).
Animals that remained alive after 28 days were held under
similar conditions for 7 days longer and fed brine shrimp
that contained control concentrations of Cd. In all
surviving shrimp, Cd residues were less than those attained
during the feeding exposure. This characteristic of
residues was discussed by Branson, et al. (1975) as a
kinetic relationship among exposure concentration, body
burden, and rate of depuration.

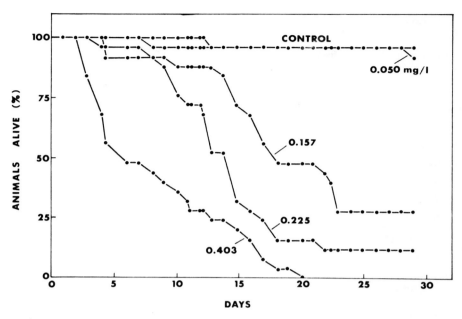

Fig. 3.3. Mortality curves of P. vulgaris in 29-day bio-
           assay at various measured concentrations of Cd in
           water. 29-day LC50 = 0.12 mg Cd/ℓ.

    No grass shrimp died during the feeding tests,
indicating that high concentrations of Cd in food were not
lethal.

Comparison of Accumulation from Food and Water

    Comparison of Cd accumulation from different sources is
aided by defining several terms. Cd accumulated directly
from water by a shrimp results in residues in tissues that
exceed the concentration in water. By comparing the tissue
residues in PPM (μg/g) to water concentrations in PPM (mg/ℓ),
the 'concentration factor' can be calculated. This term is
used extensively to describe both organic and inorganic
toxicant accumulation data. Several terms could be used to
describe Cd transfer from food to secondary consumer. Bioconcentration, biomagnification, and accumulation factor are
terms that could serve this purpose, however, in this study
Cd was transferred from Artemia to Palaemonetes, but by
factors of less than 1. Since it is difficult to describe
this transfer of Cd as bioconcentration, 'transfer potential'
is used to define the maximum transfer of Cd (under defined

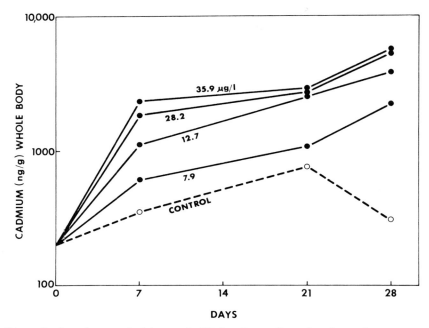

Fig. 3.4.   Accumulation of Cd by P. vulgaris from low
            concentrations in flowing seawater. Measured
            concentrations of Cd in water are shown.

test conditions) from food to the next trophic level. The
term 'food-chain potential' describes the maximum overall
accumulation and transfer of a toxicant within a food chain
under defined test conditions. The terms 'transfer
potential' and 'food-chain potential' can serve another
purpose beyond the scope of this report. These terms can be
used to compare various toxicants in specific food chains,
or could aid in describing the accumulation and transfer of
a specific toxicant in various food chains.

    The accumulation of Cd from water by factors of 18 to
84 times the initial concentration in the water indicates
that the brine shrimp accumulated Cd from water more
efficiently as exposure concentrations decreased (Table 3.2).
Cadmium was transferred to grass shrimp consistently (0.018
to 0.027 times) for all concentrations of Cd in food. We
believe these transfer potentials represent the maximum
transfer of Cd from bring shrimp to grass shrimp that could
occur under these conditions.

    The food chain potential is shown in Row 6 of Table 3.2.
This ratio represents the Cd transfer from water to

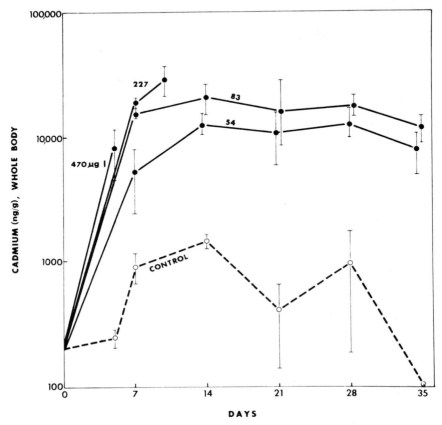

Fig. 3.5. Accumulation of Cd by P. pugio from low
concentrations in flowing seawater. Measured
concentrations of Cd in water are shown.
Vertical bars depict $\bar{X} \pm 2$ S.E.M.

secondary consumer, and is a function of accumulation of Cd
from water by bring shrimp and transfer of Cd from food
(brine shrimp) to grass shrimp. The food-chain potential
reflects the small transfer of Cd from the original source
to the secondary consumer.

Of particular interest is the comparison of Cd
accumulation from water and from food. In Table 3.3 the
numbers along the diagonal represent Cd residues (mg/kg) in
grass shrimp that were exposed to Cd in seawater or residues
in live food. Since exact residues were not available from
either the feeding test or from the low-level accumulation

Table 3.1 Whole-body Cd residues (µg/kg) accumulated during a 14-day flow-through feeding test of Palaemonetes pugio (25° ± 2°C, 20 °/oo ± 2 °/oo).

Time	Control (0.76 mg/kg)	Cadmium concentration in food			
		27.02 mg/kg	41.15 mg/kg	87.50 mg/kg	181.87 mg/kg
7 days (individual samples)	200	95	261	1123	2053
	150	236	292	912	2868
	246	401	235	1161	2070
	162	65		293	2288
	222	279			
	$\overline{X}$= 196	$\overline{X}$= 215	$\overline{X}$= 263	$\overline{X}$- 910	$\overline{X}$= 2564
14 days (composite samples)	$\overline{x}$= 155	$\overline{X}$= 352	$\overline{X}$= 994	$\overline{X}$= 1704	$\overline{X}$= 3926

172

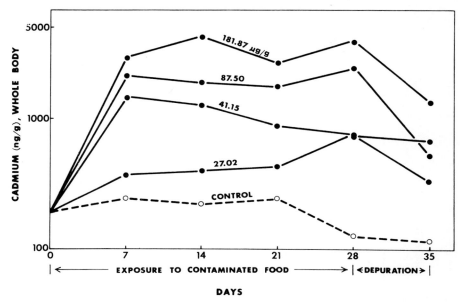

Fig. 3.6.   Accumulation of Cd by P. pugio from A. salina
with high whole-body concentrations of Cd.
Measured concentrations in Artemia are shown.

test, we plotted the available residue data and estimated the
concentrations of Cd in water or in food that would produce
comparable residues.   If these concentrations in food and in
water that produce similar residues in grass shrimp are
compared, the ratios of Cd in food to Cd in water range from
11,750 to 16,900 (Table 3.4).   We interpret this to mean
that to produce equivalent whole-body Cd residues in grass
shrimp, the concentration in food must be about 15,000 times
more than that in water.   Similar data for an organic
compound, DDT, in lake trout (Salvelinus namaycush) exposed
for 25 days (Reinert, Stone, and Bergman, 1974) permit a
comparison of accumulation of an organic compound with that
of a metal, Cd.   The DDT accumulation ratio in that study
was 15 times that of the maximum Cd accumulation ratio in
our study.

Histological Changes

     In contrast to the healthy gill structure of control
shrimp (Fig. 3.7), gill lamellae of P. vulgaris exposed to
75 µg Cd/ℓ for 10 days were melanized, necrotic, and

Table 3.2. Cadmium Transfer in Food Chain

	Control				
(1) Cd in Brine Shrimp Media (mg/ℓ)	Control	0.32 ppm	1.0 ppm	3.2 ppm	10.0 ppm
(2) Cd residues in Brine Shrimp (mg/kg)	Control	27.02 ppm	41.15 ppm	87.50 ppm	181.87 ppm
(3) Concentration Factor from Water ((2)/(1))	--	84	41	27	18
(4) Cd Residues in Grass Shrimp (mg/kg)	Control	0.49 *ppm	1.22 *ppm	2.07 *ppm	3.46 *ppm
(5) Transfer Potential from Food ((4)/(2))	--	0.018	0.027	0.024	0.019
(6) Food Chain Potential ((4)/(1)) or (3)x(5)	--	1.53	1.12	0.65	0.35

*Mean values for 7-, 14-, 21-, & 28-day samples.

174

Table 3.3. Comparison of accumulation of cadmium from water and food by grass shrimp (Measured Cd residues mg/kg of grass shrimp on diagonal).

Cd in Water (μg/ℓ)	Control	27,020	41,150	87,500	120,000**	181,870	215,000**
			Cd in Food (μg/kg)				
Control	.21*						
Control	.47***	.49*					
3.5**			1.12*				
6.9**				2.07*			
7.9					2.30***		
11.1**						3.46*	
12.7							3.80***

*Mean values for 7-, 14-, 21-, & 28-day samples.
**Calculated from measured data.
***Values from 28-day accumulation test (water).

Table 3.4. Comparison of cadmium bioaccumulation ratios in grass and
a DDT ratio in fish. (Measured concentrations in food and water
that gave equivalent residues).

Toxicant	Concentration in Food (μg/kg)	Concentration in Water (μg/ℓ)	Ratio	Time
Cd	27,020	Control	--	28 days
Cd	41,150	3.5**	11,757	" "
Cd	87,500	6.9**	12,681	" "
Cd	120,000**	7.9	15,189	" "
Cd	181,870	11.1**	16,385	" "
Cd	215,000**	12.7	16,929	" "
DDT*	2,000	0.008	250,000	25 days

* Reinert, Stone, and Bergman, 1974. (Regression analysis of original data.)

**Calculated from measured data.

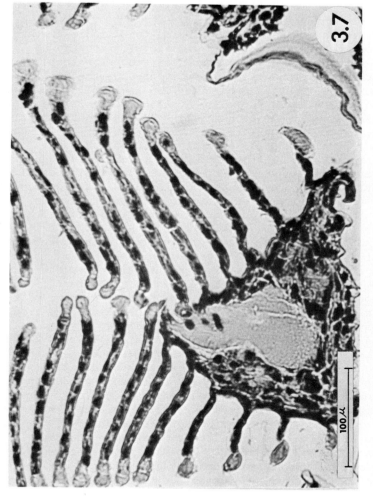

Fig. 3.7. Normal gill from grass shrimp, Palaemonetes vulgaris.

100 µ

3.7

distended due to congestion with large numbers of hemocytes
(Fig. 3.8). With the exception of distension, gill lamellae
of exposed P. vulgaris were grossly similar to those of
exposed Penaeus duorarum described in Part 2. The degree of
gill blackening appeared to be dose-related since blackened
gills were not noticed on shrimp that were exposed to sublethal Cd concentrations.

Three responses of grass shrimp to Cd indicate that at
least one target organ is the gill: (1) Exposure to sublethal concentrations in seawater or food show that Cd is
bioaccumulated from water (probably through the gills or gut)
much more readily than from live food (gut only), indicating
that unless Cd is in a readily available form in the food, Cd
is accumulated mainly by the gills; (2) Residues in grass
shrimp, exposed to sublethal concentrations of Cd in seawater
or food, level off after several days, indicating an
equilibrated kinetic state in which the gills can function in
respiration and physiological ion balance; (3) Lethal
concentrations of Cd produce a qualitative dose-response of
gill blackening, which may lead to disruption of normal
respiration (due to gross necrosis of gill lamellae) and
death.

Palaemonetes spp. are easy to maintain and test in the
laboratory, making them a preferred bioassay animal. Larval
stages are generally more sensitive to toxicants than are
adults. Entire-life-cycle bioassays of the effect of Cd on
Palaemonetes would permit more effective evaluation of the
chronic toxicity of Cd to this estuarine animal and feeding
bioassays could also be conducted simultaneously using
Artemia nauplii as food for most life stages. This method
could be modified for use with other toxicants. In any
feeding bioassay care must be taken that concentrations of
toxicant in exposure water do not rise above concentrations
in natural water and thereby obscure residues of toxicant
that accumulated solely from the food. In future bioassays
with shrimp the combined effect of accumulation of toxicants
from water, food, and sediments should be considered.

Rates at which toxicants depurate may be as important to
the animal as the toxicant concentration in the environment.
Therefore, to better understand the transfer of toxicants
through an estuarine trophic level, detailed depuration
tests should be conducted as an integral part of
accumulation experiments.

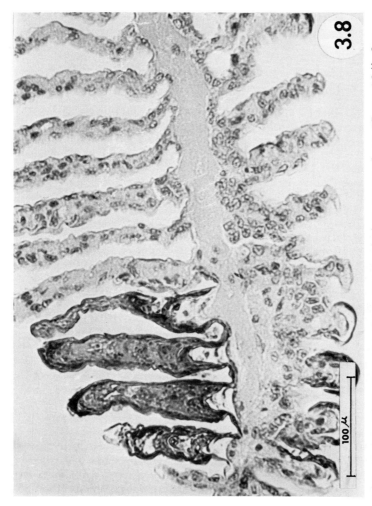

Fig. 3.8. Gill process from grass shrimp exposed to 75 μg Cd/ℓ for 10 days.

179

OVERALL CONCLUSIONS

1. Grass shrimp, Palaemonetes vulgaris, were acutely and
    chronically more sensitive to Cd than are pink shrimp,
    Penaeus duorarum. The LC50 at 96 hrs was 0.76 mg Cd/ℓ
    for grass shrimp whereas the 96-hr LC50 for pink shrimp
    was 3.5 mg Cd/ℓ (Bahner and Nimmo, 1975). The 29-day
    LC50 for grass shrimp was 0.12 mg Cd/ℓ and for pink
    shrimp, the 30-day LC50 was 0.718 mg Cd/ℓ (Nimmo and
    Bahner, in press).
2. Bioaccumulation of Cd from water occurred at concentra
    tions as low as 2 µg/ℓ in P. duorarum and 7.9 µg/ℓ in P.
    vulgaris. Cadmium in the tissues of P. duorarum did not
    plateau at low concentrations (<30 µg/ℓ), but Cd in the
    tissues of P. pugio did plateau at concentrations
    >54 µg/ℓ within 7 days. Cadmium in tissues of P. duorarum
    varied with concentrations in hepatopancreas > gills >
    exoskeleton > muscle > serum. Concentrations of Cd
    increased in muscle after Cd-exposed shrimp were trans-
    ferred to Cd-free water. Natural levels of Cd in shrimp
    were reduced by holding feral animals in flowing water
    from the estuary.
3. Pink shrimp, exposed to cadmium concentrations near LC50's,
    consistently developed blackened foci or blackened
    lamellae in the branchia. Occasionally, blackened
    cuticular lesions on the appendages and general body
    surface were observed. It is possible that Cd was
    collected by hemocytes and accumulated in the gills which
    were then walled off and later sloughed. When shrimp
    that survived exposures to Cd were placed in Cd-free
    water, they sloughed blackened portions of the branchia
    and appeared normal within 14 days. Likewise, the gill
    lamellae of P. vulgaris were blackened, necrotic, and
    distended due to congestion with large numbers of hemo-
    cytes after the shrimp were exposed to 75 µg Cd/ℓ for 10
    days.
4. When the brine shrimp, Artemia, containing Cd were used
    as food, the transfer of Cd to grass shrimp was much less
    efficient than transfer of Cd directly from the water.
    To produce equivalent whole-body residues in the shrimp,
    about 15,000 times more Cd must be introduced in food
    than could be obtained from seawater.

ACKNOWLEDGMENTS

     We thank Steven S. Foss for preparing the figures and
Imogene Sanderson for preparing the histological materials.

Part of this manuscript is taken from a thesis (L.H.B.) presented to the faculty of the Biology Department, The University of West Florida, in partial fulfillment of the requirements for the Master of Science degree. Another part was accomplished while the author (D.V.L.) was employed by the National Marine Fisheries Service, Gulf Coast Fisheries Center, Galveston Laboratory, Galveston, Texas.

LITERATURE CITED

Bahner, L. H., C. D. Craft, and D. R. Nimmo. 1975. A salt-water flow-through bioassay method with controlled temperature and salinity. Prog. Fish-Cult. 37: 126-129.
_____ and D. R. Nimmo. 1976. A precision live-feeder for flow-through larval culture or food chain bioassays. Prog. Fish-Cult. 38: 51-52.
_____ and _____. 1975. Methods to assess effects of combinations of toxicants, salinity, and temperature on estuarine animals. pp. 169-177. In: Trace Substances in Environmental Health - IX. University of Missouri at Columbia.
Benayoun, G., S. W. Fowler, and B. Oregioni. 1974. Flux of cadmium through euphausiids. Mar. Biol. 27: 205-212.
Branson, D. R., G. E. Blau, H. C. Alexander, and W. B. Neely. 1975. Bioconcentration of 2,2', 4,4'-tetrachloro-biphenyl in rainbow trout as measured by an accelerated test. Trans. Am. Fish. Soc. 104: 785-792.
Bryan, G. 1971. The effects of heavy metals (other than mercury) on marine and estuarine organisms. Proc. R. Soc. Lon. B 177: 389-410.
Cunningham, P. A. and M. R. Tripp. 1975a. Accumulation, tissue distribution and elimination of $^{203}HgCl_2$ and CH_3 $^{203}HgCl$ in tissues of the American oyster, Crassostrea virginica. Mar. Biol. 31: 321-334.
_____ and _____. 1975b. Factors affecting the accumulation and removal of mercury from tissues of the American oyster, Crassostrea virginica. Mar. Biol. 31: 311-319.
Egusa, S. and T. Ueda. 1972. A Fusarium sp. associated with black gill disease of the Kuruma prawn, Penaeus japonicus Bate. Bull. Japan. Soc. Sci. Fisher. 38: 1253-1260.
Eisler, R. 1971. Cadmium poisoning in Fundulus heteroclitus (Pisces: Cyprinodontidae) and other marine organisms. J. Fish. Res. Board Can. 28: 1225-1234.

_____ and G. R. Gardner. 1973. Acute toxicology to an estuarine teleost of mixtures of cadmium, copper, and zinc salts. J. Fish. Biol. 5: 131-142.

_____, G. E. Zaroogian, and R. J. Hennekey. 1972. Cadmium uptake by marine organisms. J. Fish. Res. Board Can. 29: 1367-1369.

Fontaine, C. T. and D. V. Lightner. 1974. Observations on the phagocytosis and elimination of carmine particles injected into the abdominal musculature of the white shrimp, Penaeus setiferus. J. Invert. Pathol. 24: 141-148.

Fowler, S. W. and G. Benayoun. 1974. Experimental studies on cadmium flux through marine biota. pp. 158-178. In: Comparative Studies of Food and Environmental Contamination. Vienna: IAEA.

Gardner, G. R. and P. P. Yevich. 1969. Toxicological effects of cadmium on Fundulus heteroclitus under various oxygen, pH, salinity and temperature regimes. Am. Zool. 9: 1096.

_____ and _____. 1970. Histological and hematological responses of an estuarine teleost to cadmium. J. Fish. Res. Board Can. 27: 2185-2196.

Holmes, C. W., E. A. Slade, and C. J. McLerran. 1974. Migration and redistribution of zinc and cadmium in marine estuarine systems. Environ. Sci. Technol. 8: 255-259.

Hood, M. A. and S. P. Meyers. 1974. Distribution of chitinoclastic bacteria in natural estuarine waters and aquarial systems. pp. 115-121. In: Proc. Gulf Coast Regional Symposium on Diseases of Aquatic Animals. Publ. No. LSU-SG-74-05, Louisiana State Univ.

Johnson, S. K. 1974. Fusarium sp. in laboratory-held pink shrimp. Texas A&M Univ. Fish Disease Diagnostic Laboratory, FDDL-S1.

Lightner, D. V. 1975. Some potentially serious disease problems in the culture of penaeid shrimp in North America. pp. 75-97. In: Proc. U. S.-Japan Natural Resources Program. Symposium on Aquaculture Diseases. Tokyo, 1974.

_____ and C. T. Fontaine. 1975. A mycosis of the American lobster, Homarus americanus, caused by Fusarium sp. J. Invert. Pathol. 25: 239-245.

_____, _____, and K. Hanks. 1975. Some forms of gill disease in penaeid shrimp reared in intensive culture. pp. 347-365. In: Proc. 6th Ann. Workshop, World Mariculture Soc. Seattle, Wash.

Mount, D. I. and W. A. Brungs. 1967. A simplified dosing apparatus for fish toxicology studies. Water Res. 1: 21-29.

Newman, M. and S. A. MacLean. 1974. Physiological response of the cunner, _Tautogolabrus adspersus_ to cadmium. VI. Histopathology. U. S. Dept. of Commerce NOAA Technical Report NMSF SSRF-68, pp. 27-33.

Nimmo, D. R. and L. H. Bahner. (In press). Metals, pesticides and PCB's: Toxicities to shrimp singly and in combination. _Proc. 3rd Ann. Internat. Estuarine Res., Fed. Conf._, Galveston, Texas. 1975.

Reinert, R. E., L. J. Stone, and H. L. Bergman. 1974. Dieldrin and DDT: Accumulation from water and food by lake trout (_Salvelinus namaycush_) in the laboratory. pp. 52-58. In: _Proc. 17th Conf. Great Lakes Res._, Int. Assoc. Great Lakes.

Segar, D. A. 1971. The use of the heated graphite atomizer in marine sciences. pp. 523-532. In: _Proc. 3rd International Congress of Atomic Absorption and Atomic Fluorescence Spectroscopy._ Adam Hilger, London.

_____ and J. L. Gilio. 1973. The determination of trace transition elements in biological tissues using flameless atom reservoir atomic absorption. _Int. J. Environ. Anal. Chem._ 2: 291-301.

Shaw, B. L. and H. I. Battle. 1957. The gross and microscopic anatomy of the digestive tract of the oyster, _Crassostrea virginica_ (Gmelin), _Can. J. Zool._ 35: 325-347.

Toussoun, T. A. and P. E. Nelson. 1968. A pictorial guide to the identification of _Fusarium_ species according to the taxonomic system of Synder and Hansen. Penn. State Univ., University Park, Pa.

Response of the Lobster, *Homarus americanus,* to Sublethal Levels of Cadmium and Mercury

F. P. THURBERG[1], A. CALABRESE[1], E. GOULD[1], R. A. GREIG[1],
M. A. DAWSON[1], and R. K. TUCKER[2]

National Marine Fisheries Service
Middle Atlantic Coastal Fisheries Center
[1]Milford Laboratory
Milford, Connecticut 06460

[2]Sandy Hook Laboratory
Highlands, New Jersey 07732

Although a considerable amount of data has been
published on various aspects of lobster physiology (Lewis,
1970), very little attention has been devoted to the sublethal effects of heavy metal pollutants on this animal
(Eisler, 1973). The American lobster, Homarus americanus,
one of the most commercially important marine crustaceans
in the United States, is found in waters adjacent to the
heavily industrialized northeastern United States. Hence,
many of its numbers are exposed to metal pollutants from
coastal manufacturing areas. Cadmium and mercury, the two
metals examined in this study, are of immediate concern in
these coastal waters. Both have been implicated in mortality and sublethal physiological alterations in various
marine species (Thurberg and Dawson, 1974; Waldichuk, 1974;
Calabrese et al., 1975).

The present study was undertaken to evaluate the sublethal physiological effects of cadmium and mercury on the
American lobster. The parameters chosen for examination
included changes in oxygen consumption, serum osmolality,
heart aminotransferase level, gill ATPase activity, and
tissue uptake of the two metals. This report is an attempt

to describe the findings, both positive and negative, in each area. It is not a definitive report, but rather a starting point for future work on the sublethal effects of metals on various aspects of lobster physiology.

METHODS AND MATERIALS

Exposure

Lobsters were trap-collected in Long Island Sound near Milford, Connecticut and held in the laboratory in flowing, sand-filtered sea water for one to two weeks prior to mercury or cadmium exposure. They were fed chopped clams (Spisula solidissima) daily during holding and throughout the exposure period. Test animals were exposed in 285-liter fiberglass tanks filled to 228 liters with sand-filtered sea water from Long Island Sound (24-26 o/oo salinity) by a proportional-dilution apparatus (Mount and Brungs, 1967). This dilutor controlled the intermittent delivery of toxicant-containing water and control water at a flow rate of 1.5 liters to each tank every 2.5 min throughout the test period. This flow provided approximately four complete daily exchanges of water in each tank. Cadmium, as cadmium chloride ($CdCl_2.2-1/2H_2O$), and mercury, as mercuric chloride ($HgCl_2$), were added at concentrations of 3 and 6 parts per billion (ppb). These concentrations are realistic in terms of cadmium present in polluted waters. Certain areas of Raritan Bay, New Jersey, contain up to 13 ppb cadmium (Tucker, unpublished data). To compare the effects of these two metals, the same concentrations were used for mercury. Background levels of these two metals in the incoming sea water were less than 1 ppb. Each experiment consisted of 18 lobsters per concentration of one metal, 6 lobsters per tank, averaging 290 g in weight (65% of the test animals were males). Thirty-day-exposure experiments were conducted in duplicate for cadmium and for mercury. A single 60-day test was completed for each metal. Water temperature ranged from 18° to $20^\circ C$ during the 30-day exposures and 10° to $15^\circ C$ during the 60-day tests. After each test lobsters were removed and tissues dissected for study. All data in this study were analyzed by Student's t distribution.

Respiration

Two gills were dissected from each lobster and placed in 15-ml Warburg-type flasks, each containing 5 ml of metal-treated or control water from the tank in which the lobster

had been exposed. Gill-tissue oxygen consumption was
monitored over a 4-hr period at 20°C using a Gilson
Differential Respirometer. The flasks were shaken at a rate
of 80 cycles per min. Oxygen consumption rates were calculated as microliters of oxygen consumed per hour per gram
dry weight of gill tissue ($\mu\ell$/hr/g) corrected to microliters
of dry gas at standard temperature and pressure.

Serum Osmolality

Blood samples were taken from each lobster by puncturing
the membrane at the base of a walking leg and drawing a
sample from the sinus with a disposable Pasteur pipette. The
blood was centrifuged at 10,000 g for 20 min and 0.2-ml
serum samples were read on an Advanced 3L Osmometer.

Biochemical Studies

Lobster hearts were packaged individually in small
plastic pouches from which as much air as possible was
excluded and frozen-stored at -29°C until testing. A 10%
homogenate, w/v in iced glass-distilled H_2O, was made from
each heart in a small glass homogenizer containing 25-μm
glass powder to facilitate grinding. The homogenates were
centrifuged for 45 min at 15,000 g at 4°C. At least 1.5 ml
supernate was diluted precisely 1:1 with iced H_2O and a
smaller portion (ca 0.50 ml) diluted precisely 1:1 with iced
220 m$\underline{M}$ $MgCl_2 \cdot 6H_2O$. One-ml aliquots of the water dilution
were used for protein determinations by the biuret method
(Layne, 1957).

The assay for aspartate aminotransferase (E.C.2.6.1.1;
AAT) was based on the standard coupled reaction as described
by Bergmeyer and Bernt (1963) except for the proportions and
concentrations of some of the reagent solutions. No malic
dehydrogenase (MdH) was added, as there was ample endogenous
MdH in the crude enzyme preparation. Water used in preparing
all solutions was doubly glass-distilled; solutions of
reduced nicotinamide adenine dinucleotide (NADH) were
prepared fresh daily; α-ketoglutarate (KG) was prepared in
200-ml volumes and frozen-stored in small amounts (5-7 ml).
A double-beam ratio-recording spectrophotometer and a linear-
log recorder were used to follow reaction rates, measured by
the decreasing absorbance at 340 nm as NADH was oxidized.

The assay medium contained 2.60 ml buffer-amino acid
solution (0.1 $\underline{M}$ phosphate buffer, pH 7.5, containing 0.25 $\underline{M}$
L-aspartic acid, neutralized with KOH before addition); 0.10
ml NADH (9 mg/ml/H_2O); 0.05 ml enzyme preparation; and, to
start the reaction, 0.20 ml KG (0.1 $\underline{M}$ neutralized with KOH).

The solutions were pipetted into an optical cuvette, 1-cm
pathlength, and allowed to stand for 10 min at room
temperature. Absorbance was read at 340 nm against a
reference cuvette (H_2O) adjusted with small amounts of NADH
to a differential reading of 0.800-1.000. When the
absorbance was stable and no further oxidation of NADH
could be detected, the reaction was started by the addition
of KG. The recorder's log mode was used with a linear chart,
and the slope was drawn from the fastest linear portion,
some 40-90 sec after the reaction's start. The unit of
activity was the change in absorbance of 0.001/min/mg
protein.
     A modification of the Bonting (1970) ATPase assay was
conducted on freeze-dried gills from a limited number of
lobsters exposed to 3 and 6 ppb cadmium for 30 and 60 days.

Chemical Analyses

     Cadmium was analyzed by a modification of the method of
Bertine et al. (1972). In the modified procedure up to 10 g
of sample were weighed into 250-ml glass beakers, concentrated nitric acid (70% reagent grade) was added to cover the
sample, and the mixture was heated on a hot plate at a gentle
boil to evaporate the acid to dryness. This process was
repeated several times until frothing or foaming of the
sample ceased. The temperature was then increased, and
acid was added in small volumes (under 2 ml) until a whitish
residue remained in the beaker. The residue was filtered
through Whatman No. 2 paper and the filtrate diluted to 25 ml
with 10% nitric acid. The samples were then analyzed
directly on an atomic absorption spectrophotometer (Perkin-
Elmer Model 403). Mercury analyses were conducted by the
cold-vapor, atomic absorption method described by Greig
et al. (1975).

RESULTS AND DISCUSSION

Respiration

     Changes in gill-tissue oxygen consumption are valuable
indicators of stress and are frequently used to evaluate
changes in metabolism due to an environmental alteration.
Lobsters exhibited significantly elevated ($P<.01$) gill-
tissue oxygen consumption values after 30 days' exposure to
3 and 6 ppb cadmium (Fig. 1). Lobsters exposed for 60 days
also showed elevated rates, but only 6 animals were tested
per concentration, a number too low for valid statistical

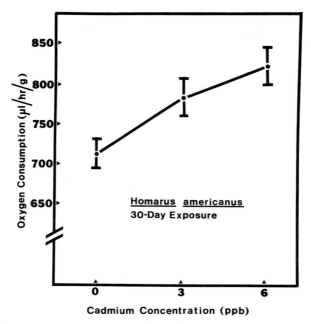

Fig. 1.   Homarus americanus.   Effects of cadmium on
           oxygen consumption.   Each point comprises
           the mean gill-tissue oxygen consumption
           value of 24 animals.   Bars: standard
           errors.   Temperature: 20°C.

analysis.   No sex-related differences were noted.   This
elevation in oxygen consumption contrasts with the
depression in gill-tissue oxygen consumption noted in
cadmium-exposed green crabs, Carcinus maenas (Thurberg
et al., 1973).   This contrast may be due to species
differences or to the different exposure regimes used; the
crabs were exposed for 48 hours to high levels of cadmium
(ppm), while the lobsters were exposed for 30 days to lower
levels (ppb).
     No differences in gill-tissue oxygen consumption were
noted between controls and any mercury-exposed group of
lobsters.   This was true for both 30- and 60-day exposed
animals.   In contrast, Vernberg and Vernberg (1972) reported
mercury-induced changes in the respiration of fiddler crabs,
Uca pugilator, exposed to 0.18 ppm mercury for 21 days.   This
may indicate that the lobsters were exposed to levels of
mercury too low to provoke detectable physiological damage.

Osmoregulation

The lobsters used in this study were exposed to
estuarine waters with a salinity range from 24-26 o/oo. Dall
(1970) reported that lobsters possess some osmoregulatory
ability and noted that blood is maintained hyperosmotic to
the sea water over this salinity range. Thurberg et al.
(1973) reported osmoregulatory disruption in the green crab,
Carcinus maenas, after 48-hr exposures to high levels (0.5 to
8.0 ppm) of cadmium, and more recently Dawson (unpublished
results) noted similar disruption in the rock crab, Cancer
irroratus, after 30-day exposures to low concentrations (25
ppb) of cadmium. No changes in serum osmolality were noted,
however, in lobsters exposed to cadmium or mercury in this
study. The serum osmolality of experimental and control
animals was maintained at about 40 milliosmoles above that
of the sea water.

Biochemical Studies

Biochemical systems involved in the mobilization of
energy are also appropriate targets for pollutant studies.
In invertebrates especially, amino acid metabolism has been
directly implicated in energy production (McAllan and
Chefurka, 1961; Schoffeniels and Gilles, 1970). Of the
enzymes driving that metabolism, one of the most important
is aspartate aminotransferase. Because we found that
crustacean heart contained large amounts of this enzyme
(Gould et al., in press), we measured heart AAT in this
study and found that there was no significant difference
between control and exposed animals.

AAT, however, is subject to regulation by modulators
that generally act, by adjusting enzyme-substrate affinities,
to compensate for environmental variables, such as temperature,
pressure, or salinity (Hochachka et al., 1970; Hochachka and
Somero, 1973). One such biochemical modulator is magnesium,
whose local free-ion concentration is greatly influenced by
the relative amounts of the adenine nucleotides, and these,
in turn, are directly concerned with the regulation of
energy metabolism (Noda, 1973). Because ATP binds Mg^{2+} more
tightly than does ADP or AMP, one might reasonably speculate
that a drain on energy reserves (i.e., hydrolysis of ATP)
may be accompanied by a rise in free Mg^{2+}. In decapod
crustaceans an inverse correlation has been observed
between blood Mg^{2+} and metabolic rate (Robertson, 1960;
Wolvekamp and Waterman, 1960), and elevated serum Mg^{2+} has
been observed in crabs stressed by disease (Drilhon, 1936)
and by short-term exposure to cadmium (Gould et al., in

press). For this study, therefore, we monitored lobster
heart AAT, not only under standard assay conditions, but also
in the presence of added magnesium chloride, at a calculated
tenfold increase over the normal concentration in lobster
blood, as reported by Cole (1940).

In heart preparations containing excess magnesium (110
mM) one-fourth of the normal AAT activity was lost in the
controls, whereas only one-tenth was lost in lobsters
exposed to 3 ppb cadmium for 30 days and none at all in
those exposed to 6 ppb (Table 1). To a lesser extent, the
same phenomenon was observed in lobsters exposed to sublethal amounts of mercury (Table 1).

Serum magnesium levels in the metal-exposed lobsters,
however, did not differ significantly from those in the
control animals. The difference in AAT activity in magne-
sium-stressed heart preparations, therefore, should not be
attributed to a stress- or metal-induced elevation of serum
magnesium in vivo, but rather to a structural or functional
alteration brought about by the animals' exposure to cadmium
or to mercury. The sensitivity of the normal enzyme to the
negative modulation of a tenfold increase in Mg^{2+} is
considerably less in animals exposed to 3 ppb Cd and dis-
appears entirely in animals exposed to 6 ppb Cd. At least
two reasonable explanations present themselves: the blocking of the Mg^{2+}-modulator site on the transaminase molecule
or the production of an isoenzyme lacking the normal sensitivity to divalent metal cations, such as magnesium. As no
electrophoretic studies were made, any interpretation at
this point can be only conjecture. The difference in catalytic efficiency is real, however, and suggests a cadmium-
induced reduction of sensitivity to physiological modulators.

Increased metabolism in the lobsters in this study, as
evidenced by the enhanced gill-tissue respiration, might
well be accompanied by higher enzyme levels in the gills.
Of the gill-tissue enzymes one of the most important is
sodium-potassium ATPase (ATP phosphohydrolase, E.C.3.6.13).
Although we were unable, in the course of this study, to
complete a thorough examination of this enzyme, preliminary
work indicates a possible increase in its activity in cadmium-exposed lobsters. The enzyme concentration, however,
varies greatly among individual animals, and a more produc-
tive approach may be the investigation of this enzyme's
kinetics.

Chemical Analyses

Crustaceans exposed to cadmium and mercury frequently
concentrate these metals in the gills and digestive gland

Table 1
Aspartate aminotransferase in heart muscle of lobsters
(<u>Homarus</u> <u>americanus</u>) exposed to cadmium and mercury for 30
days.

Exposure Level (ppb)	No. Animals	Std. H$_2$0 Prepns.*	Mg Prepns., 110 mM (% Std. Activity
		Cadmium	
0	16	1048 + 43	75.9 + 4.4
3	16	1006 + 45	90.8 + 3.6
6	17	1044 + 69	105.6 + 3.9 (P<.001)
		Mercury	
0	14	1182 + 57	71.7 + 3.6
3	9	1021 + 21	71.7 + 3.6
6	10	1003 + 79	83.7 + 3.5 (P<.05)

*Unit AAT Activity: Change in absorbance at 340 nm of 0.001/
min/mg protein.

(Vernberg and Vernberg, 1972; O'Hara, 1973; Hutcheson, 1974).
In this study no difference in digestive-gland cadmium
concentration was noted among control and exposed lobsters,
although levels above 12 ppm were found in all groups.
Lobsters exposed to cadmium accumulated significantly higher
levels of this metal in gill tissue than did control animals
(Table 2). Lobsters exposed to 6 ppb for 30 days accumulated
twice as much (3.4 ppm vs 1.5 ppm), while those exposed for
60 days accumulated slightly less than twice the amount (2.6
ppm vs 1.6 ppm). Eisler <u>et</u> <u>al</u>. (1972) found that the level
of cadmium in lobster gills doubled after exposure to 10 ppb
for 21 days. Such high levels of cadmium in the gills may be
a factor in gill-tissue respiration changes noted in this
study. Tail muscle samples examined for cadmium were below
the limits of detection (<0.12 ppm). Mercury was accumulated to a much greater degree than was cadmium (Table 3).

Table 2
Uptake of cadmium by lobsters (Homarus americanus) exposed to cadmium in sea water for 30 and 60 days.

Exposure Level (ppb)	Cadmium Concentrations Mean ppm ± s.e. (wet weight)	
	30 Days*	60 Days**
	Digestive Gland	
0.0	20.3 ± 4.2	13.7 ± 3.4
3.0	22.5 ± 3.8	12.9 ± 1.5
6.0	23.5 ± 4.5	13.9 ± 2.0
	Gills	
0.0	1.5 ± 0.09	1.63 ± 0.14
3.0	1.8 ± 0.19	2.32 ± 0.30
6.0	3.4 ± 0.29	2.63 ± 0.27
	Muscle	
0.0	<0.12	<0.12
3.0	<0.12	<0.12
6.0	<0.12	<0.12

 *Data from 12 lobsters.
**Data from 9 lobsters.

Lobsters exposed to 6 ppb mercury for 30 days contained an average of 1 ppm in the tail muscle, 15 ppm in the digestive gland, and 85 ppm in the gill tissues. After 60 days' exposure the mercury level in the gills rose to an average of 119 ppm.

SUMMARY

     Lobsters chronically exposed to sublethal amounts of cadmium had elevated rates of gill-tissue oxygen consumption,

Table 3
Uptake of mercury by lobsters (<u>Homarus</u> <u>americanus</u>) exposed
to mercury in sea water for 30 and 60 days.

Exposure Level (ppb)	Mercury Concentrations Mean ppm ± s.e. (wet weight)	
	30 Days	60 Days
	Digestive Gland[a]	
0	0.12 ± 0.018	---
3	4.12 ± 0.860	---
6	15.20 ± 3.200	---
	Gills[b]	
0	0.14 ± 0.025	0.05 ± ---
3	40.90 ± 4.060	43.80 ± 2.2
6	85.30 ± 7.000	119.50 ± 4.8
	Tail Muscle[c]	
0	0.23 ± 0.018	---
3	0.54 ± 0.076	---
6	1.00 ± 0.101	---

[a]Data from 10 lobsters.
[b]Data from 16 lobsters (30-day) and 10 lobsters (60-day).
[c]Data from 10 lobsters.

as well as a loss of magnesium sensitivity in heart muscle
transaminase. In the gills of cadmium-exposed lobsters
levels of that metal were significantly higher than in
control animals; the gills also showed some indication of
increased ATPase activity. Exposure to similar levels of
mercury had a lesser effect on the magnesium sensitivity of
heart AAT, and none at all on gill-tissue respiration,
although very high levels of this metal were detected in
gills, digestive gland, and tail muscle. Neither metal
affected serum osmolality, and very little difference was
found in any test when data from 30- and 60-day exposures
were compared.

Note: The use of trade names is to facilitate description and does not imply endorsement by the National Marine
Fisheries Service.

ACKNOWLEDGMENT

The authors thank Miss Rita S. Riccio for her critical
reading and typing of this manuscript.

LITERATURE CITED

Bergmeyer, H. U. and E. Bernt. 1963. Glutamateoxaloacetate
    transaminase. In: Methods of Enzymatic Analysis,
    p. 839, ed. by H. U. Bergmeyer. New York: Academic
    Press.
Bertine, K., R. Carpenter, N. Cutshall, D. E. Robertson, and
    H. L. Windom. 1972. Marine pollution monitoring:
    Strategies for a national program. Deliberation of a
    workshop held at the Allan Hancock Foundation,
    University of Southern California, Los Angeles,
    California.
Bonting, S. L. 1970. Sodium-potassium activated adenosine-
    triphosphatase and cation transport. In: Membranes
    and Ion Transport. Vol. 1, pp. 257-363, ed. by E. E.
    Bittar. New York: Wiley-Interscience.
Calabrese, A., F. P. Thurberg, M. A. Dawson, and D. R.
    Wenzloff. 1975. Sublethal physiological stress
    induced by cadmium and mercury in the winter flounder,
    Pseudopleuronectes americanus. In: Sublethal Effects
    of Toxic Chemicals on Aquatic Life, pp. 15-21, ed. by
    J. H. Koeman and J. J. T. W. A. Strik. Amsterdam:
    Elsevier.
Cole, W. H. 1940. The composition of fluids and sera of
    some marine animals and of the sea water in which they
    live. J. Gen. Physiol. 23: 575-584.
Dall, W. 1970. Osmoregulation in the lobster Homarus
    americanus. J. Fish. Res. Bd. Can. 27: 1123-1130.
Drilhon, A. 1936. Cited in M. Florkin. 1960. Ecology and
    Metabolism. In: The Physiology of Crustacea, p. 405,
    ed. by T. H. Waterman. New York: Academic Press.
Eisler, R. 1973. Annotated bibliography on biological
    effects of metals in aquatic environments. Ecological
    Research Series EPA-R3-73-007. Washington, D. C.:
    U. S. Govt. Printing Office.
_____, G. E. Zaroogian, and R. J. Hennekey. 1972.
    Cadmium uptake by marine organisms. J. Fish. Res. Bd.
    Can. 29: 1367-1369.

Gould, E., R. S. Collier, J. J. Karolus, and S. A. Givens. Heart transaminase in the rock crab, Cancer irroratus, exposed to cadmium salts. Bull. Environ. Contam. Toxicol. (In press)

Greig, R. A., D. R. Wenzloff, and C. Shelpuk. 1975. Mercury concentrations in fish, North Atlantic offshore waters - 1971. Pesticides Monitoring J. 9: 15-20.

Hochachka, P. W., D. E. Schneider, and A. Kuznetsov. 1970. Interacting pressure and temperature effects on enzymes of marine poikilotherms: Catalytic and regulatory properties of FDPase from deep and shallow-water fishes. Mar. Biol. 7: 285-293.

_____ and G. W. Somero. 1973. Strategies of Biochemical Adaptation. Phila.: W. B. Saunders Co.

Hutcheson, M. S. 1974. The effect of temperature and salinity on cadmium uptake by the blue crab, Callinectes sapidus. Ches. Sci. 15: 237-241.

Layne, E. 1957. Spectrophotometric and turbidimetric methods for measuring proteins. III. Biuret method. In: Methods in Enzymology. Vol. III, pp. 456-461, ed. by S. P. Colowick and N. O. Kaplan. New York: Academic Press.

Lewis, R. D. 1970. A bibliography of the lobsters, genus Homarus. U. S. Fish Wildl. Serv. SSRF-591.

McAllan, J. A. and W. Chefurka. 1961. Some physiological aspects of glutamate-aspartate transamination in insects. Comp. Biochem. Physiol. 2: 290-299.

Mount, D. I. and W. A. Brungs. 1967. A simplified dosing apparatus for fish toxicology studies. Water Res. 1: 21-29.

Noda, L. 1973. Adenylate kinase. In: The Enzymes, pp. 279-305, ed. by P. D. Boyer. New York: Academic Press.

O'Hara, J. 1973. The influence of temperature and salinity on the toxicity of cadmium to the fiddler crab, Uca pugilator. Fish. Bull. 71: 149-153.

Robertson, J. S. 1960. Osmotic and ionic regulation. In: The Physiology of Crustacea, pp. 317-339, 395-410, ed. by T. H. Waterman. New York: Academic Press.

Schoffeniels, E. and R. Gilles. 1970. Nitrogenous constituents and nitrogen metabolism in arthropods. In: Chemical Zoology, Vol. V, pp. 199-227, ed. by M. Florkin and B. T. Scheer. New York: Academic Press.

Thurberg, F. P. and M. A. Dawson. 1974. Physiological response of the cunner, Tautogolabrus adspersus, to cadmium. III. Changes in osmoregulation and oxygen consumption. In: Physiological Response of the Cunner, Tautogolabrus adspersus, to Cadmium, pp. 11-13. NOAA Tech. Rep., NMFS, SSRF-681.

_____, _____, and R. Collier. 1973. Effects of copper and cadmium on osmoregulation and oxygen consumption in two species of estuarine crabs. Mar. Biol. 23: 171-175.

Vernberg, W. B. and F. J. Vernberg. 1972. The synergistic effects of temperature, salinity and mercury on survival and metabolism of the adult fiddler crab, Uca pugilator. Fish. Bull. 70: 415-420.

Waldichuk, M. 1974. Some biological concerns in heavy metal pollution. In: Pollution and Physiology of Marine Organisms, pp. 1-57, ed. by F. J. Vernberg and W. B. Vernberg. New York: Academic Press.

Wolvekamp, H. P. and T. H. Waterman. 1960. Respiration. In: The Physiology of Crustacea, Vol. 1, pp. 35-100, ed. by T. H. Waterman. New York: Academic Press.

Effects of Chromium on the Life History of *Capitella capitata* (Annelida: Polychaeta)

DONALD J. REISH

Department of Biology
California State University, Long Beach
Long Beach, California 90840

Chromium is one of the most common heavy metals present in waste discharges. Concentrations of chromium in effluents of different discharges in southern California are 0.02 to 0.86 mg/l (Mitchell and McDermott, 1975). Approximately 85 percent of the chromium is associated with particulates which settle to the bottom. The remaining 15 percent, which is dissolved, is considered the more toxic fraction (Young and Jan, 1975a). Chromium occurs in two common forms-- trivalent and hexavalent. The trivalent form constitutes 90 percent of dissolved chromium in the effluent of the City of Los Angeles (Young and Jan, 1975b). Chromium is of importance because the influence of the sewage discharges from the City and County of Los Angeles now overlap as measured by the concentration of chromium in the sediments. Since these two outfalls are located some 20 miles apart, a knowledge of the toxicity of chromium is important.

Since polychaetous annelids constitute the majority of the benthic macroinvertebrate species off southern California (So. Calif. Coastal Water Research Project, SCCWRP, 1973) it is of importance to investigate the effect of chromium on these animals. Only a few studies have been concerned with the effect of chromium on polychaetes. Raymont and Shields (1963) found a 21-day LC_{50} of 1.0 mg/l for hexavalent chromium for Nereis virens Sars (=Neanthes virens (Sars), Hartman, 1961). The 28-day LC_{50} for hexavalent chromium to Neanthes arenaceodentata (delle Chiaji) and Capitella

capitata (Fabricius) was 0.55 and 0.28 mg/l, respectively
(Reish et al., 1976). Oshida and Reish (1975) found a
56-day LC_{50} of hexavalent chromium to Neanthes arenaceodentata (Moore) to be 0.2 mg/l. No deaths to this species
were recorded at a concentration of 12.5 mg/l trivalent
chromium after 21 days. In a reproductive experiment the
presence of hexavalent chromium at an experimental level of
0.0125 mg/l caused a decrease of over one-half mean brood
size of the P_1 offspring of N. arenaceodentata.
    The objective of this study was to investigate the
effect of hexavalent chromium on the life cycle of Capitella
capitata, the most commonly encountered invertebrate
collected in the vicinity of the Los Angeles County discharge (SCCWRP, 1973). Because C. capitata is an indicator
of polluted conditions (Reish, 1955), knowledge of the concentrations of chromium which adversely effect its life
cycle would be useful in deciding levels of chromium which
could be discharged into the marine environment.
    Since sublethal amounts of copper and zinc are known to
induce abnormal development of C. capitata larvae (Reish et
al., 1974), a second objective was to determine whether
chromium would do the same and, if so, to determine if a
relationship exists between chromium concentration and the
occurrence of abnormalities.

MATERIALS AND METHODS

    Experimental animals were from an inbred laboratory
population of C. capitata originally derived from one female
collected from Los Angeles Harbor in 1968. The population
has undergone 50-70 generations in the laboratory at 19.5$\pm^{\circ}$C
and 35.0$\pm$ 1°/oo. The worms were initially fed the green
alga Entermorpha crinita (Rath) J. Agardh, but for the past
two years finely ground Tetramin (Tetra Werke, W. Germany)
has been used as the food source. The worms are cultured in
one-gallon jars with 2500 ml of continuously aerated sea
water. The worms reside on the bottom of each culture jar
within mucoid tubes that intertwine with one another to form
clumps to which food and fecal pellets readily adhere.
    To obtain trochophore larvae for experiments clumps of
C. capitata were examined under a dissecting microscope for
females incubating embryos. Such females generally occupy
the larger tubes, which being mucoid, allow the embryos to
be readily seen (Fig. 1). The embryos gradually change in
color from white to gray-green as they develop into trochophores. Females with tubes bearing trochophores were individually isolated in dishes where each tube (Fig. 2) was

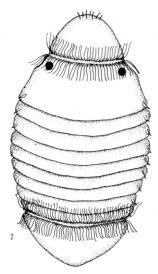

Fig. 1.   <u>Capitella</u> <u>capitata</u>. Free-swimming metatrochophore
larvae. Larvae in Figures 1 and 3 were drawn at
different scales; total length is identical.

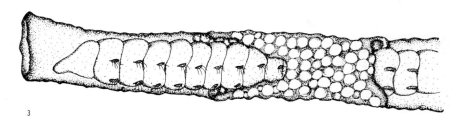

Fig. 2.   <u>Capitella</u> <u>capitata</u>. Female in tube with developing embryos.

then teased apart to allow the larvae to swim free. About
800 swimming trochophores were then pipetted into a petri
dish and served as a pool from which individuals were subsequently obtained for exposure to various concentrations of
hexavalent chromium. Twenty-five trochophores were pipetted
into a separate dish and subsequently placed in a gallon jar
containing 2500 ml sea water and toxicant. Four gallon jars,

each seeded with 25 larvae, were used with different concentrations of metal plus a sea water control. Hence, a
total of 100 larvae were exposed to each concentration of
chromium. Although the larvae are lecithotrophic, food was
provided so as to be present when they began feeding upon
settlement. Initial food consisted of a suspension of
Entermorpha crinita-Tetramin which had been previously
ground and had passed through a 0.061 mm sieve. The initial
food was a suspension of 1.0 g Entermorpha and 0.5 g
Tetramin in 100 ml sea water. Each container was provided
with 5.0 ml of the food suspension. Subsequent feedings of
5.0 ml, at weekly intervals, consisted of a suspension of
1.5 g Tetramin in 150 ml sea water.

A stock solution of hexavalent chromium was prepared
from $K_2Cr_2O_7$. Appropriate dilutions were made from this
solution to give concentrations of 0.025, 0.05, 0.1, 0.2, and
0.4 mg/l. Four jars per concentration and a sea water
control jar, each containing 2500 ml of solution were used.
Trochophores were then placed in each jar. Food was added,
and jars were aerated.

Jars were checked at 15 days and every 2-3 days thereafter until the conclusion of the experiment. Sea water was
decanted off and material from the bottom was transferred to
a petri dish where it was examined under a dissecting microscope for survival and presence of females incubating
embryos. If the brood appeared white, the tube was returned
to the sea water from the one-gallon culture jar. However,
if gray-green in color, indicating the presence of trochophores, the brood was placed with the female in a syracuse
watch glass for a larval count. The tube mass was separated
allowing larvae to escape. Larvae were counted and examined
for bifurcated posterior ends (Fig. 3). Abnormal larvae are
usually seen initially by noting aberrant swimming behavior.
Tube masses containing immature worms were returned to the
jar together with the sea water. The experiment ran for a
five-month period. No changes were made of the experimental
solutions during the course of the experiment, since it had
been determined earlier that the concentration of hexavalent
chromium did not decrease appreciably with time (Oshida et
al., 1976).

RESULTS

The results of this experiment are summarized in
Table 1. Data include (1) number of survivors, (2) number
of females which reproduced, (3) average number of offspring
produced, and (4) number and percent occurrence of abnormal

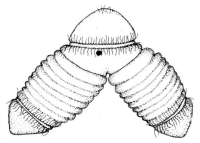

Fig. 3. <u>Capitella</u> <u>capitata</u>. Abnormal metatrochophore
larvae.

larvae. It can be seen that the number of survivors was
markedly reduced at 0.1 mg/l chromium over those at lower
concentrations. Furthermore, at this concentration a downward trend also appeared in the number of females reproducing
and the number of eggs per female. Only a few females
reproduced at the two highest concentrations. The percent
occurrence of abnormal larvae increased with increasing
concentration of chromium.

DISCUSSION

The 96-hour and 28-day LC_{50} for <u>C</u>. <u>capitata</u> exposed to
hexavalent chromium was reported as 5.0 and 0.28 mg/l,
respectively (Reish <u>et</u> <u>al</u>., 1976) which are well above the
0.1 mg/l concentration at which suppression of reproduction
was first noted (Table 1). The difference between LC_{50}
data and the lowest concentration at which abnormal larvae
were observed is even greater--a magnitude of one when
compared to the 28-day LC_{50} or over two orders of magnitude
when compared to the 96-hour LC_{50}. Since very few females
reproduced at 0.4 mg/l, presumably, no abnormal larvae would
occur at concentrations greater than this because of either
complete suppression of reproduction or death of experimental
animals. Since abnormalities occurred in 0.44% of the larvae
at the lowest experimental concentration tested, presumably,
this is not the minimum concentration of chromium needed to
induce abnormal larvae.

The LC_{50} levels of <u>C</u>. <u>capitata</u> are below those observed
by Raymont and Shields (1963) for <u>Neanthes</u> <u>virens</u> and Reish
<u>et</u> <u>al</u>. (1976) for <u>N</u>. <u>arenaceodentata</u>. However, chromium
apparently has a more detrimental effect on reproduction
activities in <u>N</u>. <u>arenaceodentata</u> than in <u>C</u>. <u>capitata</u>.
Suppression of reproduction in <u>N</u>. <u>arenaceodentata</u>, as

Table 1
Effects of Hexavalent Chromium on the Life Cycle of Capitella capitata.

Concentration in mg/l	Initial Number of Worms	Number of Survivors	Number of Females That Reproduced	Average Number of Offspring	Number of Abnormal Larvae	Percent Occurrence of Abnormal Larvae
Control	100	86	33	243	0	0
0.025	100	82	38	269	28	0.44
0.05	100	95	40	279	41	0.88
0.1	100	58	31	174*	23	1.3
0.2	100	57	3	144*	5	1.2
0.4	100	21	4	167*	8	1.7

*Significant at the 95% level, Student T-test.

measured by number of offspring, was observed at 0.0125 mg/l
(Oshida and Reish, 1975), the lowest test concentration used
or one order of magnitude less than the suppression level
observed for C. capitata. Furthermore, N. arenaceodentata
failed to reproduce at either 0.1 or 0.2 mg/l, whereas, at
least some C. capitata reproduced in reduced numbers at
0.4 mg/l.

The State of California water quality standard (1972)
states that the level of total chromium should not exceed
0.005 mg/l 50 percent of the time or 0.01 mg/l 10 percent
of the time. The concentration of chromium in the final
effluents in southern California ranges from 0.02 to 0.86
mg/l with the higher figures (0.21 - 0.86 mg/l) from those
sanitation districts with the largest discharges (SCCWRP,
1972). If the 10 percent concentration of hexavalent
chromium present in the effluent of the City of Los Angeles
(Young and Jan, 1975b) can be used as a typical amount of
this element in waste discharges, then the highest value
(0.086 mg/l hexavalent chromium) is near the 0.1 mg/l
concentration where reproductive suppression occurred
(Table 1). While no suppression of reproduction occurred at
0.025 mg/l, there was an induction of abnormal larvae at
this level which falls within the calculated percent hexavalent chromium concentration (0.021 - 0.086 mg/l) present in
the larger effluents. While the effect of 0.01 mg/l hexavalent chromium was not tested, in light of these data, it
may be near the lowest level which induces abnormal larvae.
If the 10 percent concentration of hexavalent chromium is
applied to the California water quality standards, then the
allowable amount of this form of chromium becomes 0.001 mg/l
which is one order of magnitude below the probable lowest
concentration which induces abnormal larvae. Since C.
capitata is an indicator of domestic pollution, this species
is undoubtedly more tolerant to chromium than other species.
It is apparent that the standard for allowable amounts of
chromium should be re-evaluated in light of these data. The
induction of abnormal larvae by still another toxicant,
which now stands at four (detergent Foret, 1972) copper and
zinc (Reish et al., 1974)) indicates that this may even be a
more sensitive measure of the effects of a toxicant than is
suppressed reproduction, as measured by the number of offspring produced.

ACKNOWLEDGMENT

This research was supported by a grant from the United
States Environmental Protection Agency (R-800962). The
author wishes to express his thanks to Diana Vermillion and
Philip Oshida for their assistance in this research.

LITERATURE CITED

Foret, J. O. 1972. Etude des effets à long terme de quelques détérgents (issue de la pétrolechimie) sur le séquence due développement de deux espèces de Poly-chètes Sedentaires: Scolelepis fuliginosa (Claparède) et Capitella capitata (Fabricius). Thesis, Univ. Marseille-Luminy. 125 p.

Hartman, O. 1961. Polychaetous annelids from California. Allan Hancock Foundation Pacific Exped. 25: 1-225.

Mitchell, F. K. and D. J. McDermott. 1975. Characteristics of municipal wastewater discharges, 1974. So. Calif. Coastal Water Research Project, Annual Rept. 1975, pp. 163-165.

Oshida, P. and D. J. Reish. 1975. Effects of chromium on reproduction in polychaetes. So. Calif. Coastal Water Research Project, Annual Rept. 1975, pp. 55-60.

_____, A. J. Mearns, D. J. Reish, and C. S. Word. 1976. The effects of hexavalent and trivalent chromium on Neanthes arenaceodentata (Polychaeta: Annelids). So. Calif. Coastal Water Research Project Publ. T M 225, 58 p.

Raymont, J. E. G. and J. Shields. 1963. Toxicity of copper and chromium in the marine environment. Int. J. Water Poll. 7: 435-443.

Reish, D. J. 1955. The relations of polychaetous annelids to harbor pollution. U. S. Public Health Repts. 70: 1168-1174.

_____, J. M. Martin, F. Piltz, and J. Q. Word. 1974. Induction of abnormal polychaete larvae by heavy metals. Mar. Poll. Bull. 5: 125-126.

_____, F. Piltz, N. M. Martin, and J. Q. Word. 1976. The effect of heavy metals on laboratory populations of two species of polychaetes with comparisons to the water quality conditions and standards in southern California marine waters. Water Res. 10: 299-302.

Southern California Coastal Water Research Project. 1973. The ecology of the southern California bight. Impli-cations for water quality management. So. Calif. Coastal Water Res. Project Publ. TR 104: 531 pp.

State of California, State Water Resources Control Board. 1972. Water Quality Control Plan, Ocean Waters of California. Sacramento. 13 p.

Young, D. J. and T. K. Jan. 1975a. Trace metals in near-shore sea water. So. Calif. Coastal Water Research Project, Annual Rept. 1975. pp. 143-146.

_____ and _____. 1975b. Chromium in municipal wastewater and sea water. So. Calif. Coastal Water

Calabrese et al. (1974, 1975), among others, have
repeatedly emphasized the need to perform chronic-exposure
studies with marine animals and sublethal levels of pollutants,
as well as the more usual bioassay tests. Following the course
of a disease or physiological challenge in its early stages can
be more enlightening than mortality statistics or even post-
mortem pathology reports. Persuaded of this approach, this
laboratory has performed several chronic-exposure experimental
series with previously determined sublethal concentrations of
various heavy metal salts, using several different marine
animals, among them the winter flounder (Calabrese et al.,
1975); the lobster, Homarus americanus (Thurberg et al., in
press, and the striped bass, Morone saxatilis (Dawson et al.,
in prepn.).

The biochemistry of marine animals, as compared to
terrestrial vertebrates, is largely uncharted territory. In
preparation for a multidisciplinary series of experiments
exposing winter flounder to sublethal levels of cadmium, we
examined a variety of tissues from unexposed fish, looking
especially at metalloenzymes and metal-activated enzymes. As
a result, we chose for particular attention the discrete mass
of tissue comprising the kidney and hematopoietic material
for three reasons: a practical one, that it was large
enough to provide material for more than a few enzyme assays;
a logical one, that the kidney is second only to the gills as
an effector organ in ionic regulation; and a compelling one,
that the tissue preparations contained a very active
magnesium-linked shunt enzyme, glucose-6-phosphate dehydrogenase (D-glucose-6-phosphate: NADP oxidoreductase,
E.C.1.1.1.49; G6PdH). In cold-acclimated fish, the hexose
monophosphate shunt is more active than the Embden-Meyerhof
pathway of glycolysis, providing the first example of what
Hochachka and Somero (1973) have called "seasonal metabolic
reorganization"; and the flounder for our study were
collected and exposed during the winter months.

This particular enzyme, G6PdH, because of its well-
documented sensitivity to biochemical modulation by free
magnesium ions, provided a possible means of learning some-
thing of the mechanisms by which sublethal amounts of heavy
metals produce biochemical and, as a result, physiological
stress. If an enzyme's catalytic efficiency is normally
enhanced by a divalent metal cation, such as magnesium, it
might reasonably be expected to undergo some structural
perturbation in the presence of other divalent metal cations,
such as cadmium. Any configurational change in an enzyme at
or near ligand sites will affect enzyme-ligand affinities,
which control reaction rates. Ligand affinities have been
shown to be radically altered, not only by such changes in

the physical environment as pH (Atkinson, 1965; 1966), temperature (Behrisch and Hochachka, 1969; Somero, 1969), and pressure (Hochachka et al., 1970), but also by the availability of such physiological modulators as magnesium (Gould, 1969; Gould and Karolus, 1974). A major part of this study, therefore, was the investigation of G6PdH sensitivity to magnesium activation.

In addition, we examined the activity of two metallo-enzymes in whose molecular structure zinc is incorporated, and for whose function zinc is essential: carbonic anhydrase (carbonate hydro-lyase, E.C.4.2.1.1: CA) and leucine amino-peptidase (1-leucyl-peptide hydrolase, E.C.3.4.1.1: LAP). Both enzymes, widely distributed in nature, have been puri-fied and analyzed from a variety of sources. CA has been found in the tissues of marine as well as terrestrial vertebrates, and also in almost all of the major invertebrate groups. Analyses from every source thus far have shown it to contain zinc at the active site (Vallee, 1960; Lindskog et al., 1971); and although most such analyses have been on enzymes from mammalian tissues, CA in tuna erythrocytes was found to contain zinc (Leiner et al., 1962), and Addink (1968) found molluscan CA also to be a zinc enzyme. A similar case can be made for LAP (Smith and Hill, 1960; Vallee and Wacker, 1970; Delange and Smith, 1971), although apparently no metal analysis has been carried out as yet on LAP in marine animals. For this study, then, we proceeded on the assumption that both CA and LAP are zinc enzymes in flounder tissues also, and are probable targets for cadmium attack, according to the Irving-Williams binding order for divalent metal cations (cf. Williams, 1971).

Cadmium has been implicated in the poisoning of testi-cular carbonic anhydrase in rats during both in vivo and in vitro studies (Pařízek, 1956; Hodgen et al., 1969), by what is considered to be competitive inhibition at the zinc site. These mammalian studies, however, were short-term, acute exposures. Nordberg (1972) has reported that although testicular impairment is the most prominent pathological change during acute exposure to cadmium, the kidney is the organ most affected during chronic exposure, an observation corroborating Kendrey and Roe's report (1969) on cadmium accumulation in human tissues, in which the kidney was the major cadmium sink.

In the work considered here with the flounder kidney and its intimately associated hematopoietic tissue, we offer two possible explanations for biochemical malfunction and subsequent physiological stress in an animal chronically exposed to sublethal amounts of cadmium.

METHODS AND MATERIALS

Animal Exposure

The flounder were collected in November by otter trawl within 2 km of Milford Harbor, Connecticut, and held in flowing, sand-filtered sea water for 1 to 2 weeks prior to cadmium exposure. The fish were fed chopped clams (Spisula solidissima) during the time of acclimation and throughout the exposure period. The proportional diluter apparatus (Mount and Brungs, 1967), which delivered calibrated amounts of the metal to the fiberglass tanks holding the fish, and the flow-through facility itself have been described in detail by Calabrese et al. (1975). Cadmium was added as $CdCl_2.2-1/2H_2O$ at the concentrations of 5 and 10 ppb Cd. Background concentration in the incoming sea water was less than 1 ppb Cd. The fish averaged 219 g in weight (range 98-465 g) and 282 mm in total length (range 219-390 mm). The tests were conducted in duplicate, for a total of 36 fish per concentration. Water temperature ranged from 3° to $6^{\circ}C$. After exposure for 60 days, the fish were removed and samples of various organs excised for testing.

Tissue Preparation

The tissue masses comprising the flounder kidneys embedded in hematopoietic tissue (KH) were packaged individually, except for small specimens. These were pooled for analysis, as were the partial specimens obtained when tissue samples were taken for histopathological examination. Each sample was placed in a small plastic pouch from which air was excluded, sealed tightly with masking tape, and stored frozen at $-29^{\circ}C$.

On the day tissue preparations were made, the KH samples were homogenized 1:9, w/v, in ice-cold, doubly glass-distilled water, in a small glass homogenizer containing 25-µm glass powder to facilitate grinding. Centrifugation was at 17,000 g and $4^{\circ}C$ for 45 min, and the supernatant fraction served as the crude enzyme preparation. Protein was determined by the biuret test (Gornal et al., 1949), against a crystallized bovine serum albumin standard. All solutions used in the following assays were also prepared with doubly glass-distilled water.

Enzyme Assays

The esterase activity of carbonic anhydrase was measured by the spectrophotometric technique described by Pocker and Stone (1967), employing the catalytic hydrolysis of p-nitrophenyl acetate at an alkaline pH. Instead of Tris buffer, we used 0.02 $\underline{M}$ K_2HPO_4, pH 8.84. Sulfonamide inhibition, which operates powerfully on CA (Mann and Keilin, 1940; Pocker and Stone, 1965), was used to distinguish the enzyme's activity from that of other esterases, which are not similarly affected. Acetazolamide was the sulfonamide used. CA's esterase activity can be determined with greater ease and accuracy than its hydratase activity, although the latter is much stronger (Patterson et al., 1963). Enzyme was added as 0.10 ml of the 10% preparation.

Leucine aminopeptidase was measured using the reagents and procedure described in Sigma Technical Bulletin No. 251 (Sigma Chemical Co., St. Louis, Mo., 63178), with incubation temperature lowered to 30°C. The procedure was used on the grounds of greater speed and relative accuracy, respectively, than the manometric and titrimetric techniques (Davis and Smith, 1955). Because the substrate employed, L-leucyl-β-naphthylamide, can be hydrolyzed by several different kinds of enzymes, Armstrong et al. (1966) have suggested that "arylamidases" might be a more appropriate term for the enzyme activity measured by this test. On the assumption that flounder LAP is a zinc enzyme, like its mammalian analogues, and therefore probably subject to attack by cadmium in vitro as well as in vivo, we had tested preparations from unexposed fish in companion assays containing several different concentrations of cadmium chloride. This testing established that at assay concentrations below 0.25 m$\underline{M}$, cadmium inhibited a small fraction of the total activity. The inhibited fraction we considered to be a truer indication of specific LAP activity than the total activity of the standard assay. For testing the experimental samples, therefore, an analogous test series containing added $CdCl_2-1/2H_2$), 0.10 m$\underline{M}$ assay concentration, as well as the standard assays, were performed. The enzyme preparations were diluted 1:39, v/v, for a final 400-fold dilution, of which 0.20 ml was used per assay.

Glucose-6-phosphate dehydrogenase activity was measured in the presence of $MgCl_2.6H_2O$, assay concentration 2.5 m$\underline{M}$, and also without Mg. Assays were performed at 340 nm using a double-beam, ratio-recording spectrophotometer with chamber temperature 25°C, in an optical cuvette having a 10-mm pathlength and 3.00-ml final assay volume. A linear-log recorder was used to follow reaction rates, which were

read from the fastest portion of each recording. For the
standard G6PdH assay, each cuvette contained 2.55 ml tris
(hydroxymethyl) aminomethane buffer, 0.10 $\underline{M}$, pH 8.00;
0.10 ml nicotinamide adenine dinucleotide phosphate, 8 mg/ml
H_2O; 0.10 ml glucose-6-phosphate, sodium dihydrate, 15 m$\underline{M}$;
0.15 ml $MgCl_2 \cdot 6H_2O$, 50 m$\underline{M}$; and 0.10 ml of the 10% enzyme
preparations. For the assay without Mg, 0.15 ml H_2O was
substituted. As with the other enzyme assays, the water used
in preparing all solutions was iced and doubly glass-
distilled. The enzyme's kinetics were also examined with
and without assay Mg; K_m and V_{max} for each preparation were
determined by the classic double-reciprocal plot (Lineweaver
and Burk, 1934), with a minimum of 4 points per slope.
Preliminary testing on KH from unexposed fish established
the most suitable substrate concentrations, whose reciprocals
provided the necessary point spread.

RESULTS AND DISCUSSION

Compared to the controls, there was no apparent change
in arylamidase activity in the kidney-hematopoietic tissue
from cadmium-exposed flounder, as measured by the rate of
hydrolysis of L-leucyl-β-naphthylamide (Fig. 1). Activity
was undoubtedly present in both tissues, as the widely-
distributed LAP has been found in both kidney and spleen of
mammals (Smith and Hill, 1960). Because several kinds of
enzymes hydrolyse this substrate, the overall activity
actually represents an indeterminate number of hydrolases,
including the zinc enzyme LAP. Enzymes of the general class
of N-terminal exopeptidases, to which LAP belongs, have
numerous overlapping substrate specificities, and peptide
hydrolysis by LAP is not restricted to leucyl compounds
(Davis and Smith, 1955; Smith and Hill, 1960; Delange and
Smith, 1971). Davis and Smith (1955) recommended using
several different substrates for the identification of a
purified peptidase; using only one substrate to measure
specific LAP activity in a crude preparation, then, will
produce questionable results. LAP, however, is the only
metal-containing terminal aminopeptidase listed by Vallee
and Wacker (1970), and Smith and Hill (1960) reported that
in vitro cadmium completely inhibited the hydrolysis of L-
leucylamide by purified LAP, not a surprising observation
with a zinc enzyme. Moreover, in our initial study, we
found that adding cadmium to the standard assay medium did
indeed inhibit a small and variable portion of the standard
arylamidase activity in the KH tissue from unexposed fish.
The inhibited fraction, therefore, we considered to offer

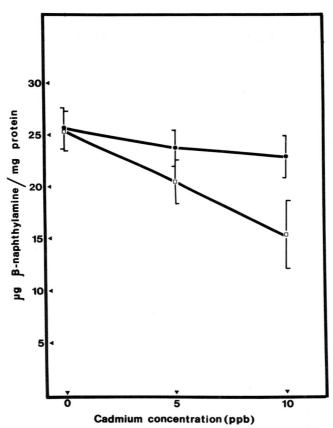

Fig. 1. Leucine aminopeptidase activity in the
kidney and hematopoietic tissue of flounder
chronically exposed to sublethal amounts of
cadmium chloride. The upper curve represents
the standard activity as measured by the
hydrolysis of L-leucyl-β-naphthylamide.
The lower curve represents the activity in
assays containing cadmium chloride, 0.10 mM
assay concn. The difference between the 2
curves is cadmium-inhibited activity, and is
considered to be a truer measure of LAP
activity than the standard assay (see text).
Significance level in fish exposed to 10 ppb
Cd is <.01.

evidence that flounder LAP, like mammalian LAP, is a zinc
enzyme, and that the degree of cadmium inhibition is a more
accurate measure of LAP activity (or of another metallo-
arylamidase) than is the standard assay. A parallel series
of arylamidase assays containing added cadmium chloride
(0.10 mM assay concentration) is represented by the lower
curve in Figure 1. The upper curve represents the standard
activity. Very little cadmium-inhibited activity was
present in the controls, but a significant increase in
cadmium inhibition (P<.01) can be seen in fish exposed to
10 ppb Cd. Exposure to sublethal amounts of cadmium, therefore, appears to have induced an increased biosynthesis of a
cadmium-inhibited enzyme.

The phenomenon of stimulated enzyme activity in cadmium-
exposed fish is seen again in carbonic anhydrase, the second
presumptive-zinc enzyme examined (Table 1). CA activity
increased very significantly (P<.001) in preparations from
flounder exposed to 10 ppb Cd. Rawls and Maren (1962) found
that acetazolamide does not inhibit acidification of the
urine in winter flounder, an observation suggesting that
renal CA, if present, is non-functional. Because all fish, as
is true of all known vertebrates, have erythrocyte CA (Maren,
1967), the activity measured in these preparations probably
stems from the hematopoietic tissue, rather than the kidney
proper.

Induction of this zinc enzyme in animals exposed to
sublethal amounts of cadmium has also been reported for
mammalian and avian tissues. Johnson and Walker (1970)
introduced 0.03 mM cadmium chloride by intraperitoneal
injection into rats and domestic fowl, and noted in each case
an initial inhibition of testicular CA, followed by an
increase in the enzyme's activity 10 min after injection.
Recent work with winter flounder (Manen et al., in ms.)
showed a similar pattern in an enzyme of liver and kidney,
following a single i.v. injection of methyl mercury; that is,
there was a sevenfold evaluation of activity after an initial
short-term decrease. The enzyme under study was ornithine
decarboxylase, a pyridoxal-phosphate protein, whose synthesis
requires a divalent metal ion (Meister, 1955).

The toxicity of most metal ions lies in their capacity
to form complexes obeying the rules of Lewis's hard and soft
acids and bases, as definitively set forth by Pearson in 1968,
and as discussed in relation to enzyme catalysis by Mildvan
(1970). For example, cadmium and mercury (soft acids) seek
out sulfhydryl groups (soft bases), displacing metal ions
like zinc (moderately soft acid) with a weaker capacity to
complex. The increased activity of cadmium-inhibited enzymes
in the cadmium-exposed flounder, therefore, may reasonably be

Table 1
Carbonic anhydrase activity in the kidney and hematopoietic
tissue of winter flounder, Pseudopleuronectes americanus,
chronically exposed to sublethal levels of cadmium chloride.

Experimental Conditions	Number of Samples	CA Activity[a] $\bar{x}$+S.E.		Range[a]	Level of Significance
Control	19	70.3	5.5	23-100	
5 ppb Cd	20	84.2	5.5	50-127	
10 ppb Cd	16	100.2	4.8	72-128	P<.001

[a] Unit of CA activity = change in absorbance at 400 nm of 0.001/min/mg protein.

interpreted as an inductive response to cadmium poisoning at the molecular level, a physiological mechanism that compensates for at least partial enzyme inhibition.

Glucose-6-phosphate dehydrogenase, which catalyzes the first step of the pentose phosphate shunt, is not a metalloenzyme, but a metal-activated one. It does not require magnesium for activity, but is stimulated by that cation. Glaser and Brown (1955) discovered that free magnesium ion increased the activity of purified yeast G6PdH by lowering the K_m's for both substrate and coenzyme; that is, magnesium enabled the enzyme to function optimally at lower concentrations of both ligands than was possible when magnesium was not present. In the flounder preparations of this study, G6PdH activity measured in the presence of magnesium did not vary significantly from control to cadmium-exposed fish (Fig. 2). With no magnesium in the assay medium, the activity was lower in all cases, but slightly higher in the cadmium-exposed fish, although not to a statistically significant degree. When the magnesium-induced activity was calculated as percent of activity without magnesium, however, the data became highly significant (p<.001). Sensitivity to magnesium activation is clearly lower in the metal-exposed fish. Even partial blocking of such physiological controls points to some loss of the metabolic flexibility necessary for adaptation and survival in changing environmental conditions.

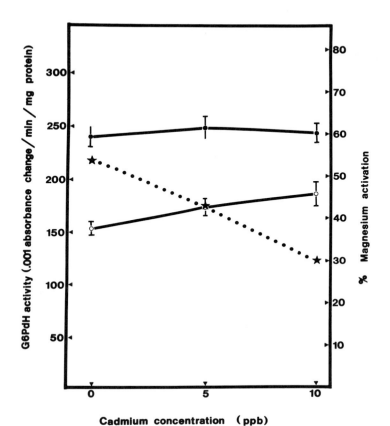

Fig. 2. Glucose-6-phosphate dehydrogenase activity
in the kidney and hematopoietic tissue of
flounder chronically exposed to sublethal
amounts of cadmium chloride. The lower
curve represents the standard activity without magnesium, and the upper curve, the
activity in the presence of $MgCl_2.6H_2O$,
2.5 m$\underline{M}$ assay concn. The dotted line
represents % magnesium activation. In fish
exposed to 10 ppb Cd, the significance
level is <.001.

In the kinetics study of G6PdH (Table 2), data obtained
from the control fish agree with the findings of Glaser and
Brown (1955) with purified yeast enzyme, namely: magnesium
increases the efficiency of G6PdH by lowering its K_m for
substrate (P<.05) without changing the maximal velocity
(V_{max}). On the other hand, the same enzyme in preparations

Table 2
Glucose-6-phosphate dehydrogenase in kidney and hematopoietic tissue of winter flounder, Pseudopleuronectes americanus, chronically exposed to cadmium chloride.

Experimental Conditions	K_m (G6P) [a] $\bar{x}$ [c] $\pm$ S.E.		V_{max} [b] $\bar{x}$ [c] $\pm$ S.E.	
Control: no Mg	65.4	3.7	176	11
with Mg[d]	52.7	3.2 (P<.05)	171	10
10 ppb Cd: no Mg	64.3	2.6	177	8
with Mg[d]	64.3	2.3	200	8 (P<.05)

[a] G6P assay concentration, μM.

[b] Unit of activity = change in absorbance at 340 nm of 0.001/min/gm protein.

[c] Arithmetic mean for 6 samples.

[d] Assay concentration, 2.5 mM $MgCl_2.6H_2O$.

from cadmium-exposed fish shows no such receptivity to magnesium's stimulus, but does have an increased V_{max} (P<.05). In other words, there seems to be a higher concentration of a less efficient enzyme in the cadmium-exposed fish. Figure 2 may be interpreted, therefore, as showing greater amounts of an impaired enzyme functioning at a slower rate in the cadmium-exposed fish than the enzyme in the control fish.

No remarkable histopathology was noted in any of the tissues, including the kidney and hematopoietic mass (Murchelano, 1975), from the fish in this experimental series, nor did hematological tests show any significant changes between control and exposed flounder (Calabrese et al., 1975). Only one physiological indication of stress was observed, a decreased rate of gill-tissue respiration (Calabrese et al., 1975).

In this study, then, chronic exposure to sublethal levels of cadmium chloride produced more subtly toxic effects than discernible tissue pathology. In the case of the presumptive zinc enzymes, CA and LAP, it stimulated an increased production of the enzymes under attack, so that their functions could proceed at near-normal rates. Synthesis of enzymes requires energy. Another significant effect was the loss of G6PdH's sensitivity to modulation by magnesium, the attenuation of an important metabolic control.

The observations reported here reinforce earlier work with cadmium-exposed marine animals: the cunner, Tautogolabrus adspersus (Gould and Karolus, 1974); the rock crab, Cancer irroratus (Gould et al., in press); and the lobster, Homarus americanus (Thurberg et al., in press). On the whole, they suggest that sublethal amounts of cadmium affect both metalloenzymes and enzymes whose catalytic efficiency is largely dependent upon allosteric mechanisms. The implications are that exposure to sublethal amounts of cadmium chloride causes: 1) a drain of metabolic energy by reason of increased enzyme biosynthesis, and 2) a loss of metabolic flexibility by reason of reduced sensitivity to magnesium modulation.

ACKNOWLEDGMENTS

The author thanks Nancy C. Rawson and Gayle Whittaker for their expert technical assistance; Drs. William B. Kinter and David S. Miller, of the Mount Desert Island Biological Laboratory, for helpful discussions on flounder tissues in general and carbonic anhydrase in particular; and Dr. J. M. Smith, Jr., of the Lederle Laboratories (American Cyanamid Company), for his gift of Diamox, the acetazolamide used in this study. Reference to trade names is to facilitate identification. It does not imply endorsement by the National Marine Fisheries Service.

LITERATURE CITED

Addink, A. D. F. 1968. Some aspects of carbonic anhydrase of Sepia officinalis (L.). Lindskog et al. (1971) (v.i.).
Armstrong, J. McD., D. V. Myers, J. A. Verpoorte, and J. T. Edsall. 1966. Purification and properties of human erythrocyte carbonic anhydrases. J. Biol. Chem. 241: 5137-5149.
Atkinson, D. E. 1965. Biological feedback control at the molecular level. Science 150: 851-857.
_____ 1966. Regulation of enzyme activity. Ann. Rev. Biochem. 35: 85-124.

Behrisch, H. W. and P. W. Hochachka. 1969. Temperature and the regulation of enzyme activity in poikilotherms. Properties of rainbow trout fructose diphosphatase. Biochem. J. 111: 287-295.

Bigelow, H. B. and W. C. Schroeder. 1953. Fishes of the Gulf of Maine. U. S. Fish Wildl. Serv. Fish. Bull. 74, 53: 276-283.

Calabrese, A., R. S. Collier, and J. E. Miller. 1974. Physiological response of the cunner, Tautogolabrus adspersus, to cadmium. I. Introduction and experimental design. In: Physiological Response of the Cunner, Tautogolabrus adspersus, to Cadmium, pp. 1-3. NOAA Tech. Rep., NMFS, SSRF-681.

_____, _____, D. A. Nelson, and J. R. MacInnes. 1973. The toxicity of heavy metals to embryos of the American oyster Crassostrea virginica. Mar. Biol. 18: 162-166.

_____, F. P. Thurberg, M. A. Dawson, and D. R. Wenzloff. 1975. Sublethal physiological stress induced by cadmium and mercury in the winter flounder, Pseudopleuronectes americanus. In: Sublethal Effects of Toxic Chemicals on Aquatic Animals, pp. 15-21, eds. J. H. Koeman and J. J. Strik. Amsterdam: Elsevier Scientific Publ. Co.

Chan, J. P., M. T. Cheung, and F. P. Li. 1974. Trace metals in Hong Kong waters. Mar. Pollut. Bull. 5: 171-174.

Davis, N. C. and E. L. Smith. 1955. Assay of proteolytic enzymes. In: Methods of Biochemical Analysis, Vol. 2, pp. 215-257, ed. by D. Glick. New York: Interscience Pub.

Dawson, M. A., E. Gould, R. A. Greig, F. P. Thurberg, and A. Calabrese. Physiological response of juvenile striped bass, Morone saxatilis, to low levels of cadmium and mercury. In prepn.

Delange, R. J. and E. L. Smith. 1971. Leucine aminopeptidase and other N-terminal exopeptidases. In: The Enzymes, 3rd ed., Vol. III, pp. 81-118, ed. by P. D. Boyer. New York: Academic Press.

Eisler, R. 1971. Cadmium poisoning in Fundulus heteroclitus (Pisces: Cyprinodontidae) and other marine organisms. J. Fish. Res. Bd. Canada 28: 1225-1234.

Gardner, G. R. and P. P. Yevich. 1970. Histological and hematological responses of an estuarine teleost to cadmium. J. Fish. Res. Bd. Canada 27: 2185-2196.

Glaser, L. and D. H. Brown. 1955. Purification and properties of D-glucose-6-phosphate dehydrogenase. J. Biol. Chem. 216: 67-79.

Gornal, A. G., C. S. Bardawill, and M. M. David. 1949. Cited in E. Layne, 1957. Spectrophotometric and turbidi-

metric methods for measuring proteins. III. Biuret
method. In: Methods in Enzymology, Vol. III, pp. 450-
451, ed. by S. P. Colowick and N. O. Kaplan. New York:
Academic Press.

Gould, E. 1969. Alpha-glycerophosphate dehydrogenase as an
index of iced-storage age of fresh, gutted haddock. J.
Fish. Res. Bd. Canada 26: 3175-3181.

_____ and J. J. Karolus. 1974. Physiological response
of the cunner, Tautogolabrus adspersus, to cadmium. V.
Observations on the biochemistry. In: Physiological
Response of the Cunner, Tautogolabrus adspersus, to
Cadmium, pp. 21-25. NOAA Tech. Rep., NMFS, SSRF-681.

_____, R. S. Collier, J. J. Karolus, and S. A. Givens.
Heart transaminase in the rock crab, Cancer irroratus,
exposed to cadmium salts. Bull. Environ. Contam. Toxicol.
In press.

Greig, R. A. 1975. Personal communication. NOAA, NMFS,
MACFC, Milford Laboratory, Milford, Conn. 06460.

Hochachka, P. W. and G. Somero. 1973. Strategies of Biochemical Adaptation, p. 220. Philadelphia: W. B.
Saunders Co.

_____, D. E. Schneider, and A. Kuznetsov. 1970. Inter-
acting pressure and temperature effects on enzymes of
marine poikilotherms: catalytic and regulatory properties
of FDPase from deep- and shallow-water fishes. Mar. Biol.
7: 285-293.

Hodgen, G. D., W. R. Butler, and W. R. Gomes. 1969. In vivo
and in vitro effects of cadmium chloride on carbonic
anhydrase. J. Reprod. Fert. 18: 156-157.

Johnson, A. D. and G. P. Walker. 1970. Early actions of
cadmium in the rat and domestic fowl testis. V. Inhibition of carbonic anhydrase. J. Reprod. Fert. 23: 463-
468.

Kendrey, G. and F. J. C. Roe. 1969. Cadmium toxicology.
Lancet 2: 1206-1207.

Leiner, M., H. Beck, and H. Eckert. 1962. Über die kohlen-
saure-dehydratase in den einzelnen wirbeltierk-lassen. Z.
Physiol. Chem. 327: 144-165.

Lindskog, S., L. E. Henderson, K. K. Kannan, A. Liljas, P. O.
Nyman, and O. Strandberg. 1971. Carbonic anhydrase.
In: The Enzymes, 3rd ed., Vol. V, pp. 587-665, ed. by
P. D. Boyer. New York: Academic Press.

Lineweaver, H. and D. Burk. 1934. Cited in J. M. Reiner.
1959. Behavior of Enzyme Systems, p. 27. Minneapolis:
Burgess Pub. Co.

Manen, C.-A, B. Schmidt-Nielsen, and D. H. Russell. Alterations of polyamine synthesis in liver and kidney of the
winter flounder in response to methylmercury. In prepn.

Mann, T. and D. Keilin. 1940. Sulphanilamide as a specific inhibitor of carbonic anhydrase. Nature 146: 164-165.

Maren, T. H. 1967. Carbonic anhydrase: chemistry, physiology, and inhibition. Physiol. Rev. 47: 595-781.

Meister, A. 1955. Transamination. In: Advances in Enzymology, Vol. 16, pp. 185-246, ed. by F. F. Nord. New York: Interscience Pub.

Mildvan, A. S. 1970. Metals in enzyme catalysis. In: The Enzymes, 3rd ed., Vol. II, pp. 445-536, ed. by P. D. Boyer. New York: Academic Press.

Mount, D. I. and W. A. Brungs. 1967. A simplified dosing apparatus for fish toxicology studies. Wat. Res. 1: 21-29.

Murchelano, R. A. 1975. Personal communication. NOAA, NMFS, MACFC, Pathobiology Investigations, Oxford, Md. 21654.

Nordberg, G. F. 1972. Cadmium metabolism and toxicity. Environ. Physiol. Biochem. 2: 7-36.

Parizek, J. 1956. Effect of cadmium salts on testicular tissue. Nature 177: 1036-1037.

Patterson, E. K., H. Shu-Hsi, and A. Keppel. 1963. Studies on dipeptidases and aminopeptidases. I. Distinction between leucine aminopeptidase and enzymes that hydrolyze L-leucyl-β-naphthylamide. J. Biol. Chem. 238: 3611-3623.

Pearson, R. G. 1968. Hard and soft acids and bases, HSAB, Part I. Fundamental principles. J. Chem. Educ. 45: 581-587.

Pocker, Y. and J. T. Stone. 1965. The catalytic versatility of erythrocyte carbonic anhydrase. The enzyme-catalyzed hydrolysis of p-nitrophenyl acetate. J. Am. Chem. Soc. 87: 5497-5498.

_____ and J. T. Stone. 1967. The catalytic versatility of carbonic anhydrase. III. Kinetic studies of the enzyme-catalyzed hydrolysis of p-nitrophenyl acetate. Biochem. J. 6: 668-678.

Rawls, J. A. and T. H. Maren. 1962. Effect of carbonic anhydrase inhibition on urine of a marine teleost, Pseudopleuronectes americanus. Bull. Mt. Desert Island Biol. Lab. IV: 57-58.

Smith, E. L. and R. L. Hill. 1960. Leucine aminopeptidase. In: The Enzymes, 2nd ed., Vol. 4, pp. 37-62, ed. by P. D. Boyer, H. Lardy, and K. Myrback. New York: Academic Press.

Somero, G. 1969. Pyruvate kinase variants of the Alaska king crab. Evidence for a temperature-dependent interconversion between two forms having distinct and adaptive kinetic properties. Biochem. J. 114: 237-241.

Thurberg, F. P., A. Calabrese, E. Gould, R. A. Greig, M. A. Dawson, and R. K. Tucker. Response of the lobster, Homarus americanus, to sublethal levels of cadmium and mercury. In: Physiological Responses of Marine Biota to Pollutants, ed. by F. J. Vernberg, A. Calabrese, F. P. Thurberg, and W. B. Vernberg. New York: Academic Press. This volume.

_____, M. A. Dawson, and R. S. Collier. 1973. Effects of copper and cadmium on osmoregulation and oxygen consumption in two species of estuarine crabs. Mar. Biol. 23: 171-175.

Tucker, R. K. 1975. Personal communication. NOAA, NMFS, MACFC, Sandy Hook Laboratory, Highlands, N. J. 07732.

Vallee, B. L. 1960. Metal and enzyme interactions: correlation of composition, function, and structure. In: The Enzymes, 2nd ed., Vol. 3, pp. 225-276, ed. by P. D. Boyer, H. Lardy, and K. Myrback. New York: Academic Press.

_____ and W. E. C. Wacker. 1970. Metalloenzymes. In: Handbook of Biochemistry, 2nd ed., pp. C51-C53, ed. by H. A. Sober. Cleveland: CRC Press.

Williams, D. R. 1971. The Metals of Life. London: Van Nostrand Reinhold.

The Effect of Chromium on Filtration Rates and Metabolic Activity of *Mytilis edulis* L. and *Mya arenaria* L.

JUDITH McDOWELL CAPUZZO[1] and JOHN J. SASNER, JR.[2]

[1]Woods Hole Oceanographic Institution
Woods Hole, Massachusetts   02543

[2]University of New Hampshire
Zoology Department
Durham, New Hampshire   03824

The input of trace metals from industrial wastes is a problem of increasing concern in the marine environment. Sedimentary accumulation of metals above natural levels, resulting in drastic changes in benthic diversity, have been attributed to such discharges in polluted coastal environ-ments (Pearce, 1969, 1972; Mackay et al., 1972). Based on recent evidence, it has been suggested that marine bivalves are important in further concentrating these metals through particulate uptake. Raymont (1972) and Romeril (1971) compared the concentrations of iron, zinc, and copper in estuarine sediments and tissues of the quahog Mercenaria mercenaria. Ayling (1974) correlated the concentrations of chromium, cadmium, copper, lead, and zinc in the oyster Crassostrea gigas with the surrounding muds. In both studies bivalve tissue had consistently higher levels of these trace metals than the sediments. Whereas particulate ingestion has been shown to be a mechanism of metal uptake in bivalves, no information is available on the physiological effects of this uptake in polluted areas.

The object of this study was to determine the effects of particulate uptake of chromium by two filter-feeding bivalves-Mytilus edulis L., the blue mussel, and Mya arenaria L., the soft shelled clam. Recent sediments in the Great Bay estuary of New Hampshire were found to contain high concentrations of chromium associated with the fine (clay-silt) fraction of sediment due to the input of chromium from tannery wastes (Capuzzo and Anderson, 1973). Because clay particles are within the size range of particles filtered and retained by filter-feeding bivalves (Moore, 1971), accumulation of fine sediment and the associated chromium concentrations may cause subtle physiological changes in exposed organisms. The rates of filtering activity and oxygen consumption were used as indicators of physiological stress after exposure to chromium contaminated sediments and clay suspensions.

MATERIALS AND METHODS

Specimens of _Mytilus_ _edulis_ L. and _Mya_ _arenaria_ L., ranging in size from 30 mm to 80 mm in shell length, were collected from coastal areas of New Hampshire during the summer and early fall. Animals were maintained at $15^{\circ}C$ and 27.5 o/oo for a period of 4 weeks before exposure to either natural sediments or clay suspensions. Sediment samples (60 cm x 30 cm x 5 cm) were collected from the Great Bay estuary and varied in chromium content with the distance from the site of chromium discharge (Capuzzo and Anderson, 1973). The sampling sites were chosen to be representative of sediments with natural chromium levels (Adam's Point) and high chromium levels (Pomeroy's Cove and Fresh Creek). Sediment samples and specimens of _Mytilus_ were placed in 20 l plexiglass aquaria for a period of 24 weeks in continuously flowing sea water. In order to maintain a suspension of sediment with a minimum of loss, sea water and aeration were supplied to the bottom of each aquarium; the average liquid turnover time in each aquarium was ∿6 hours. Temperature, salinity, and food concentration varied with the ambient seasonal (summer-fall) conditions of the Great Bay estuary.

Artificial sediments were prepared from clay suspensions of bentonite and kaolinite (Ward Scientific Co., Rochester, New York), aerated in distilled water with 1.4 mg Cr (as $CrCl_3$)/g clay for a period of 1 week. The suspensions were then filtered (Millipore ®, 0.45 µm) and dried to constant weight at $60^{\circ}C$. Clay suspensions with natural chromium levels were prepared in a similar manner. These minerals were chosen because of differences in their ion exchange capacity. The bentonite minerals have a high cation exchange capacity through lattice substitution, whereas the kaolinite minerals have relatively low cation exchange capacity through surface absorption (Kelley, 1955). Temporarily high exchange capacity may occur in the kaolinite minerals; however, the weak bonds formed are not stable and cations may be released to the water column at the sediment-water interface.

Artificial sediment samples (30 cm x 30 cm x 2.5 cm) and specimens of _Mytilus_ or _Mya_ were placed in 10 l aquaria for a period of 4-6 weeks. Filtered (Millipore ®, 0.45 μm) sea water was continuously supplied to each aquarium and salinity and temperature were maintained at 27.5 o/oo and 15°C; the average liquid turnover time in each aquarium was ∿6 hours. Artificial sediments were kept suspended with a minimum of loss as in the experiments with natural sediments, with sea water and aeration being supplied to the bottom of each aquarium.

Control bivalves were maintained in filtered sea water (27.5 o/oo, 15°C) with no artificial sediment. Exposure of bivalves to 1.0 mg Cr (as $CrCl_3$)/l sea water was used as an indicator of sublethal chromium toxicity. The pH of the sea water (7.9-8.1) was not altered by the addition of the test suspensions or $CrCl_3$. All bivalves were fed a daily ration (100 ml - 1 x 10^6 cells/ml) of the diatom _Phaeodactylum tricornutum_ Bohlin during the test period.

After exposure, filtration rates were determined by an indirect method, monitoring the reduction in particle concentration per unit volume during a time interval. _Phaeodactylum tricornutum_ was used as the suspended particle for filtration rate determinations. Individual bivalves were placed in 1.5 l aquaria with 1 l of media having a diatom concentration of 1 x 10^4 cells/ml and were acclimated for a period of 30 minutes. Filtering rates were determined at 30 minute intervals for 90 minutes and an average filtration rate was recorded. Samples of 20 ml were collected and duplicate portions of 0.5 ml were counted using a Coulter Counter, Model F_N, and filtration rates were calculated from a modified version of the formula derived by Jorgensen (1943):

$$Filtration\ Rate = (\log C_o - \log C_t) \cdot V/ \log e \cdot t/ dry\ wt.$$

where, C_o = initial concentration of cells in suspension;
C_t = concentration of cells in suspension after time t;
V = volume of media (milliliters);
t = time (hours);
dry weight = soft tissues dried to constant weight at 110°C (milligrams).

Oxygen consumption rates of excised gill tissue were measured as an indicator of ciliary activity. Specimens of _Mytilus_ were dissected after exposure to the different test conditions and oxygen uptake of whole gill tissue at 22°C was

measured using a Gilson Differential Respirometer. Gill
tissue was placed in a 15 ml reaction flask with 5 ml of
filtered sea water. Filter paper soaked with 20% KOH was
placed in the center well of each flask as the CO_2 absorbent.
Rates of oxygen consumption are reported as μl O_2/h/mg dry
weight.

Chromium concentrations of sediments and clay suspensions were determined by atomic absorption spectrophotometry
(Perkin-Elmer Model 303; Capuzzo and Anderson, 1973). The
clay fraction of sediments was determined by the pipette
analysis method of Folk (1965); chromium concentrations are
reported as mg Cr/g clay. Chromium concentrations of whole
tissue and dissected parts (muscle, mantle, gills, and
viscera) were determined by atomic absorption spectrophotometry according to the method described by Riley and Segar
(1970); concentrations are reported as μg Cr/g dry weight.

The relationship of filtration rates (ml/h/mg) or
oxygen consumption rates (μl/h/mg) to dry weight of control
animals was determined by log-log linear regression analysis.
All experimental data were compared to the regression line
for control data of _Mytilus_ or _Mya_ and presented on double
logarithmic plots. The allometric equation ($y = a \cdot x^b$ or
$\log y = \log a + b \log x$) quantitatively describes the
dependency of metabolic activity on body size (vonBertalanffy,
1964); for these experiments: _y_ = filtration rate (ml/h);
_x_ = dry weight of soft tissue (mg); and _a_ and _b_ = the
regression coefficients for each set of experimental data.
The constant _a_ determines the absolute filtration rate and
_b_, the slope of the regression line, describes the dependency
of filtration rate on body size. The regression equation for
each experiment was compared with the control experiment by
analysis of covariance.

RESULTS

Filtration Rate Experiments

The results indicate that chromium, in dissolved and
particulate forms, has an effect on filtration rates in
Mytilus and, to a much lesser extent, in _Mya_. Exposure of
Mytilus to sediments from Pomeroy's Cove (0.15 mg Cr/g clay)
and Fresh Creek (0.99 mg Cr/g clay) caused a reduction in
filtration rates, the degree of toxicity being in accord
with the concentration of chromium in the sediments. No
significant difference from control data was observed with
exposure to sediments from Adam's Point (0.01 mg Cr/g clay;
Fig. 1).

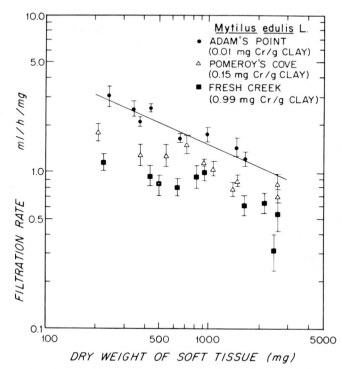

Fig. 1.   Filtration rates of Mytilus edulis L.
exposed to natural sediments, compared
to regression line of control experiments
(r = -0.77).   Each point represents the
mean filtration rate of 10 experiments ± 1
standard error.

Exposure to the artificial sediments with high chromium
concentrations resulted in a varying degree of toxicity in
Mytilus and Mya dependent on the cation exchange capacity of
each mineral and the feeding behavior of each bivalve.
Uptake of kaolinite with a chromium concentration of 1.2 mg
Cr/g clay caused a reduction in filtration rates in both
Mytilus (Fig. 2) and Mya(Fig. 3).   Because of the weak bonds
associated with chromium uptake by the kaolinite minerals
and the possible release of chromium to the water column at
the sediment-water interface, it is assumed that uptake of
chromium by both species was due mainly to diffusion of
dissolved chromium.   Through analysis of kaolinite suspensions during the experimental period, a reduction of chromium
concentration to 0.05 mg Cr/g clay was noted; this reduction

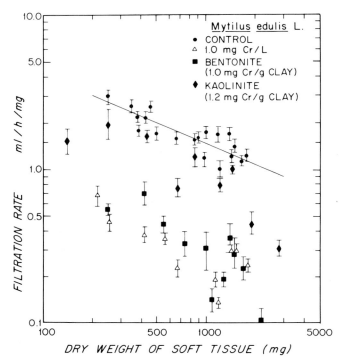

Fig. 2.   Filtration rates of <u>Mytilus</u> <u>edulis</u> L.
exposed to artificial sediments with
high chromium concentrations and 1 mg
Cr/l sea water, compared to regression
line of control experiments (r = -0.77).
Each point represents the mean filtration
rate of 10 experiments ± 1 standard error.

was not detected in bentonite suspensions. Uptake of bentonite with a chromium concentration of 1.0 mg Cr/g clay
resulted in a significant reduction of filtration rates in
<u>Mytilus</u> (Fig. 2), but little change was observed in <u>Mya</u>
(Fig. 3). Differences in feeding time, filtering rates, and
filtration efficiency may explain this variation, with
<u>Mytilus</u> accumulating more particulate matter than <u>Mya</u>.
      Exposure to 1 mg Cr/l sea water also resulted in signi-
ficant reductions in filtering activity in both <u>Mytilus</u>
(Fig. 2) and <u>Mya</u> (Fig. 3). It is concluded that diffusion
from sea water and particulate uptake are important pathways
for chromium exposure in <u>Mytilus</u> but only diffusion is
important in <u>Mya</u>. No significant reduction in filtration
rates of either species was observed in the absence of

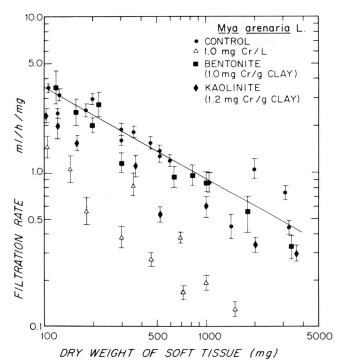

Fig. 3.   Filtration rates of <u>Mya arenaria</u> L.
          exposed to artificial sediments with
          high chromium concentrations and 1 mg
          Cr/l sea water, compared to regression
          line of control experiments (r = -0.94).
          Each point represents the mean filtration
          rate of 10 experiments ± 1 standard error.

chromium. Analysis of the allometric relationship of each
set of experimental data is presented in Table 1, indicating
significant changes in filtration rates with exposure to
chromium. The weight dependency of the filtration rate was
not changed significantly by any treatment, except exposure
of <u>Mya</u> to 1 mg Cr/l with greater reductions in filtration
rates being observed in larger specimens.

Ciliary Activity Experiments

     Direct observation of <u>Mytilus</u> gills exposed to 1 mg
Cr/l sea water and bentonite with a chromium concentration
of 1.0 mg Cr/g clay revealed a slower, more erratic movement
of the latero-frontal cilia when compared with gills from

Table 1

Analysis of Covariance of Filtration Rate Experiments of
Mytilus edulis L. and _Mya arenaria_ L.

Experiment	log a	B	F Ratio of Slopes	F Ratio of Means
Mytilus edulis L.				
Control	1.516	0.551	–	–
Adam's Point	1.267	0.658	NS	NS
Pomeroy's Cove	1.012	0.672	NS	10.257**
Fresh Creek	1.129	0.577	NS	47.606**
Bentonite	1.447	0.575	NS	NS
Bentonite + Cr	1.794	0.205	NS	98.473**
Kaolinite	1.635	0.518	NS	NS
Kaolinite + Cr	1.324	0.555	NS	5.853*
1 mg Cr/l sea water	0.827	0.529	NS	227.500**
Mya arenaria L.				
Control	1.698	0.421	–	–
Bentonite	1.836	0.352	NS	NS
Bentonite + Cr	1.873	0.332	NS	NS
Kaolinite	1.853	0.354	NS	NS
Kaolinite + Cr	1.742	0.334	NS	12.030**
1 mg Cr/l sea water	2.043	0.056	6.763*	67.710**

  * $P < 0.05$
** $P < 0.01$
NS – not significant

control animals; the erratic movement of the cilia resulted
in inefficient retention of particles. The rate of ciliary
activity is correlated with the rate of oxygen consumption
(Gray, 1928). Reduction of oxygen consumption of excised
gill tissue was observed with exposure to chromium,
indicating an inhibition of ciliary activity (Fig. 4).

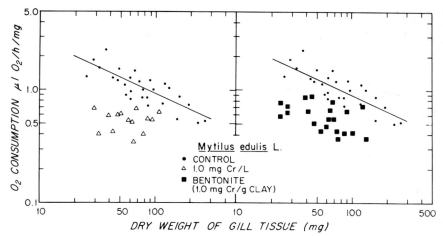

Fig. 4. Oxygen consumption rates of excised gill tissue of
Mytilus edulis L. exposed to bentonite + 1.0 mg
Cr/g clay and 1.0 mg Cr/l sea water, compared to
regression line of control experiments (r = -0.81).

Chromium Accumulation in Tissues

    Data of chromium concentrations in tissues of Mytilus
and Mya after exposure to test conditions are presented in
Table 2. Uptake of chromium from particulate bound chromium
was lower than uptake from dissolved chromium. However,
toxic action on the filtering mechanism of the bivalves
studied was observed with both mechanisms of exposure.
Cation exchange between particulate matter and bivalve
tissue might be a complex process causing a deterioration of
tissue with initial uptake and a reversible exchange taking
place after the damage has occurred. Changes in pH that
occur during digestion and at the gill surface during
aerobic and anaerobic respiration might explain such cation
exchange.

DISCUSSION

    The toxicity of chromium to marine organisms has been
determined in standard bioassays by several workers (Raymont
and Shields, 1963; Calabrese et al., 1973). However, concentrations of dissolved chromium used in such studies far
exceeded amounts detected in the water column of polluted
coastal environments. Whereas these studies have provided

Table 2
Chromium Concentrations of Soft Tissue of Mytilus edulis L. and Mya arenaria L.

Experiment	Sediment mg Cr/g Clay	Soft Tissue µg Cr/g Dry Weight			
		Whole	% Muscle	% Mantle and Gills	% Viscera
Mytilus edulis L.					
Control	–	8	25.0%	17.5%	57.5%
Adam's Point	0.01	8	–	–	–
Pomeroy's Cove	0.15	10	–	–	–
Fresh Creek	0.99	13	–	–	–
Bentonite	<0.01	8	–	–	–
Bentonite + Cr	1.00	14	20.6%	20.6%	58.8%
Kaolinite	<0.01	8	–	–	–
Kaolinite + Cr	1.20	40	–	–	–
1 mg Cr/1 sea water	–	430	26.1%	34.9%	39.0%
Mya arenaria L.					
Control	–	4	–	–	–
Bentonite	<0.01	2	–	–	–
Bentonite + Cr	1.00	7	–	–	–
Kaolinite	<0.01	4	–	–	–
Kaolinite + Cr	1.20	22	–	–	–
1 mg Cr/1 sea water	–	765	–	–	–

information on the relative toxicity of chromium in comparison with other metals, an accurate assessment of the actual toxic action of chromium in a polluted environment has not been made. The reactivity of chromium and other trace metals with particulate mineral and organic material can lead to accumulation of these metals in the sedimentary portion of an ecosystem; thus, additional mechanisms besides diffusion of dissolved chromium are important pathways for the distribution and toxicity of chromium in marine organisms.

Reduction of filtration rates and disturbed ciliary activity were observed in response to uptake of dissolved chromium by Mya arenaria L. and uptake of dissolved and particulate chromium by Mytilus edulis L. The decline in oxygen consumption of excised gill tissue suggests that chromium interferes with an energy supplying metabolic process, resulting in an inhibition of ciliary activity, or vice versa. The physiological effects of other trace metals in marine bivalves have been the subject of several recent studies. Exposure of Mytilus to copper (Brown and Newell, 1972; Scott and Major, 1972) resulted in reductions of oxygen consumption, whereas exposure of Mytilus and Mya to silver (Thurberg et al.,1974) resulted in increased oxygen consumption. These studies have indicated that such physiological responses in marine bivalves are dependent on the specific metal toxicant; additional information on the biochemical action of trace metal toxicants is needed.

Reduction of filtration rates in response to stress conditions is a more serious problem for a population of bivalves than indicated by lowered metabolic activity in an individual organism. Other studies have shown that physiological stress in marine bivalves has resulted in decreased growth rates (Galstoff et al.,1947), losses of carbohydrate and protein reserves (Bayne and Thompson, 1970), interference with spawning time (Roberts, 1972), and greater production of abnormal offspring (Bayne, 1975). It is concluded, therefore, that chromium accumulation in areas affected by industrial wastes results in serious consequences to filter-feeding bivalves.

ACKNOWLEDGMENTS

This study was part of a dissertation submitted to the University of New Hampshire by J. Capuzzo in partial fulfillment of the requirements for the Doctor of Philosophy Degree. The authors wish to thank Dr. R. Ukeles, National Marine Fisheries Service, Milford, Connecticut, for supplying cultures of Phaeodactylum tricornutum Bohlin; and to Mr. Paul

Pelton, Center for Industrial and Institutional Development, University of New Hampshire, for assistance in the analysis of samples by atomic absorption spectrophotometry. This study was supported by a National Defense Education Act pre-doctoral fellowship and a University of New Hampshire dissertation fellowship to J. Capuzzo. All experiments were conducted at the University of New Hampshire, Jackson Estuarine Laboratory and Zoology Department facilities.

LITERATURE CITED

Ayling, G. M. 1974. Uptake of cadmium, zinc, copper, lead and chromium in the Pacific oyster, Crassostrea gigas, grown in the Tamar River, Tasmania. Water Research, 8: 729-738.

Bayne, B. 1975. Reproduction in bivalve molluscs under environmental stress. In: Physiological Ecology of Estuarine Organisms, pp. 259-277, ed. by F. J. Vernberg. Columbia, S. C.: University of South Carolina Press.

Bayne, B. and R. J. Thompson. 1970. Some physiological consequences of keeping Mytilus edulis in the laboratory. Helgoländer wiss. Meeresunters, 20: 526-552.

Brown, B. E. and R. C. Newell. 1972. The effect of copper and zinc on the metabolism of the mussel Mytilus edulis. Mar. Biol. 16: 108-118.

Calabrese, A., R. S. Collier, D. A. Nelson, and J. R. MacInnes. 1973. The toxicity of heavy metals to embryos of the American oyster Crassostrea virginica. Mar. Biol. 18: 162-166.

Capuzzo, J. M. and F. E. Anderson. 1973. The use of modern chromium accumulations to determine estuarine sedimen-tation rates. Mar. Geol. 14: 225-235.

Folk, R. L. 1965. Petrology of Sedimentary Rocks. Austin, Texas: Hemphill's.

Galstoff, P. S., W. A. Chipman, J. B. Engle, and H. H. Calderwood. 1947. Ecological and physiological studies on the effect of sulphate pulp mill wastes on oysters in the York River, Virginia. Fish. Bull. 51: 58-186.

Gray, J. 1928. Ciliary Movement. Cambridge: Cambridge University Press.

Jorgensen, C. B. 1943. On the water transport through the gills of bivalves. Acta Physiol. scand. 5: 297-304.

Kelley, W. P. 1955. Base exchange in relation to sediments. In: Recent Marine Sediments, pp. 454-465, ed. by P. D. Trask. New York: Dover Publications.

Mackay, D. W., W. Halcrow, and I. Thornton. 1972. Sludge dumping in the Firth of Clyde. Mar. Poll. Bull. 3: 7-10.

Moore, H. J. 1971. The structure of the latero-frontal cirri on the gills of certain lamellibranch molluscs and their role in suspension feeding. Mar. Biol. 11: 23-27.

Pearce, J. B. 1969. The effects of waste disposal in the New York Bight. Interim Report, Sandy Hook Mar. Lab., U. S. Bur. Sport Fish. Wildl. Dec. 1969.

_____. 1972. The effects of solid waste disposal on benthic communities in the New York Bight. In: Marine Pollution and Sea Life, pp. 404-411, ed. by M. Ruivo. London: Fishing News Ltd.

Raymont, J. E. G. 1972. Some aspects of pollution in Southampton water. Proc. Roy. Soc. London, Ser. B, 180: 451-468.

_____ and J. Shields. 1963. Toxicity of copper and chromium in the marine environment. Int. J. Air and Water Poll. 7: 435-443.

Riley, J. P. and D. A. Segar. 1970. The distribution of the major and some minor elements in marine animals. I. Echinoderms and coelenterates. J. mar. biol. Ass. U. K. 50: 721-730.

Roberts, D. 1972. The assimilation and chronic effects of sublethal concentrations of endosulfan on condition and spawning in the common mussel Mytilus edulis. Mar. Biol. 16: 119-125.

Romeril, M. G. 1971. Central Electricity Research Laboratories, Laboratory Note No. RD/L/N31/71. Leatherhead, England.

Scott, D. and C. W. Major. 1972. The effect of copper (II) on survival, respiration, and heart rate in the common blue mussel, Mytilus edulis. Biol. Bull. 143: 679-688.

Thurberg, F. P., A. Calabrese, and M. A. Dawson. 1974. Effects of silver on oxygen consumption of bivalves at various salinities. In: Pollution and Physiology of Marine Organisms, pp. 67-78, ed. by F. J. Vernberg and W. B. Vernberg. New York: Academic Press.

VonBertalanffy, L. 1964. Basic concepts in quantitative biology of metabolism. Helgolander wiss. Meeresunters 9: 5-37.

Part III.
Petroleum Hydrocarbons

Effects of Petroleum Hydrocarbons on the Rate of Heart Beat and Hatching Success of Estuarine Fish Embryos

J. W. ANDERSON[1], D. B. DIXIT[2], G. S. WARD[3], and R. S. FOSTER[1]

[1]Department of Biology
Texas A&M University
College Station, Texas   77840

[2]Department of Biology
Northern Virginia Community College
Sterling, Virginia   22170

[3]Bionomics Marine Laboratory
Pensacola, Florida   32507

There are numerous reports on the effects of salinity
and temperature, singly or in combination, on hatching
success and/or the rate of embryonic development of fishes.
Many of these studies and the methods for analyzing the data
have been reviewed by Alderdice (1972). More recently,
investigators have utilized similar methods of fertilizing
and rearing embryos and fry of several species of fish to
test the toxicity of a given pollutant. While the literature
contains many such reports, only a few of the more recent
studies will be noted.

Von Westernhagen et al. (1974) examined the combined
effects of cadmium and salinity on the development and
survival of herring eggs. The Baltic herring (Clupea
harangus L.) eggs were more sensitive to Cd when exposed at
low salinity (5 and 16%), and increasing Cd content shortened
incubation time. The length of hatched larvae was inversely
related to Cd concentration of the medium, and the Cd content
of eggs increased with decreasing salinity. A rather
interesting finding was that embryonic activity (revolving or

trembling) decreased with low salinity and high Cd concentration. Also, with increasing Cd concentration the yolk sac
volume of hatched larvae increased. It was suggested that Cd
decreased activity, resulting in a lower embryonic utilization
of yolks by the organisms during development.

Long-term studies on the effects of chlorinated biphenyls
(PCB's), Aroclor 1242, 1248, and 1254, on fathead minnows
(Pimephales promelas) and flagfish (Jordanella floridae) were
collected by Nebeker et al. (1974). Extensive data on tissue
content, survival, and several reproductive parameters were
reported. The effects of Aroclor 1254 on embryos and fry of
sheepshead minnows, Cyprinodon variegatus, have also been
tested (Schimmel et al., 1974). Fry were shown to be more
susceptible to this toxicant than embryos, juveniles, or
adults.

There are several reports of the response of fish eggs to
solutions containing oil dispersants and/or oils. Linden
(1974) tested the effects of two oil dispersants (BP1100X and
Finasol SC) on the development of Baltic herring and found
inhibition of cell differentiation, crippled embryos (50 ppm
of Finasol SC) and a slowing of embryonic heart rate.
Exposure to Finasol SC at concentrations between 0 and 50 ppm
resulted in embryonic heart beat rates which decreased
consistently with increasing concentration. The response to
BP1100X at much higher concentrations was not clearly related
to concentration nor was it as dramatic. It was also shown
that hatched larvae were more sensitive to the dispersants
than the eggs. In a recent study Linden (1975) exposed
freshly hatched larvae of the Baltic herring to either
Venezuelan crude oil alone or in combination with Finasol OSR-
2 or BP1100X. Larvae were 50 to 100 times more sensitive to
both oil-dispersant combinations than to the oil alone.
Abnormal responses included hyperactive spiral swimming and
quick movements to the surface followed by sinking.
Deterioration of fins and vertebral column were also noted.
The duration of oil toxicity was lengthened by dispersants,
which presumably held the oil in suspension for a greater
time period.

The effects of crude oil extracts on the eggs of the cod,
Gadus morhua L., was reported by Kuhnhold (1974). He
reported significant differences in the toxicity of the three
crude oil extracts utilized and the rates at which they
produced abnormal responses. Several types of morphogenic
abnormalities were reported as well as loss of pigmentation
and an attempt was made to relate the effects to specific
hydrocarbons in the extracts. Rice et al. (1975) studied the
effect of Prudhoe Bay crude oil on the eggs, alevins and fry
of pink salmon (Oncorhynchus gorbuscha). Growth of alevins

exposed to sublethal doses of oil for 10 days varied in relation to dose and age at time of exposure, such that
susceptibility increased with increasing age. As reported in
other studies, eggs were the most resistant. In a review
paper on the effects of oil on the development of eggs and
larvae of marine organisms, Kühnhold (1975) emphasized the
lack of analytical data on oils and their water extracts in
most of the earlier studies. He pointed out the various
types of investigations on the reproduction and development
of organisms that should be conducted to clearly evaluate the
effects of hydrocarbons. To determine the impact of hydrocarbons on the productivity of fish species one must consider
the specific hydrocarbons involved, the duration of exposure,
and the stage of the life history exposed.

Studies reported here are concerned with the influence
of water-soluble fractions (WSF) of a refined (No. 2 Fuel
Oil) and a crude oil (South Louisiana) on the development
rate, heart beat rate, and hatching success of embryos from
three species of estuarine fish. The specific hydrocarbons
present in the water extracts have been well characterized
(Anderson et al.,1974a) and analyses of water and tissues
were conducted during the study.

MATERIALS AND METHODS

In the laboratory sperm and eggs of Cyprinodon
variegatus, Fundulus heteroclitus, and F. similus were
obtained by stripping the adults (Trinkaus, 1967). Embryos
resulting from the fertilized ova of each species were maintained in 6" glass fingerbowls at room temperature (21 $\pm$
1°C) prior to use in experiments. The period of development
at which exposure to hydrocarbons was initiated varied from
immediately after fertilization to 130 hours of development.
In general no aeration was supplied to the bowls, which
contained either 80 or 100 ml of water. In one case aeration
was supplied daily for 5 minutes immediately before that
solution was decanted off and replaced with "fresh" solution.
Some tests were conducted without renewing the media over the
entire period of development. Even in the presence of hydrocarbons the dissolved oxygen concentration of the water,
containing as many as 25 eggs, remained near saturation
(approximately 7 ppm O_2). Seawater used for all controls and
in the preparation of the water-soluble extracts was 20 o/oo
S Instant Ocean. In each test the hatching rate of embryos
and the percent of successful hatching were recorded. At
the stage of development where the heart was visible and

active, daily measurements of heart beat rate were
determined for control and hydrocarbon-exposed embryos.

To determine the permeability of the chorionic membrane
of the embryos two different tests were conducted. First,
the chorion of a portion of the Cyprinodon variegatus embryos
was surgically removed using watchmaker forceps. The
tolerance of these embryos was compared to those embryos of
the same age with intact chorionic membranes. In addition,
the concentration of specific hydrocarbons (naphthalenes)
present in F. similus eggs exposed to three concentrations
of WSF from South Louisiana crude oil was determined by the
method of Neff and Anderson (1975a).

Extensive analyses of the hydrocarbons in four reference
oils and their water extracts have already been reported
(Anderson et al., 1974a). Naphthalenes (naphthalene, methylnaphthalenes and dimethylnaphthalenes) were significant
contributors to the aromatic fraction of these extracts and
represented approximately one third of the total water
solubles extracted from No. 2 Fuel Oil. The toxicity,
accumulation and retention time of these compounds (Anderson,
1975; Anderson et al., 1974b) are justification for monitoring
their presence in the water and tissues during these tests
with fish embryos. The WSF's used in this study were
prepared, as described in earlier reports, by mixing one
part oil over 9 parts water for 20 hours. In several cases
the total hydrocarbon concentrations in the 100% WSF (water
phase after mixing) was determined by infrared spectrophotometry (API, 1958). In addition, the content of naphthalenes
in the 100% WSF's was measured by the method of Neff and
Anderson (1975a), utilizing an ultraviolet spectrophotometer.
Dilutions of the 100% WSF were prepared for each test and in
some cases they were renewed daily.

Since it is often more convenient to refer to a percent
of the water-soluble fraction (%WSF) of a given oil than to
parts per million of total hydrocarbons (ppm TH) or ppm total
naphthalenes (ppm TN) the conversion of these percentages
should be explained. During these studies numerous analyses
of water were conducted and the 100% WSF of So. La. crude oil
was shown to produce total naphthalene concentrations of 0.24
to 0.28 ppm. Analyses of the No. 2 Fuel Oil 100% WSF resulted
in means and standard deviation of 11.55 $\pm$ 3.22 ppm TH and a
mean of 2.3 ppm TN. These slightly higher than normal levels
of hydrocarbons from No. 2 Fuel Oil are likely the result of
mixing smaller quantities of oil in a container with different
surface to volume characteristics (1 gallon jar instead of the
normal 5 gallon carboy). Considering earlier analyses and
those of the present study, the following conversion factors
should be used:

100% WSF No. 2 Fuel Oil = 10 ppm TH = 2 ppm TN
100% WSF South Louisiana Crude Oil = 20 ppm TH = 0.3 ppm
     TN

Accurate dilution of these solutions have been shown to be
possible since droplets or surface films are not present.

RESULTS

     The results of research on hatching success of embryos
of all three species are summarized in Table 1. Some
interesting comparisons can be made between the results of
tests involving different treatment conditions. In those
studies where the media were not renewed during the course
of exposure, which was generally about 8 days, the 100% WSF
of No. 2 Fuel Oil was the only concentration producing 100%
mortality. Survival to hatching for both F. heteroclitus
and Cyprinodon variegatus was approximately 50% at the 50%
WSF concentration. The cyprinodons survived to levels of 60
to 100% at a WSF concentration of 25%. It should be noted
that embryos of this species, which lacked the chorionic
membrane during a 6-day exposure (from an age of 5 days to 11
days) survived at about the same level as normal embryos.
     The effect of daily renewals of WSF's is evident from
the decreased survival of cyprinodon embryos at 30% WSF of
No. 2 Fuel Oil. Without renewal, survival and hatching of
Cyprinodon variegatus was at 40 to 62% in 50% WSF and 80 to
88% in 25% WSF. Since survival was reduced to 0% at a
concentration of 30% WSF, the replenishment of toxic hydrocarbons daily significantly reduced hatching success. When
the South Louisiana crude oil WSF's were renewed daily in
tests with F. similus, the critical concentration was shown
to lie between 50 and 70% WSF. While one cannot be certain
from these tests, it is apparent that the tolerances of
embryos from all three species of estuarine fish are
approximately equal. Adult F. similus and Cyprinodon
variegatus were shown to be very similar in tolerance to
WSF's of No. 2 Fuel Oil and Venezuelan Bunker C (Anderson et
al. 1974a). It is, therefore, likely that the significantly
greater sensitivity exhibited by cyprinodon embryos in 30%
WSF of No. 2 Fuel Oil (100% mortality), as compared to F.
similus in 30% WSF of South Louisiana (96% hatching), is the
result of differences in the hydrocarbon content of the
extracts. Since tests 1 and 2 (with chorion) on cyprinodons
and the test for F. heteroclitus were conducted in exactly
the same fashion, it would appear that embryos from these
two species are very similar in tolerance to hydrocarbons.

As noted above, those embryos of Cyprinodon variegatus
which lacked the chorionic membrane responded to the various
hydrocarbon solutions very much like normal embryos. These
findings would indicate that this outer membrane is not a
barrier to hydrocarbons, but analyses of embryos for hydrocarbon content would be more direct evidence. In a later
experiment with F. similus those embryos which were obviously
deformed, making hatching impossible, were sacrificed for
hydrocarbon analyses by the method of Neff and Anderson
(1975a). Table 2 shows the results of these analyses for
embryonic content of naphthalene, methylnaphthalenes, and
dimethylnaphthalenes. Dimethylnaphthalenes were the highest
in concentration in all embryos tested and the levels of
total naphthalenes ranged from about 5 to 16 ppm. It should
be noted that 4% of the embryos exposed to 70% WSF of South
Louisiana crude oil survived to hatching (Table 1), although
they apparently contained approximately 5 ppm of total
naphthalenes.

The rate of hatching by F. similus embryos was also
recorded, and the results are quite interesting. As shown in
Figure 1, exposure to hydrocarbons apparently stimulated
hatching, since those embryos in 10% WSF of South Louisiana
crude oil began hatching on the 17th day. By the time
control embryos began to hatch (day 24), oil-exposed
organisms exhibited hatching percentages of 80%, 56%, and
40% at 30, 10, and 50% WSF's respectively. The final
hatching percentages at these latter concentrations and for
control animals ranged from 72 to 96%. While the hatching
rate of oil exposed embryos was rather irregular and took
place over a period of about 10 days, control organisms all
hatched within a period of 4 days.

The final parameter measured in these studies was the
rate of embryonic heart beat. In Cyprinodon variegatus the
heart developed sufficiently to observe pulsations by 130
hours while heart beats of F. heteroclitus were not recorded
until the 11th day of development. Figures 2A and 2B
illustrate the effects of WSF's of No. 2 Fuel Oil on the
heart beat rates of F. heteroclitus (Fig. 2A) and Cyprinodon
variegatus (Fig. 2B). The mean heart beat rates for control
F. heteroclitus embryos and those exposed to a single dose
of 25% WSF of No. 2 Fuel Oil was approximately 135 beats per
minute, while cyprinodon embryos under the same conditions
exhibited a mean of about 100 beats per minute. In both
species the 50% WSF produced a rate of heart beat which was
significantly lower, and in the case of the cyprinodons this
suppression increased gradually with time. An even more
drastic reduction in heart beat rate was exhibited by
embryos of both species when subjected to the 100% WSF of

Table 1. The hatching success of embryos of three fish species exposed to WSFs of No. 2 Fuel Oil (#2 FO) and South Louisiana (So. La.) crude oil

Species	Oil	Condition	0	10	25	30	50	70	75	90	100
Fundulus heteroclitus	#2FO	No. Dead	0		0		2				5
		No. Hatched	5		5		3				0
		% Hatched	100		100		60				0
Cyprinodon variegatus (with chorion)	#2FO Test 1	No. Dead	0		1		3				8
		No. Hatched	7		7		5				0
		% Hatched	100		88		62				0
	Test 2	No. Dead	0		1		3				5
		No. Hatched	5		4		2				0
		% Hatched	100		80		40				0
(Chorion removed)	#2FO	No. Dead	0		4		4				10
		No. Hatched	9		6		6				0
		% Hatched	100		60		60				0
(with chorion)	#2FO Renewed Daily	No. Dead	3	1		25	25	25		25	
		No. Hatched	22	24		0	0	0		0	
		% Hatched	88	82		0	0	0		0	
Fundulus similis	So. La. Crude Renewed Daily	No. Dead	2	7		1	7	24		25	25
		No. Hatched	23	18		24	18	1		0	0
		% Hatched	92	72		96	72	4		0	0

Table 2

Naphthalene (N), methylnaphthalenes (MN), and dimethylnaphthalenes (DMN), and total naphthalenes (TN) concentrations in eggs exposed to 70, 90, and 100% WSF's of South Louisiana crude oil. Standard deviations were approximately 10% of the means.

Sampling Time	Exposure Concentration	Concentration (ppm)			
		N	MN	DMN	TN
28 days	70%	0.88	1.59	2.50	4.97
27 days	90%	4.01	4.47	7.60	16.08
25 days	100%	2.02	3.21	4.22	9.45

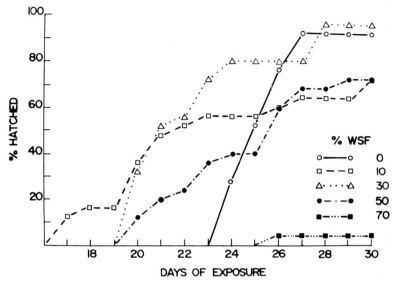

Fig. 1.   Relationship between the percent hatch of Fundulus
similus exposed to WSFs of South Louisiana crude oil
and the time of exposure and development (20 eggs at
each concentration, with 10 in each fingerbowl).

No. 2 Fuel Oil. In both species the suppression began
immediately after exposure and the decrease continued until
death of the embryos. Further studies with cyprinodons
verified the above findings.
     The studies on the effects of WSF's of South Louisiana
crude oil on F. similus embryonic heart beat are summarized
in Figure 3. Since there were daily variations in the mean
rates of heart beat in control, 10%, 30%, and 50% WSF's, the
zone of variation has been enclosed by shading. There was no
apparent relationship between heart beat rate and concentration of WSF at concentrations of 0 to 50%. However, at
concentrations of 70, 90, and 100% WSF the heart beat rates
were lower from day 4 to hatching on day 20. In the 90 and
100% WSF concentrations embryonic heart beat steadily
decreased in rate during the course of exposure. The
embryos exposed to 70% WSF decreased to a low on day 12, but
then increased and finally leveled off at approximately the
12 day heart beat rate. As shown in Table 1, none of the
embryos hatched in concentrations of 90 or 100% WSF, but 4%
in the 70% WSF did hatch.

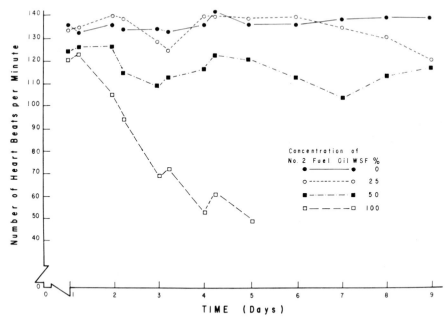

Fig. 2A. Relationship between the mean heart beat rates of
fish embryos exposed to concentrations of No. 2 Fuel
Oil WSF's and time of exposure. Sample size was
only 5 (F. heteroclitus) or 8 (C. variegatus) per
concentration in these preliminary experiments, but
the heart beat rates were surprisingly uniform
(standard deviations generally about 10% of the
mean).
A. Fundulus heteroclitus embryos

DISCUSSION

Since analyses were performed during the course of
these experiments, it is possible to relate the reduction
in hatching success to the content of total hydrocarbons (TH)
and total naphthalenes (TN) in the exposure media. As
discussed in earlier publications (Anderson, 1975; Anderson
et al., 1974a, b) the concentration of naphthalenes in the
WSF of No. 2 Fuel Oil is substantially higher (approximately
6 times) than that of South Louisiana crude oil. It is,
therefore, not surprising that when both water-soluble
fractions were renewed daily, percent hatching was zero for
C. variegatus exposed to 30 and 50% WSF of Fuel Oil, while
there was 96 and 72% hatching of F. similus embryos exposed

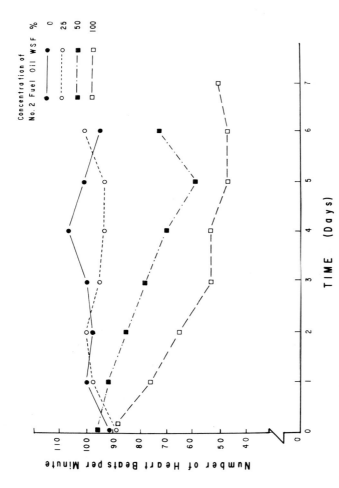

Fig. 2B. Relationship between the mean heart beat rates of fish embryos exposed to concentrations of No. 2 Fuel Oil WSF's and time of exposure. Sample size was only 5 (F. heteroclitus) or 8 (C. variegatus) per concentration in these preliminary experiments, but the heart beat rates were surprisingly uniform (standard deviations generally about 10% of the mean).

B. Cyprinodon variegatus embryos

251

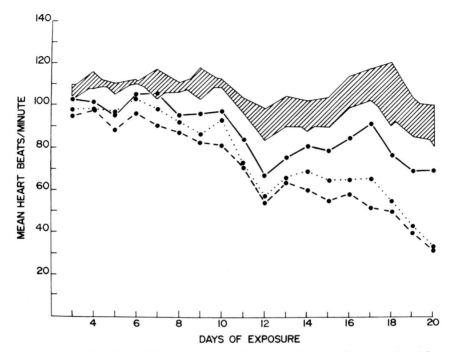

Fig. 3.   Relationships between the mean heart beat rate of
F. *similus* embryos and the time of exposure to
WSF's of South Louisiana crude oil and development.
Standard deviations from the means were within 10%.
Five embryos were selected at random from the 25
organisms at each concentration for daily rate
determinations. The shaded zone represents the
range of all means derived from embryos exposed to
concentrations from 0 to 50% WSF. Solid line = 70%
WSF; dotted line = 90% WSF; and dashed line = 100%
WSF.

to these concentrations of South Louisiana WSF. Judging from
all other studies in our laboratory with the three species of
fish and their embryos, the observed differences in hatching
are not likely to be the result of species variability.
    An attempt should be made to determine the critical
level of hydrocarbons resulting in a significant reduction of
hatching success. It could be stated that any reduction in
the production of offspring is a significant event, but the
normal procedure is to determine the concentration that
reduces survival by 50%. It is recognized that a considerable
portion of the offspring from a given hatch are lost from

natural factors, such as predation, disease, and harsh
environmental conditions. Perhaps the combined influence of
natural factors and pollutants should be addressed, but the
former factors are extremely difficult to evaluate. From the
data on heart beat rates and hatching success, the embryos
(F. heteroclitus and C. variegatus) exposed to a single dose
of hydrocarbons from No. 2 Fuel Oil appear to be significantly
affected by the 50% WSF. Since the survival fluctuates
between 40 and 60% and heart beat was not drastically
suppressed, it is possible that a 60% WSF would be closer to
a median tolerance level. There are approximately 5 to 6 ppm
of total hydrocarbons and 1 to 1.2 ppm of total naphthalenes
in the WSF's of No. 2 Fuel Oil producing the effects. It
should be emphasized that these are initial concentrations
and, without renewal, the levels in the fingerbowls likely
dropped to very low values after 3 to 4 days. Unfortunately
analyses of eggs were not conducted in these earlier tests,
but data to be discussed below indicate that accumulation by
eggs could have taken place during the first stages of
exposure.

Those studies in which the media were changed daily more
clearly indicated the difference between No. 2 Fuel Oil and
South Louisiana crude oil WSF's. The median tolerance level
for Cyprinodon variegatus embryos exposed to Fuel Oil is
apparently near 25% WSF, or 2.5 ppm TH and 0.5 ppm TN, under
these conditions. It should be noted that the continued
replenishment of toxic hydrocarbons at this level more
closely approximates a heavily contaminated chronically-
exposed environment, while the one-dose system discussed
above might simulate a spill situation. The concentration of
South Louisiana crude oil, which would produce a similar 50%
reduction in hatching of Fundulus similus embryos, is
approximately 60% WSF. This fraction contains 12 ppm TH and
about 0.2 ppm of total naphthalenes. One is immediately
impressed by the similarity between the levels of total
naphthalenes (0.2 - 0.5 ppm) in both WSF's which would
produce a 50% reduction in hatching success. As noted in
earlier studies (Anderson, 1975; Anderson et al., 1974b, and
Cox, 1974) the content of naphthalenes appears to be
responsible for a major portion of the toxicity produced by
the WSF of No. 2 Fuel Oil and may be a significant contributor
to toxic effects produced by South Louisiana crude oil. Since
the WSF's of South Louisiana were prepared daily and
analytical data have shown that there are significant amounts
of the lighter and more volatile aromatic compounds, such as
toluene, benzenes and xylenes, these no doubt contributed to
the toxicity of the solutions.

The absence of the chorionic membrane in one group of tests with cyprinodons apparently had little effect on the hatching success of the embryos. There has been speculation that this membrane may be responsible for the relative hardiness of fish eggs. From these studies it seems that it does not serve a protective function and that it must be permeable, at least to naphthalenes. The concentrations found associated with the eggs of F. similus were quite high (5-16 ppm), considering the fact that levels of naphthalenes in the exposure media ranged from 0.21 to 0.28 ppm. Since the embryos were rinsed and blotted before extraction in hexane for analyses, the concentration of naphthalenes is not a reflection of surface adsorption, but internal accumulation as we have reported earlier for several adult organisms. One of the Fundulus similus embryos exposed to 70% WSF hatched; the remainder of the group (24) contained approximately 5 ppm TN. This may indicate that tissue levels near this concentration severely decrease survival of the embryos. The fry surviving the above exposure lived one week longer under the same experimental conditions. Using larger groups of eggs at each concentration, it should be possible to determine more closely the levels of naphthalenes in the tissue associated with a 50% reduction in hatching success.

The effect of exposure to sublethal levels of hydro-carbons on the rate of hatching was a general stimulation. Those concentrations of South Louisiana WSF, which did not significantly alter percent hatching, did induce hatching several days before the process was initiated in controls. The hatching rate of control embryos was much more rapid and consistent. The influence of hydrocarbons on hatching rate may be considered an example of the concept of "sufficient challenge" (Smyth, 1967). There is considerable evidence which indicates that the development and metabolism of mammals may be stimulated by various types of natural environmental factors, drugs, and pollutants. Data accumulated in this laboratory support this hypothesis, particularly the unpublished data by Foster, which shows that after 24 hours exposure to 10 and 50% WSF's of South Louisiana crude oil F. heteroclitus embryos respire at rates of 50 to 80% higher than controls.

Heart beat rates of embryos were measured in an attempt to evaluate this parameter as a meaningful sublethal response. Many behavioral and physiological parameters have been examined in a search for an indicator of organismic "well-being" (Anderson et al., 1974c). Caution must be taken in interpreting laboratory studies associated with sublethal responses, as abnormal responses may either be short-term or not be exhibited until toxicant levels are very near lethal

levels. The reduction in heart beat rate exhibited by all
three fish species was produced by hydrocarbon levels which
resulted in either partial or complete mortality. This
parameter therefore has limited usefulness as an indicator of
pollution related stress. Since the reduction in heart beat
rate is exhibited very early in the development of embryos
which may take one month to hatch, it does serve as an
indicator of the final hatching success. There was little
evidence of increased sensitivity of fry over that of
embryos since fry kept under the same conditions after
emerging from the eggs did not suffer significant mortality.
After 2 weeks there were no fry mortalities in the 0-30% WSF
of South Louisiana (renewed daily), but the one fry from 70%
and 2 in 50% died after one week.

The levels of WSF's which significantly suppressed heart
beat rate were the same as those resulting in 50% or greater
reduction in hatching success (Fuel Oil: 2.5 ppm TH, 0.5 ppm
TN; So. La.: 12 ppm TH, 0.2 ppm TN). It should be noted
that these concentrations pertain to exposures to solutions
which are renewed on a daily basis, while in the single-dose
system of exposure the values increase to approximately
twice the concentration of hydrocarbons from the WSF of No. 2
Fuel Oil. It is not known whether or not the embryos have
the capability, as do adult fishes (Lee et al., 1972a), of
metabolizing naphthalenes and other petroleum hydrocarbons.
In a single-dose exposure system, embryos have the opportunity
to release hydrocarbons, and generally there is an equilibrium
established in fish constantly exposed, presumably via
metabolic and excretory processes. In subsequent studies the
retention time of hydrocarbons in eggs should be examined and
freshly hatched larvae from exposed eggs should be analyzed
for hydrocarbon content. The hydrocarbon release rates of
numerous adult marine organisms has been investigated
(Anderson, 1975; Neff and Anderson, 1975b; Anderson et al.,
1974b,c; Cox, 1974; Lee et al., 1972b; Stegeman and Teal,
1973), but no data are available on the accumulation or
release of hydrocarbons by eggs or recently hatched larvae
of marine invertebrates or fish.

The sequence of events leading to the death of embryos
can be described in terms of morphological abnormalities, but
the molecular nature of hydrocarbon influence on development
is virtually unknown. It may be that these lipid soluble
compounds are interfering with the normal construction of the
lipoprotein portion of the cell membrane. There is also
apparently an inhibition of melanin production; in this study
and that of Kühnhold (1974), loss of pigmentation was noted
with increasing hydrocarbon exposure. The mechanism of
toxicant attack is always difficult to determine, but

hopefully this study will lead to investigations designed to answer questions regarding the mode of chemical influence. Finally, one should attempt to relate the level of hydrocarbons found in this and other similar studies to effect normal development to those reported for the marine environment. Natural concentrations of petroleum hydrocarbons in the open ocean have been reviewed by Gordon et al. (1974) and in general values are on the order of 10 to 50 ppb for total hydrocarbons and 5 ppb or less for aromatic hydrocarbons. Some reports from highly contaminated areas, including spill sites, may be as high as 200 ppb total dissolved hydrocarbons (Gordon and Prouse, 1973; McAuliffe et al., 1975). It would appear that rather drastic conditions of contamination are required to produce the abnormal responses discussed in this paper.

ACKNOWLEDGMENTS

The authors wish to thank all those in the laboratory lending aid in these studies. The support for this research was provided by a contract from the American Petroleum Institute and grant #IDO75-04890 from the IDOE Office of NSF.

LITERATURE CITED

Alderdice, D. F. 1972. Responses of marine poikilotherms to environmental factors acting in concert, Chapter 12. Factor combinations. In: Marine Ecology 1(3): 1659-1722, ed. by O. Kinne. London: Wiley-Interscience.

American Petroleum Institute. 1958. Determination of volatile and non-volatile oily material. Infrared Spectrophotometric Method, API Pub. No. 733-748.

Anderson, J. W. 1975. Laboratory studies on the effects of oil on marine organisms: an overview. API Pub. No. 4249.

_____, J. M. Neff, B. A. Cox, H. E. Tatem, and G. M. Hightower. 1974a. Characteristics of dispersions and water-soluble extracts of crude and refined oils and their toxicity on estuarine crustaceans and fish. Mar. Biol. 27(1): 75-88.

_____, _____, _____, _____, and _____ . 1974b. The effects of oil on estuarine animals: toxicity, uptake and depuration, respiration. In: Pollution and Physiology of Marine Organisms, pp. 285-310, ed. by F. J. Vernberg and W. B. Vernberg. New York: Academic Press.

_____, _____, and S. R. Petrocelli. 1974c. Sub-
lethal effects of oil, heavy metals and PCBs on marine
organisms. In: Survival in Toxic Environments, pp. 83-
121, ed. by M. A. Q. Khan and J. P. Bederka, Jr. New
York: Academic Press.

Cox, B. A. 1974. Responses of the marine crustacean
Mysidopsis almyra Bowman, Penaeus aztecus Ives, and
Penaeus setiferus (Linn.) to petroleum hydrocarbons.
A Dissertation, Texas A&M University, College Station,
Texas. 167 pp.

Gordon, D. C., Jr., P. D. Keizer, and Jacqueline Dale. 1974.
Estimates using fluorescent spectroscopy of the present
state of petroleum hydrocarbon contamination in the
water column of the Northwest Atlantic Ocean. Marine
Chem. 2: 251-261.

_____ and N. J. Prouse. 1973. The effects of three
oils on marine phytoplankton photosynthesis. Marine
Biol. 22: 329-333.

Kuhnhold, Walther W. 1974. Investigations on the toxicity
of seawater-extracts of three crude oils on eggs of
cod (Gadus morhua L.). Ber. dt. wiss. Komm.
Meeresforsch. 23.

_____. 1975. The effect of mineral oils on the
development of eggs and larvae of marine species.
Presented to ICES-Workshop: Petroleum Hydrocarbons in
the Marine Environment, (In press).

Lee, R. F., R. Sauerheber, and A. A. Benson. 1972a.
Petroleum hydrocarbons: uptake and discharge by the
marine mussel Mytilus edulis. Sci. 177: 344-346.

_____, _____, and G. H. Dobbs. 1972b. Uptake,
metabolism and discharge of polycyclic aromatic hydrocarbons by marine fish. Mar. Biol. 17: 201-208.

Linden, O. 1974. Effects of oil spill dispersants on the
early development of Baltic herring. Ann. Zool.
Fennici. 11: 141-148.

_____. 1975. Acute effects of oil and oil/dispersant
mixture on larvae of Baltic herring. Ambio. 4: 130-
133.

McAuliffe, C. D., A. E. Smalley, R. D. Groover, W. M. Welsh,
W. S. Pickle, and G. E. Jones. 1975. Chevron Main
Pass Block 41 Oil Spill: Chemical and biological
investigations. In: Proceedings of 1975 Conf. on
Prevention and Control of Oil Pollution, pp. 555-566.
Wash. D. C. American Petroleum Institute.

Nebeker, Alan V., Frank A. Puglisi, and David L. DeFoe. 1974.
Effect of polychlorinated biphenyl compounds on
survival and reproduction of the fathead minnow and
flagfish. Trans. Amer. Fish. Soc. 3: 562-568.

Neff, J. M. and J. W. Anderson. 1975a. Ultraviolet spectro-
    photometric method for the determination of naphthalene
    and alkylnaphthalenes in the tissues of oil-contaminated
    marine animals. Bull. Env. Contam. Toxicol. 14: 122-
    128.
_____ and _____. 1975b. Accumulation, release and
    body distribution of benz-a-pyrene-C^{14} in the clam
    Rangia cuneata. In: Proceedings of 1975 Conference on
    Prevention and Control of Oil Pollution, pp. 469-471.
    Wash. D. C.: American Petroleum Institute.
Rice, Stanley D., D. Adam Moles, and Jeffrey W. Short. 1975.
    The effect of Prudhoe Bay crude oil on survival and
    growth of eggs, alevins and fry of pink salmon
    Oncorhynchus gorbuscha. 1975. In: Proceedings of
    1975 Conference on Prevention and Control of Oil
    Pollution, pp. 503-507, Wash. D. C.: American
    Petroleum Institute.
Schimmel, Steven C., David J. Hansen, and Jerrold Forester.
    1974. Effects of Aroclor 1254 on laboratory-reared
    embryos and fry of sheepshead minnow (Cyprinodon
    variegatus). Trans. Amer. Fish. Soc. 103: 582-586.
Smyth, H. F., Jr. 1967. Sufficient challenge. Fd. Cosmet.
    Toxicol. 5: 51-68.
Stegeman, J. J. and J. M. Teal. 1973. Accumulation,
    release and retention of petroleum hydrocarbons by the
    oyster Crassostrea virginica. Mar. Biol. 2: 37-44.
Trinkaus, J. P. 1967. Fundulus. In: Methods in Develop-
    mental Biology, pp. 113-122, ed. by F. H. Wilt and
    N. K. Wessels. New York: T. Y. Goweel Co.
von Westernhagen, H., H. Rosenthal, and K. R. Sperling. 1974.
    Combined effects of cadmium and salinity on development
    and survival of herring eggs. Helgolander wiss.
    Meeresunters 26: 416-433.

Effect of Petroleum Hydrocarbons on Breathing and Coughing Rates and Hydrocarbon Uptake-Depuration in Pink Salmon Fry

STANLEY D. RICE[1], ROBERT E. THOMAS[2], and JEFFREY W. SHORT[1]

[1]Northwest Fisheries Center, Auke Bay Fisheries Laboratory
National Marine Fisheries Service, NOAA
Post Office Box 155, Auke Bay, AK 99821

[2]Chico State University, Chico, CA 95926

Changes in respiratory activity have been used as sensitive indicators of stress in fish exposed to pollutants such
as DDT, kraft pulp mill effluent (Schaumberg et al., 1967),
bleached kraft mill effluent (Walden et al., 1970; Davis,1973),
zinc (Sparks et al., 1972), copper (Drummond et al., 1973),
combinations of copper and zinc (Sellers et al., 1975), benzene
(Brocksen and Bailey, 1973), and crude and refined oils
(Anderson et al., 1974a; Thomas and Rice, 1975). Anderson et
al. (1974a) observed increased oxygen consumption rates in
sheepshead minnows, Cyprinodon variegatus,after 24 hrs exposure
to refined and crude oils. Thomas and Rice (1975) observed an
immediate increase in opercular breathing rates of pink salmon
fry exposed to Prudhoe Bay crude oil. The increased breathing
rate response observed by Thomas and Rice was related to the
dosage level but dropped to near control levels during a 24-
hr exposure. Because of the absence of precise analytical
measurement of oil concentrations during the exposures, Thomas
and Rice could not explain the reduction of breathing rates to
near control levels. The decrease in breathing rates during
oil exposure may have been due to a change in the effective
concentration of oil or to a physiological response of the pink
salmon fry such as adaptation or narcosis.

In this paper we report the changes in breathing rates
during the extended exposure of pink salmon, Oncorhynchus
gorbuscha, fry to oil, and examine the reasons for return to

near normal rates during extended exposure. We determined:
(1) the acute toxicity of Cook Inlet and Prudhoe Bay crude
oils and No. 2 fuel oil to pink salmon fry, so that sublethal
dosage exposures and responses could be compared to a lethal
dosage exposure; (2) changes in opercular breathing and coughing rates of fish exposed to a variety of sublethal concentrations of WSF (water-soluble fractions) of the three oils;
(3) the breathing and coughing rates of fish during exposures
to a constant oil concentration for three days; and (4) the
hydrocarbon uptake and depuration by fish exposed to the WSF
of Cook Inlet crude oil for varying periods.

MATERIALS AND METHODS

We conducted the experiments at the Northwest Fisheries
Center Auke Bay Fisheries Laboratory using pink salmon fry
raised in gravel incubators (Bailey and Taylor, 1974). The
fry emerged in April 1975 and were kept in running seawater
aquaria until used in the study. Pink salmon fry normally
migrate to sea as soon as they emerge from the incubator gravel
and they readily acclimate to sea water. Temperatures in the
aquaria during the experiments, June-August 1975, ranged from
10-12.5°C. The fry, which were fed Oregon Moist Pellets daily,
appeared normal in every respect.

Preparation of the Water-soluble Fraction

Water-soluble fractions were prepared with Prudhoe Bay
and Cook Inlet crude oils and No. 2 fuel oil. One percent oil
in sea water (1 liter oil/100 liters sea water) was mixed
slowly for 20 hrs at ambient water temperatures (10-12°C). The
mixture was allowed to separate for 3 hrs before the WSF was
siphoned from under the slick. We determined the concentration
of oil in the water by UV (ultraviolet) spectroscopy, and
diluted the WSF to the desired concentrations used in the
experiments. Detailed analyses of the WSF by GC (gas chromatography) are available (Rice et al., Unpubl. manuscript).

Oil Analyses

We determined the concentrations of the WSF several times
during the exposures by both UV and IR (infrared) spectroscopy,
so that we would know what changes occurred. We used the IR
method of Gruenfeld (1973), and the UV method of Neff and
Anderson (1975). For IR analysis, samples of WSF were
extracted with 1,1,2-trichloro-1,2,2-triflouroethane (Burdick
and Jackson, Muskegon, Mich.[1]) and the absorbance of the

---

[1]Reference to trade names does not imply endorsement by
the National Marine Fisheries Service.

extract was measured on a Beckman Acculab 1 IR spectrophotometer at 2930 cm^{-1}. This method is particulary sensitive to
paraffins, since absorbance at 2930 cm^{-1} is due mainly to
methyl and methylene CH stretch (Silverstein and Bassler, 1966).
The method was calibrated by measuring absorbances of known
dilutions of the crude oil. Results are expressed in ppm
(parts per million) of oil.

For UV analysis, water samples were extracted with n-
hexane (Burdick and Jackson, Muskegon, Mich.) and the absorbance of the extract was measured at 221 nm on a Beckman model
25 scanning UV spectrophotometer. Absorbance at 221 nm is
mainly due to naphthalene and aliphatic substituted naphthalenes (Neff and Anderson, 1975). The absorbance measurement
was adjusted to correct for the extraction step where unequal
proportions of water sample to extracting solvent were used.
Results were calibrated against known concentrations of
naphthalene and expressed as ppb (parts per billion) of naphthalene equivalents. Calibrations against crude oil dilutions
would result in misleading concentrations since many high
molecular weight compounds absorb in the UV but are quite
insoluble in water and do not appear in the WSF.

Bioassays

We conducted static bioassays in aquaria (19-liter glass
jars) aerated at approximately 100 bubbles per minute and maintained at ambient seawater temperatures (10-12°C). Tissue-to-
volume ratios in the aquaria never exceeded 1 g tissue (wet
weight) per liter of sea water. Ten fish were exposed to each
dose level for 96 hrs and the response statistics were analyzed
by a computerized probit analysis.

Measurements of Opercular Breathing and Coughing Rates

We measured breathing and coughing rates by electronically
recording the opercular movements of individual fish swimming
free in small compartments (Thomas and Rice, 1975). The fish
had not undergone any anesthesia or surgery. The breathing
and coughing rate experiments were conducted in a heat and
sound insulated room to protect the fry from extraneous
stimuli--no one was in the room and the door was never opened
during the recording periods.

One fish (50 $\pm$ 5 mm long) was placed in each of 38 chambers at 1430, and the breathing rate was first recorded at
0630 the following day. This first recorded rate was averaged
for each group of fry and was used as the basal or normal
breathing rate. The use of the initial recorded rate as the
basal rate was validated in preliminary studies where no
significant diurnal rhythm was detected in nonexposed fry and

no significant changes occurred in breathing rates during
72 hrs of confinement. The average breathing rates were determined for each fry during a 3 min segment of each recording
period. In some cases, a fish may have been against an
electrode, actively swimming, or in a position within the
chamber where 3 consecutive minutes of clear recordings could
not be obtained. When this happened, that fish was not used
for that particular recording period, so sample size at different recording times varies within a group of fry. After
the basal breathing rates were recorded, the test solution
was introduced to 31 of the fish chambers at 0800; breathing
rates were recorded at that time and 3, 6, 9, 12, and 22 hrs
later. The breathing rates of the unexposed fry (controls)
in the other seven chambers were recorded at the same time
intervals. Coughing, a brief reversal of water flow in the
opercular cavity, was detected in some fish as a spike superimposed on the recordings of opercular breathing movements.
Our recording was not sensitive enough to pick up low intensity coughs in most of these small fish. However, when the
coughs could be detected in individual fish, the rates were
constant during that recording period. Coughs could be
detected in exposed fish more frequently than in controls,
apparently because the intensity of the cough was greater.
The differences in breathing and cough rates were tested by
analysis of variance and Student's t-test.

During each test, the water in the test chamber was nearly constant in oxygen, water flow, and temperature. We monitored the temperature continuously as it entered the test
aquarium, which was $11.5 \pm 0.8^\circ$C for all tests. Oxygen
(measured by a Yellow Springs Instruments Polarographic Probe)
never dropped below 8.2 ppm. The flow rate was approximately
1 liter per hour through each of the recording chambers. This
flow was maintained by pumping water from reservoirs of stock
solution into small head tanks. The use of head tanks produced a constant hydrostatic pressure to the manifold which
fed the individual chambers. Most breathing rate tests lasted
22 hrs and each used one stock of WSF. The WSF was passed
through the recording chambers and discarded. Although the
test water was not recirculated, oil concentrations in the
stock reservoir decreased during the 22 hr exposures. We
required a constant dose for a 72 hr experiment and achieved
it by adding the WSF to the large reservoir of test water
every 3 hrs (concentrations were measured by UV each time).

Measurement of Tissue Hydrocarbon Uptake and Depuration

We measured hydrocarbon uptake and depuration in one lot
of fish exposed to WSF in a large tank. To measure uptake,
we sampled fish at 0, 3, 10, 33, and 96 hrs after the initial

exposure. The oil concentration of the exposure water decreased continuously (as measured both by UV and IR) and was only about 20% of the initial concentration after 96 hrs. At 96 hrs the exposure tank was converted to running sea water for depuration. To measure depuration, we sampled fish after 3, 10, 72, and 240 hrs in clean water--gill, viscera (minus heart and kidney), and muscle were dissected out of five fish each period. The tissues of five individuals were pooled to form one sample for gill, one sample for viscera, and one for muscle. The samples were then frozen in glass jars with Teflon-lined caps. Each pooled sample was extracted and ana-lyzed by GC for nonpolar paraffins and aromatic hydrocarbons (Warner, 1976). GC analysis was done by Dr. J. Scott Warner of Battelle Memorial Laboratories, Columbus, Ohio. Dr. Warner analyzed one tissue sample by GC-MS (mass spectrometry) for positive identification of major components.

RESULTS AND DISCUSSION

Toxicity

The three oils differed in toxicity to pink salmon fry, although none of the three oils killed many animals after the initial 24 hrs (Table 1). The ranking of relative toxicities depended on the method used to measure the WSF concentrations. When measured by UV (TLm's expressed in ppb of naphthalene equivalents), the WSF of No. 2 fuel oil was much less toxic than the WSF of the crude oils. In contrast, when measured by IR (TLm expressed as ppm of oil), the WSF of No. 2 fuel oil was more toxic than either of the crude oils, and of the crude oils, Prudhoe Bay crude oil was more toxic than Cook Inlet crude oil. Rice et al. (Unpubl. manuscript) determined TLm's of Cook Inlet and Prudhoe Bay crude oils and No. 2 fuel oil with several marine invertebrates and fish and found that toxicity was more closely associated with oil concentrations measured by UV than by IR. Since naphthalene and substituted naphthalenes are detected by UV, it appears that toxicity is more closely associated with naphthalene concentration. Anderson et al. (1974b) previously observed that the toxicity of WSF of oils is a function of their diaromatic and triaromatic hydrocarbon contents. UV measurements of equivalent naphthalene concen-trations are apprently more meaningful than IR in explaining toxicity, but we include IR measurements for comparison with other studies.

The TLm's of the WSF of No. 2 fuel oil (expressed as naphthalene equivalents) are about double the TLm's measured for the two crude oils. This is because the WSF of crude oil contain high concentrations of toxic mononuclear aromatics.

These mononuclear aromatics absorb at 221 nm even though they
have low molar absorptivities at 221 nm compared to diaromatic
hydrocarbons.   In the WSF of the No. 2 fueld oil we used,
mononuclear aromatics were present only in very low concentrations.

Table 1
Median tolerance levels of pink salmon fry exposed to water-
soluble fractions of Cook Inlet and Prudhoe Bay crude oils
and No. 2 fuel oil.   Concentrations of oil are expressed in
ppb of naphthalene equivalents (UV) and ppm of oil (IR).

Test Length and Confidence Interval	Cook Inlet Crude Oil	Prudhoe Bay Crude Oil	No. 2 Fuel Oil
	ppb of naphthalene equivalents (UV)		
24-hr TLm	160	112	250
95% confidence interval	137-187	101-123	229-274
96-hr TLm	112	112	232
95% confidence interval	101-124	101-123	211-256
	ppm of oil (IR)		
24-hr TLm	4.13	1.56	0.89
95% confidence interval	3.51-4.84	1.41-1.73	0.82-0.97
96-hr TLm	2.92	1.56	0.81
95% confidence interval	2.65-3.22	1.41-1.73	0.72-0.92

Effects on Breathing Rates in 22 hr Oil Exposures

The breathing rate response during a 22-hr exposure is
similar for all three oils (Fig. 1).   In all cases, the highest breathing rate occurred between 3 and 6 hrs, when the
first measurements were made during exposure.   For all the
oils, the breathing rate subsequently declined, and only at
the higher concentrations did breathing rates continue above
normal after 22 hrs.   The increase in breathing rate was
linear with increasing oil concentrations.   This is evident
in Figure 2 where breathing rates at 3 hrs increased linearly
as exposure dose increased (expressed as percents of the 96-hr
TLm).   There appears to be little difference between the oils
when oil concentrations are expressed as percents of the 96-hr

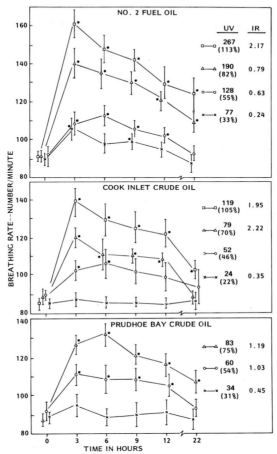

Fig. 1.   The mean opercular breathing rate ($\pm$ 95% confidence
interval) of pink salmon fry exposed to water-soluble
fractions of 3 oils.   Individual doses are given in
ppb of naphthalene equivalents (UV) percent of the
96-hr TLm (as measured by UV), and in ppm of oil (IR).
Asterisk indicates significant differences at 0.01
level between that mean and mean at 0 hrs. Mortality
was noted at the highest oil exposures at 22 hrs
(fuel oil--8 fry in 190 ppb of naphthalene (UV), 18
fry in 267 ppb naphthalene (UV); Cook Inlet--17 fry
in 119 ppb naphthalene (UV); Prudhoe Bay--10 fry in
83 ppb naphthalene (UV).   Sample sizes ranged from
15 to 32 except at 22 hrs in the above doses because
some of the fish died.

TLm. The lowest concentration that causes a significant response is estimated at about 30% of the 96-hr TLm (Fig. 2).

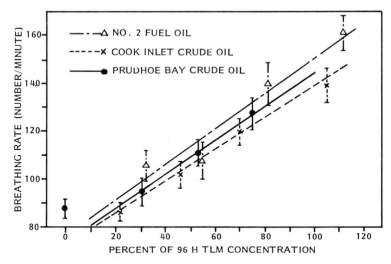

Fig. 2.   The linear relationship between mean opercular breathing rates and exposure dose
expressed as a percent of the 96-hr TLm (as
measured by UV). Mean breathing rate $\pm$ 95%
confidence interval and regression line for
breathing rates of pink salmon fry exposed
to water-soluble fractions of No. 2 fuel
oil, Cook Inlet crude oil, and Prudhoe Bay
crude oil (correlation coefficients of
0.970, 0.997, and 0.999, respectively).

The oil concentrations during the 22-hr exposures decreased significantly for all three oils (as measured by UV and IR). The average decrease in naphthalene equivalents (UV) during the 22-hr exposures was 42.1%, 53.0%, and 42.7% for No. 2 fuel oil, Cook Inlet crude oil, and Prudhoe Bay crude oil, respectively. In many cases, the final concentrations of naphthalene equivalents after 22 hrs were below the threshold concentration (30% of the 96-hr TLm) that could cause a significant increase in breathing rates.

The increased breathing rates are more closely related to increased oil concentrations in the WSF as measured by UV than measured by IR. This is best seen in results of exposures to the two highest concentrations of Cook Inlet crude oil—the breathing rate increased as the concentration of naphthalene equivalents (UV) increased and the concentration of paraffins (IR) decreased. The close association of toxicity and

breathing rate with UV measurement of naphthalene equivalents
may best be explained by the fact that naphthalene and substituted naphthalenes (measured by UV) are quite toxic, while
paraffins (measured by IR) are not very toxic. This again
agrees with observations by Rice et al. (Unpubl. manuscript)
and Anderson et al. (1974b) who found toxicity more closely
related to concentrations of aromatic hydrocarbons than to
paraffins.

Cough Response

     The pattern of cough response was the same as that of
breathing rate changes--but detected in relatively fewer fish
(Fig. 3). Sellers et al. (1975) criticized the use of either
breathing rate or cough frequency alone to indicate pollution
stress because of individual variability. They measured cough
frequency, ventilation frequency, and ventilation intensity
(buccal and opercular cavity pressures) in large trout (225
g) exposed to copper and zinc. Increased response was usually
observed with increased dose. They suggested that all three
measurements were needed for each fish, since individual
variability was large. It was not feasible to monitor ventilation intensity in our experiments with small pink salmon
fry (2 g), but we were able to get statistically significant
results at the $p = 0.01$ level with a sample size of approximately 20 per dose. Our data on cough response of sample
sizes of fish confirm our observations on breathing rate from
larger samples of fish. Although a rather large sample is
needed to obtain significant data on breathing rate responses,
large samples are easy to obtain with the system we used.

Concentrations of Hydrocarbons in Tissues

     Concentrations of both paraffinic (nonpolar) and aroma-
tic hydrocarbons were measured in samples of gill, viscera
(minus heart and kidney), and muscle taken from fish exposed
for 96 hrs to the WSF of Cook Inlet crude (25% of 96-hr TLm)
and depurated in noncontaminated sea water up to 240 hrs
(Figs. 4, 5). Only n-paraffins with carbon numbers 15-19 were
consistently present in all three tissues. Muscle tissue
showed no consistent pattern for any of these paraffins. In
the viscera paraffins generally decreased from a high level
in control samples before the exposure started to low concentrations at the end of the depuration period. In gill tissue,
4 of the 5 paraffins reached their greatest concentration
after 10 hrs of exposure, while C_{17} decreased through the
exposure and depuration periods. Paraffins $C_{12}-C_{30}$ were all
detected in gill tissue after 10 hrs exposure, but were not
detected at other times in gill tissue or in the other tissues
at any time.

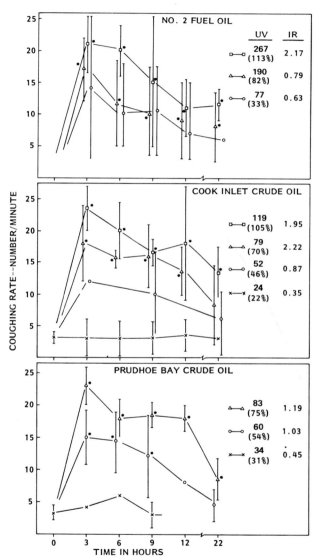

Fig. 3.   Mean coughing rate of pink salmon fry exposed up to
22 hrs to water-soluble fractions of 3 oils. Sample
sizes ranged from 1 to 14, and ± 95% interval is
indicated where sample sizes were 3 or more. Exposure doses are expressed in ppb of naphthalene equivalents (UV), percent of the 96-hr TLm (measured by
UV), and in ppm of oil (IR).

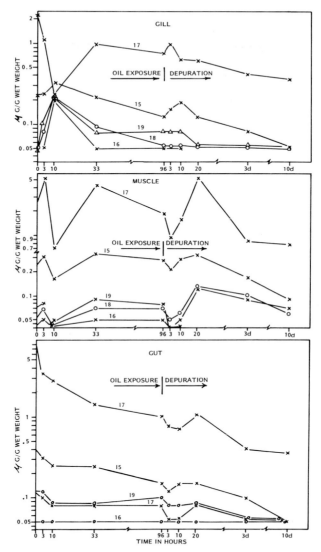

Fig. 4. Concentrations of saturated paraffinic hydrocarbons as measured by GC in gut, gill, and muscle tissue of pink salmon fry exposed up to 4 days to the WSF of Cook Inlet crude and depurated up to 10 days. Only C_{15} through C_{19} paraffins were consistently detected in all tissues, except that C_{12} through C_{30} were found in gill tissue after 10 hrs of exposure. Limits of detectability were 0.05 μg/g wet weight in gill and gut and 0.02 μg/g in muscle.

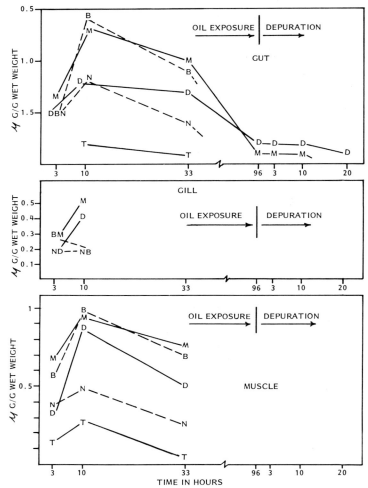

Fig. 5. Concentrations of individual aromatic hydrocarbons
as measured by GC in gut, gill, and muscle tissue of
pink salmon fry exposed up to 4 days to the WSF of
Cook Inlet crude and depurated 10 days. Samples were
taken 0, 3, 10, 33, and 96 hrs after exposure began
and after 3, 10, 20, and 72 hrs and 10 days of depuration. Missing data points represent samples that
were below the limits of detectability (0.04 μg/g
in muscle, and 0.1 μg/g in gut and gill). B=sum of
4 mononuclear aromatics, N=naphthalene, M=methyl-
naphthalene, D=dimethylnaphthalene and T=trimethyl-
naphthalene.

The peak of paraffin concentrations in the gill tissue
after 10 hrs of exposure suggests that paraffins are either
adsorbing to the surface of the gills or moving into the
fishes' gills. There is no evidence that paraffins were
absorbed into other tissues. The general decrease of paraffins in gut tissue is probably due to the lack of feeding
during the 96 hrs of oil exposure and the 24 hrs of depuration.
Paraffins of nonoil origin are normally present in the guts
of feeding fish (see controls, Fig. 4). The paraffin concentrations in muscle are little affected by the exposure, suggesting few of the paraffins reach the muscle tissue. However, in all three tissues, paraffin concentrations changed
(increased or decreased) during the first 10 hrs of exposure
and the first 10 hrs of depuration, suggesting that the fish
are affected physiologically by changes in their environment.
Paraffins are not toxic and can be metabolized. The amount of
stress, if any, caused by higher paraffin concentrations in
some tissues cannot be evaluated by these data.

Monoaromatic and diaromatic hydrocarbons were found in
the tissue samples. Methylnaphthalene was the most abundant
aromatic in the tissues, and like all other aromatics found
was most abundant after 10 hrs of exposure (Fig. 5). The
highest concentrations of all aromatics were found in the gut.
Methylnaphthalenes and dimethylnaphthalenes were the slowest
to be removed from the gut tissues. All the aromatics were
below detectable levels from the gill and muscle tissues by
the fourth day of exposure. Triaromatics were not detected
in any of the tissues.

The greatest concentration of aromatics in the tissues
occurred at 10 hrs and the concentrations of aromatics had
declined by 20 hrs. These changes in the tissues correlate
with the results in our other experiments: (1) few deaths
occurred after 24 hrs of exposure and (2) the breathing rates
declined to near control levels by 22 hrs. The stress (as
reflected by mortality and breathing rates) caused by high
concentrations of aromatics in the tissues should diminish
after 10 hrs since these compounds are disappearing from the
tissues. We have not determined whether the compounds are
being excreted directly or transformed into metabolites. The
presence and persistence of high concentrations of aromatic
hydrocarbons in the gut is consistent with other investigations (Lee et al., 1972; Neff, 1975) where the highest concentrations of naphthalenes were found in the gall bladders of
exposed fish. These observations suggest that the liver is
metabolizing some of the oil compounds and excreting the products to the gall bladder and gut. This speculation is
consistent with results obtained by Pedersen and Hershberger
(1974) who demonstrated that trout have the capability to

metabolize benzopyrene, and with Lee et al. (1972) who demonstrated metabolism and excretion of naphthalene by 3 species
of marine fish.

The return of breathing rates to near control levels during the 22-hr exposure might be explained by one of three
hypotheses: (1) the fish acclimate to the stress and decrease
their adrenal response; (2) the fish decrease their response
to the stress because the oil concentrations drop during the
exposure; (3) the fish acclimate to the stress by making
physiological adjustments that increase their ability to cope
with the stress.

The first hypothesis, adrenal response to stress by the
fish, does not explain our data. We tested the duration of
the adrenal response to stress by banging violently on the
recording chambers. Breathing rate was immediately elevated,
but dropped to near control levels within 2 hrs. The same
stress repeated during the day elicited immediate elevated
breathing rates, but the return to near control levels became
quicker.

The second hypothesis, a decreased response because of
decreased oil concentrations, does explain the observations.
Although the fish were exposed to a flow-through oil exposure
for 22 hrs, the oil concentrations decreased continuously,
apparently because of microbial degradation. After 22 hrs,
only the highest doses had oil concentrations greater than
the threshold concentration. The stress to the fish during
the 22-hr exposure was continuously lowered; both externally
in the water and internally as shown by reduction of aromatic
hydrocarbons in the tissues.

The third hypothesis, acclimation by physiological adjust-
ment, was tested by exposing fry to a constant oil concentration for 72 hrs. The oil concentration was kept at a relatively constant 65% of the 96-hr TLm for Cook Inlet crude oil--
enough freshly mixed WSF was added every 3 hrs to restore
the concentration of naphthalene equivalents (UV) to the
desired level. Both breathing and coughing rates peaked during the first 24 hrs and then dropped somewhat but remained
above the original baseline rates for the remaining 48 hrs of
exposure (Fig. 6). The elevated and sustained respiratory
demand in response to continued oil exposure suggests that a
new steady state of metabolism is maintained. We believe that
both the second and third hypotheses are true--that is, the
fish respond to dropping oil concentration by decreasing their
respiratory response, and the fish will make physiological
adjustments to cope with a continued stress of oil.

We suspect that part of the increased oxygen consumption
is needed to support increased physiological activities in
metabolism and excretion of the hydrocarbons. Part of this
increased oxygen consumption may be used to synthesize enzymes

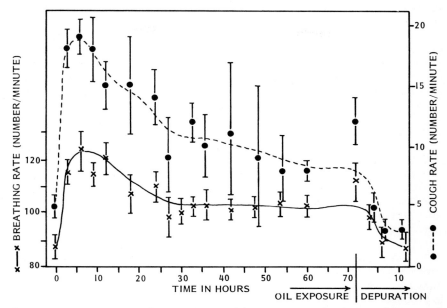

Fig. 6. Mean breathing and coughing rates of pink salmon
         exposed to a constant dose of Cook Inlet crude oil,
         water-soluble fraction for 72 hrs. Exposure dose
         was held "constant" at an average concentration of
         ppb of naphthalene equivalents (UV) by adding fresh
         water soluble fraction every 3 hrs. Significant
         differences from the 0 hr mean are indicated (aster-
         isk) for breathing rate (0.01 level) and coughing
         rate (0.05 level). 95% confidence intervals are
         given for both breathing rate (all sample sizes
         greater than 16) and for coughing rate (sample sizes
         were between 3 and 9).

since (1) the disappearance of aromatic hydrocarbons suggests
metabolism, (2) metabolism of naphthalene and benzopyrene has
been demonstrated in 3 species of marine fish (Lee et al.,
1972) and metabolism of benzopyrene by rainbow trout (Pedersen
and Hershberger, 1974), (3) induction of enzymes (aryl hydrocarbon hydroxylases) capable of breaking down aromatic hydrocarbons has been demonstrated in trout exposed to petroleum
(Payne and Penrose, 1975), and (4) synthesizing these enzymes
requires energy. Thus, we speculate that much energy is
required initially, probably to synthesize large quantities
of enzymes needed to metabolize hydrocarbons into forms that
can be excreted. The somewhat reduced but still elevated

breathing rates continuing after the initial response suggest greater than normal quantities of energy are still needed to maintain enzyme synthesis and oxidation of the hydrocarbons.

It could be argued that increased breathing rates may not mean increased oxygen consumption. However, we have observed an increase in oxygen consumption in fish exposed to an oil concentration that is 50% of a 96-hr TLm. Increased oxygen consumption after exposure to oil is also consistent with findings of Anderson et al. (1974a). The fish in our study returned to normal oxygen consumption rates after several hours.

If not overwhelmed by the initial exposure to oil, the fish can rid themselves of toxic compounds. Higher concentrations of oil are lethal, but sublethal concentrations may have substantial effects on survival. Continued exposure to sublethal concentrations results in continued elevated metabolism and energy demands. This increased energy demand requires increased food intake which puts the fish at a disadvantage in the struggle for survival. Survival rates would be reduced for a group of fish subjected to this stress for significant periods.

CONCLUSIONS

1. WSF from Cook Inlet and Prudhoe Bay crude oils and No. 2 fuel oil cause similar increases in breathing and coughing rates in pink salmon fry.

2. Breathing and coughing rates increase in proportion to oil concentrations, as measured by UV but not by IR. This suggests that naphthalenes rather than paraffins are responsible for this effect. Significant responses were detected at about 30% of the 96-hr TLm.

3. Breathing and coughing rates of pink salmon fry remained above normal during exposure to a constant dose of oil for 72 hrs.

4. Paraffinic, monoaromatic, and diaromatic hydrocarbons were found in tissues of fish exposed to the WSF of Cook Inlet oil. The fish started apparent depuration of the aromatics during the first 24 hrs of exposure, which indicates that they can cope with the stress physiologically. Our data support the concept of excretion through the liver-gall bladder-gut.

5. High breathing rates during the first 24 hrs of exposure, elimination of most aromatics by 20 hrs, and the continued high breathing rates during the constant-dose exposure for 72 hrs indicate that salmon fry can cope with a sublethal exposure to hydrocarbons, but at the cost of an increased metabolic rate. Increased metabolic rates may be detrimental to survival if the stress persists for long periods of time.

ACKNOWLEDGMENTS

This research was financed by contracts with Shell Oil Company, Standard Oil of California, Union Oil of California, Texaco Inc., Marathon Oil Company, and Phillips Petroleum Corporation and funds from the Outer Continental Shelf Energy Assessment Project. The authors also appreciate the help of Frederick Salter for design and construction of equipment as well as assistance from numerous other staff members of the Northwest Fisheries Center Auke Bay Fisheries Laboratory.

LITERATURE CITED

Anderson, J. W., J. M. Neff, B. A. Cox, H. E. Tatem, and G. M. Hightower. 1974a. The effects of oil on estuarine animals: toxicity, uptake and depuration, respiration. In: Pollution and Physiology of Marine Organisms, pp. 285-310, ed. by F. J. Vernberg and W. B. Vernberg. Academic Press, San Francisco.
_____, _____, _____, _____, and _____. 1974b. Characteristics of dispersions and water-soluble extracts of crude and refined oils and their toxicity to estuarine crustaceans and fish. Mar. Biol. 27: 75-88.
Bailey, J. E. and S. G. Taylor. 1974. Salmon fry production in a gravel incubator hatchery, Auke Creek, Alaska, 1971-72. NOAA (Nat. Oceanic Atmos. Adm.) Tech. Memo. NMFS (Natl. Mar. Fish. Serv.) ABFL-3.13 p.
Brocksen, R. W. and H. T. Bailey. 1973. Respiratory response of juvenile chinook salmon and striped bass exposed to benzene, a water-soluble component of crude oil. In: Proceedings of Joint Conference on Prevention and Control of Oil Spills, pp. 783-791. Amer. Pet. Inst., Environ. Prot. Agency, U. S. Coast Guard, Washington, D. C.
Davis, J. D. 1973. Sublethal effects of bleached kraft pulp mill effluent on respiration and circulation in sockeye salmon (Oncorhynchus nerka). J. Fish. Res. Bd. Canada 30: 369-377.
Drummond, R. A., W. A. Spoor, and G. F. Olson. 1973. Some short-term indicators of sublethal effects of copper on brook trout, Salvelinus fontinalis. J. Fish. Res. Bd. Canada 30: 698-701.
Gruenfeld, M. 1973. Extraction of dispersed oils from water for quantitative analysis by infrared spectrophotometry. Environ. Sci. Tech. 7: 636-639.
Lee, R. F., R. Sauerheber, and G. H. Dobbs. 1972. Uptake, metabolism and discharge of polycyclic aromatic hydrocarbons by marine fish. Mar. Biol. 17: 201-208.

Neff, J. M. 1975. Accumulation and release of petroleum-derived aromatic hydrocarbons by marine animals. A paper presented to Symposium on chemistry, occurrence, and measurement of polynuclear aromatic hydrocarbons.
_____ and J. W. Anderson. 1975. An ultraviolet spectrophotometric method for the determination of naphthalene and alkylnaphthalenes in the tissues of oil-contaminated marine animals. Bull. Environ. Contam. Toxicol. 14(1): 122-128.

Payne, J. F. and W. R. Penrose. 1975. Induction of aryl hydrocarbon (benzo[a] pyrene) hydroxylase in fish by petroleum. Bull. Environ. Contam. Toxicol. 14(1): 112-116.

Pedersen, M. G. and W. K. Hershberger. 1974. Metabolism of 3,4-benzpyrene in rainbow trout (Salmo gairdneri). Bull. Environ. Contam. Toxicol 12(4): 481-485.

Rice, S. D., J. W. Short, C. Brodersen, T. A. Mecklenburg, D. A. Moles, C. Misch, D. L. Cheatham, and J. F. Karinen. Unpubl. Manuscr. Final report to Shell Oil Company, Standard Oil of Calif., Union Oil of Calif., Texaco Inc., Marathon Oil Co., and Phillips Petroleum Corp. On file at Northwest Fisheries Center Auke Bay Fisheries Laboratory, National Marine Fisheries Center, NOAA, P. O. Box 155, Auke Bay, AK 99821.

Schaumburg, T., E. Howard, and C. C. Walden. 1967. A method to evaluate the effects of water pollutants on fish respiration. Water Res. 1: 731-737.

Sellers, C. M., A. G. Heath, and M. L. Bass. 1975. The effect of sublethal concentrations of copper and zinc on ventilatory activity, blood oxygen and pH in rainbow trout (Salmo gairdneri). Water Res. 9: 401-408.

Silverstein, R. M. and G. C. Bassler. 1966. Spectrometric Identification of Organic Compounds. John Wiley and Sons, Inc., New York. 177 p.

Sparks, R. E., J. Cairns, Jr., and A. G. Heath. 1972. The use of bluegill breathing rates to detect zinc. Water Res. 6: 895-911.

Thomas, R. E. and S. D. Rice. 1975. Effect of water-soluble fraction of Prudhoe Bay crude oil on opercular rates of pink salmon, Oncorhynchus gorbuscha, fry. J. Fish. Res. Bd. Canada 32: 2221-2224.

Walden, C. C., T. E. Howard, and G. C. Froud. 1970. A quantitative assay of the minimum concentrations of kraft mill effluents which affect fish respiration. Water Res. 4: 61-68.

Warner, J. S. 1976. Determination of aliphatic and aromatic hydrocarbons in marine organisms. _Anal_. _Chem_. 48: 578-586.

Willard, H. H., L. L. Merritt, Jr., and J. A. Dean. 1965. _Instrumental_ Methods _of_ _Analysis_. Van Nostrand Reinhold Co., New York. 784 p.

Some Metabolic Effects of Petroleum Hydrocarbons in Marine Fish

DENNIS J. SABO and JOHN J. STEGEMAN

Woods Hole Oceanographic Institution
Woods Hole, Massachusetts   02543

Results of both environmental and experimental studies
have demonstrated the toxicity of petroleum hydrocarbons to
fish (Zitko and Tibbo, 1971; Anderson et al., 1974), yet
there is little knowledge regarding the metabolic action of
these compounds in fish which are chronically exposed or
briefly exposed to low sublethal levels of contamination.
The particular questions that we are trying to answer are
whether or not there are identifiable metabolic and biochemical changes which result from low levels of contamination, and secondly, what is the biological significance of
these changes.

MATERIALS AND METHODS

Fundulus heteroclitus were collected from two local
areas, one of which was the Wild Harbor marsh that suffered
a major spill in 1969 of #2 fuel oil resulting in significant
mortality. The marsh still contains high levels of the oil,
but there has been some recovery, and the fish inhabiting
this marsh are considered to represent a chronically exposed
group since they presently contain up to 75 ppm petroleum
hydrocarbon in their tissues (Burns, 1975). The control
marsh, Sippewissett, has not been exposed to petroleum hydrocarbon contamination, and no petroleum hydrocarbons were
found in fish from this marsh. In the second phase of our
study F. heteroclitus were collected from an uncontaminated

area, and exposed in our laboratory tanks to 180 parts per billion of #2 fuel oil for eight days in a flow through system (Stegeman and Teal, 1973). Livers were removed from 30 fish each, pooled and immediately placed in ice cold Krebs-Ringer Bicarbonate Buffer, pH 7.4, blotted and weighed. These tissues were minced and transferred to serum bottles containing Krebs Ringer Bicarbonate Buffer, 200 mg % glucose and 0.5 microcuries of either glucose-1-^{14}C, glucose-6-^{14}C, or acetate-1-^{14}C. The vials were sealed with serum bottle stoppers, gassed with 95% oxygen-5% CO_2 for five minutes and then incubated at 25°C for two hours with gentle shaking. All incubations were done in triplicate. At the end of the incubation period liver lipids and CO_2 were extracted (Sabo et al.,1971), and the radioactivity measured by scintillation counting. Lipid classes were separated by thin layer chromatography and relative incorporation of ^{14}C into phospholipid, triglyceride, free fatty acids, cholesterol and cholesterol esters was determined in spots which were detected by staining with rhodamine G. Each spot was then scraped from the plate, and the relative incorporation determined by liquid scintillation counting. Portions of the same tissues from each group were fixed in glutaraldehyde and examined at the electron microscopic level.

RESULTS

The results in Table 1 represent data collected from approximately 200 Wild Harbor (contaminated) and Sippewissett (uncontaminated) fish and show the relative incorporation of ^{14}C into CO_2 and lipid in hepatic tissues from both groups. It is apparent that little or no difference exists between Wild Harbor and Sippewissett fish in the amount of CO_2 respired from glucose labeled in either position or from acetate. Similarly, there is little difference between the two groups representing any difference in the rate of carbon from glucose incorporated into lipid. There is, however, an apparent small effect on lipid synthesis from acetate, although the relevance of this relatively small difference to the overall scheme of lipid synthesis is not apparent at this time. It is clear that when one considers carbohydrate and acetate metabolism in the two groups there is no apparent significant difference between them. However, if we consider data concerning the lipid classes separated by thin layer chromatography (Table 2) there are differences which are quite evident. In both groups the pattern of lipid synthesis is approximately the same whether synthesized from glucose

Table 1
Glucose and Acetate Metabolism in Fundulus heteroclitus from
a Contaminated Environment.

Substrate		$^{14}CO_2$	^{14}C-Lipid	Lipid/CO_2
Glucose-1-^{14}C	Wild Harbor*	3,784	101	0.026
	Sippewissett	3,128	96	0.030
Glucose-6-^{14}-C	Wild Harbor*	1,022	109	0.110
	Sippewissett	700	129	0.180
Acetate-1-^{14}-C	Wild Harbor*	72,916	11,644	0.160
	Sippewissett	76,752	15,765	0.210

*Contaminated

labeled in the one position or in the six position. This is
not unexpected as there is little evident difference between
the two groups from the data in Table 1. However, between
groups there is an apparent difference in rates of phospholipid synthesis from all three labeled substrates and there
is a considerable decrease in the rate of triglyceride
synthesized from acetate.

The difference with regard to acetate and phospholipid
is very suggestive of an effect on cell membranes, either
cytoplasmic or intracellular, since phospholipid along with
cholesterol are the major constituents of most membranes.
The decrease in triglyceride in the Wild Harbor group is
indicative of a lack of accumulation or increase in utilization of neutral fat. This could be explained by a decrease
in accumulation and/or an increase in utilization of stored
energy for purposes that appear to be related to stress
caused by a contaminated environment. The metabolic mechanisms related to these observations are currently under
investigation.

Several differences were observed when the electron
micrographs of fish from Sippewissett and Wild Harbor were
compared. Livers from the Sippewissett fish (Fig. 1), our

Table 2
Percent Contribution to Total Hepatic Lipid by Lipid Class in Fundulus heteroclitus.

	Glucose-1-^{14}C		Glucose-6-^{14}C		Acetate-1-^{14}C	
	Sippewissett	Wild Harbor	Sippewissett	Wild Harbor	Sippewissett	Wild Harbor
Phospholipid	63.0	48.0	59.7	48.0	54.0	71.7
Cholesterol	9.7	12.5	10.7	14.0	9.7	5.7
Fatty Acid	9.0	12.0	9.7	12.6	8.0	13.3
Triglyceride	11.3	14.0	12.3	13.0	24.0	5.3
Cholesterol Ester	7.3	13.0	7.7	12.3	4.0	4.3

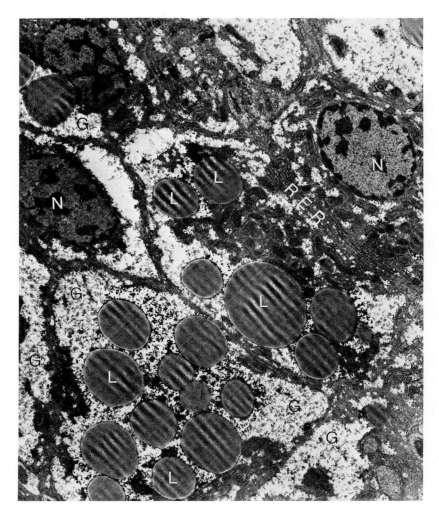

Fig. 1. Liver tissue from <u>Fundulus heteroclitus</u> from
Sippewissett March. Magnified 15,000 X

RER = rough endoplasmic reticulum
  L = lipid
  G = glycogen
  N = nucleus

control group, present normal tissue in the amount of glycogen, rough endoplasmic reticulum, free ribosomes and amount
of lipid. However, a comparison of the ultrastructure of
the hepatic tissue of fish from Wild Harbor with that of this
same tissue of fish from Sippewissett (Fig. 2) shows that
there is little glycogen in the Wild Harbor fish, there is an
increase in the rough endoplasmic reticulum as well as free
ribosomes and there is less neutral fat in the form of lipid
globules. The increase in the rough endoplasmic reticulum is
consistent with the cholesterol and phospholipid data since
there is a significant amount of phospholipid but little
cholesterol in the endoplasmic reticulum membrane. The
smaller amounts of lipid globules in the Wild Harbor group
are consistent with the decrease in accumulated triglyceride
as seen in Table 2.

　　With fish that have been experimentally exposed for
eight days to #2 fuel oil at a concentration of 180 parts
per billion (Table 3), very little difference was noted in
the amount of CO_2 and lipid synthesized from labeled glucose
between experimental and control fish. There was, however,
a noticeable difference when we used acetate. This situation
is distinct from that with fish chronically exposed in the
environment (Table 1). If we observe the various lipid
classes (Table 4) we again see differences in patterns of
lipid synthesis between control and experimental fish from
acetate. It can be seen that there is a decrease in phospholipid and an increase in cholesterol and again a decrease
in triglyceride and cholesterol ester synthesis.

　　These results, although differing from the environmental
samples, are still indicative of changes which are probably
related to structures of either cytoplasmic or intracellular
membranes. In this case, it is more likely that the changes
are related to the composition of the plasmic membrane and
particularly the outer portion, since it contains a major
portion of cellular cholesterol.

CONCLUSIONS

　　It is apparent that contamination by petroleum hydrocarbons do produce subtle biochemical changes in fish that
are both chronically and briefly exposed. Secondly, these
changes strongly suggest an altered membrane structure either
intracellularly or at the cell surface. Both types of
changes could alter some membrane functions. However, at
this stage in the development of our work we cannot make any
strong statements regarding the meaning of these changes as
they relate to membrane function or the health or

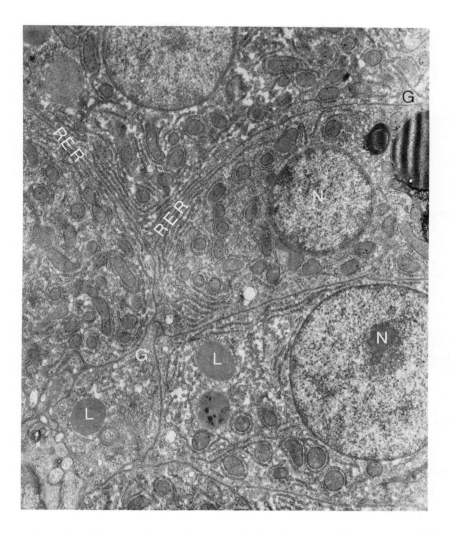

Fig. 2. Liver tissue from <u>Fundulus heteroclitus</u> from Wild
         Harbor Marsh. Magnified 15,000 X

         RER = rough endoplasmic reticulum
           L = lipid
           G = glycogen
           N = nucleus

Table 3
Glucose and Acetate Metabolism in Fundulus heteroclitus
Exposed to 180 ppb No. 2 Fuel Oil for 8 Days.

		CPM/100 mg Tissue	
Substrate	Sample	$^{14}CO_2$	^{14}C-lipid
Glucose-1-^{14}C	Control	2,441	180
	Oil	2,474	206
Acetate-1-^{14}C	Control	13,434	180
	Oil	17,645	482

Table 4
Percent Contribution to Total Lipid by Lipid Class in
Fundulus heteroclitus After Exposure to 180 ppb No. 2 Fuel
Oil for 8 Days.

	Acetate-1-^{14}C	
Lipid Class	Control	Oil
Phospholipid (PL)	27.8	19.5
Cholesterol (CH)	12.7	50.6
Fatty Acid (FA)	12.7	11.8
Triglyceride (TG)	17.2	10.2
Cholesterol Ester (CHE)	9.5	2.3

reproduction of the population of these fishes. These and related questions are presently under investigation.

ACKNOWLEDGMENTS

    We thank Albert Sherman for his excellent technical assistance and Dr. Leonard S. Gottlieb and Mr. Allen Walstrum of the Mallory Institute of Pathology, Boston, Massachusetts for preparing and assisting in the interpretation of the electron microscopic data. This work was supported by the Sea Grant #04-6-158-4416. Woods Hole Oceanographic Institution Contribution Number 3684.

LITERATURE CITED

Anderson, J. W., J. M. Neff, B. A. Cox, H. E. Tatem, and G. M. Hightower. 1974. Characteristics of dispersions and water soluble extracts of crude and refined oils and their toxicity to estuarine crustaceans and fish. Mar. Biol. 27: 75-88.
Burns, K. 1975. Doctoral Dissertation. W.H.O.I.
Sabo, D. J., R. P. Francesconi, and S. N. Gershoff. 1971. Effect of vitamin B_6 deficiency on tissue dehydrogenases and fat synthesis. J. Nutrition 101: 9.
Stegeman, J. J. and J. M. Teal. 1973. Accumulation, release and retention of petroleum hydrocarbons by the oyster Crassostrea virginica. Mar. Biol. 22: 37-44.
Zitko, V. and S. N. Tibbo. 1971. Fish kill caused by intermediate oil from coke ovens. Bull. Environ. Contam. Toxicol. 6: 24.

Effects of Natural Chronic Exposure to Petroleum Hydrocarbons on Size and Reproduction in *Mytilus californianus* Conrad

D. STRAUGHAN

Allan Hancock Foundation
University of Southern California
Los Angeles, California 90007

The mussel, Mytilus californianus, forms large beds in the lower intertidal areas in open ocean conditions along the west coast of North America from Alaska to Baja, California (Soot-Ryen, 1955). Hence, data relating to this species is of specific value over a large area. Mytilus californianus also occurs in areas chronically exposed to natural oil seepage - namely, Coal Oil Point in southern California. Therefore, data on the response of M. californianus to natural chronic exposure to petroleum is of use in predicting the influence of chronic exposure to low levels of petroleum along the west coast of North America. Naturally, care must be used in extrapolation of such data to account for differences in oil composition and in physical and ecological variables (Kanter, 1974; Straughan, 1972).

The extent of natural chronic exposure to petroleum can perhaps be best illustrated by comparison of levels of total ether extractable organics in M. californianus from the three areas. Samples were collected between May and November, 1973. The samples were chemically analyzed and chemical data interpreted by J. Scott Warner (Straughan, 1976). M. californianus from Coal Oil Point contained 130 to 435 µg/g total ether extractables; M. californianus from Santa Catalina Island contained 4 to 26 µg/g total ether extractables; M. californianus from Pismo Beach contained 2 to 7 µg/g total ether extractables. Ninety-five percent of the total ether extractables were of petroleum origin in the

Coal Oil Point samples while none were of petroleum origin in
the Pismo Beach samples. In one set of samples from Santa
Catalina Island (collected September 1973) up to 16 µg/g
total ether extractables of petroleum origin were recorded
while in the other set of samples from Santa Catalina Island
(collected November, 1973) none of the total ether extractables were of petroleum origin.

Kanter et al. (1971) and Kanter (1974) have demonstrated
experimentally that M. californianus from Coal Oil Point is
more tolerant to Santa Barbara crude oil than M.
californianus from the Santa Catalina Island and Pismo Beach.
The latter two sites are not chronically exposed to natural
oil seepage. Hence, the chronic exposure to natural oil
seepage does appear to have some influence on the populations
of M. californianus at Coal Oil Point.

This paper presents the results of field studies on
Mytilus californianus to determine if differences that could
be attributed to the natural chronic exposure to petroleum,
exist between the field population at Coal Oil Point and
field populations at two control sites, Pismo Beach and Santa
Catalina Island (Fig. 1), that are not chronically exposed to
natural oil seepage. The parameters studied were reproduction and size.

Although a population such as that at Coal Oil Point is
living and settling in an area, the species may not be
maintaining a 'healthy' population and/or breeding.
Straughan (1971) suggested a possible reduction in breeding
in 1969, but Harger and Straughan (1972) could find no change
in the 'health' of this species after the Santa Barbara oil
spill using shell length: $\sqrt{\text{dry body weight}}$ as an index.

Gonads were sectioned and examined by light microscopy
to determine if there were differences in gametogenesis as a
result of chronic exposure to oil. These sections were also
scanned for indications of abnormal growths in the light of
reports of gonadal tumors in Mya arenaria in an area exposed
to No. 2 fuel oil (La Roche, 1972).

MATERIALS AND METHODS

Ten animals were collected from Pismo Beach, Coal Oil
Point, and Santa Catalina Island at one or two monthly
intervals in 1972 and 1973. The following parameters were
measured: shell length, dry body weight, reproductive
state. The shell length was measured by calipers. The shell
was then opened and the gonad examined to determine if the
animal was in a state ready to spawn. This determination
was based on color of gonads - white in males and orange in

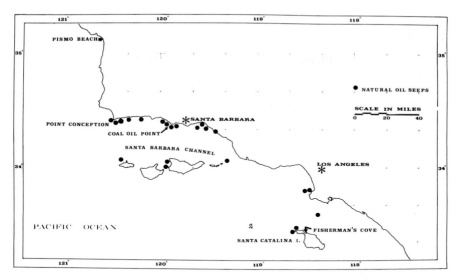

Fig. 1.   Map of Southern California to show collection sites
of M. californianus - Pismo Beach, Coal Oil Point,
and Fishermans Cove on Santa Catalina Island - and
distribution of areas of natural oil seepage.

females when ready to breed, thickness - gonads were usually
thicker prior to spawning, and presence of mature oocytes or
spermatozoa when a small amount of tissue was squashed
between a microscope slide and coverslip (Jessee, 1976).
Ideally, freshly collected animals were examined. However,
at times the animals were frozen between collection and
examination.

The wet tissue of the animal was then placed in a petri
dish in an oven (40°C) for 24 hours to determine dry tissue
weight. Tissues of larger animals were sometimes not dry
within this period, and they were left in the oven for 48
hours.

On November 16, 1973 specimens were collected especially
to study possible differences in gametogenesis by light
microscopy. Ten animals with no visible tar on the shells
were collected from each of upper and lower intertidal levels
of distribution at Coal Oil Point. A third group of ten
animals that were surrounded by tar were collected at Coal
Oil Point on the same day. A simultaneous collection of ten
animals with no visible tar on the shells was collected from
each of upper and lower intertidal levels of distribution
from Santa Catalina Island. Animals were collected from
both upper and lower intertidal levels in the M.

californianus populations at both sites following Jessee's
observations that breeding was not always synchronous at
different intertidal heights at Santa Catalina Island
(Jessee, 1976).

A section of the gonad of each animal was fixed in 10%
formalin - seawater for 24 hours. Tissues were stored in
the fixatives for 2 weeks at 8°C. The tissue was then
dehydrated in ethanol and embedded in paraplast. Sections
were cut at six microns and stained with hematoxylin and
eosin. Sections were then examined by light microscopy.

RESULTS

Examination of the ratio between shell length and dry
body weight (Table 1) at all three sites, indicates that
animals from Santa Catalina Island are relatively lighter
than the animals from Coal Oil Point and Pismo Beach, while
the relationship between Coal Oil Point and Pismo Beach
animals varies. The differences are possibly related to
factors such as water, temperature, and food supply.

A comparison of the fraction of animals breeding in the
monthly census between April and November 1973 showed an
unsynchronized breeding peak at all three sites between
April and July (Fig. 2). Fewest animals in breeding
condition were recorded at all three sites in September.
The most animals in breeding condition were recorded at
Pismo Beach and Coal Oil Point in October and at Santa
Catalina Island in November.

The microscopic examination of tissues was conducted
in November because this was the time of greatest similarity
of breeding condition of M. californianus from Coal Oil
Point and Santa Catalina Island. Examination of the stained
sections by light microscopy revealed no discernable
difference in the gonads of the control group of animals from
Santa Catalina Island and the experimental groups of animals
from Coal Oil Point. Animals from each location showed
similar interstitial cell appearance and the various stages
of gametogenesis appeared normal.

Examination of random points in the sections to
determine the relative area of gonad occupied by oocytes and
connective tissue showed that 68% of the space was occupied
by oocytes in animals from Coal Oil Point and 72% of the
space was occupied by oocytes in animals from Santa Catalina
Island. Fourteen percent of the space was occupied by
connective tissue in animals from Santa Catalina Island.
These differences coincide with the observations from the
macroscopic and squash examinations in that slightly more

Table 1
Comparison of the Ratio Between Dry Body Weight and Shell
Length.

Date	Area Comparasion
August 1972	PB = COP
October 1972	COP < PB
February 1973	CAT = PB
April 1973	CAT < PB COP
May 1973	CAT < PB = COP
June 1973	CAT < PB = COP
July 1973	CAT < PB = COP
August 1973	CAT < PB = COP
September 1973	CAT < PB = COP
October 1973	CAT <COP < PB
November 1973	CAT <COP < PB

Pismo Beach - PB
Coal Oil Point - COP
Santa Catalina Island - CAT

= indicates no significant difference at 0.05 level

< indicates a significant difference at the 0.05 level.

animals from Santa Catalina Island were in a breeding state
than those from Coal Oil Point (Fig. 2). The data do not
suggest any reduction of oocyte production at Coal Oil
Point as compared to Santa Catalina Island (550 points
examined on Coal Oil Point specimens; 300 examined on Santa
Catalina Island specimens). No abnormalities were recorded
in the gonadal tissue of the animals examined.

Oocytes, both those still attached to the follicle wall
and those free in the follicle lumen, ranged from 9.8 to
113.4 µm in diameter in animals from Coal Oil Point and 19.6
to 98.0 µm in diameter in animals from Santa Catalina Island.
The Kolmogorov-Smirnov two-sample test (Siegel, 1956:127)
rejects the Null hypothesis that there is no difference

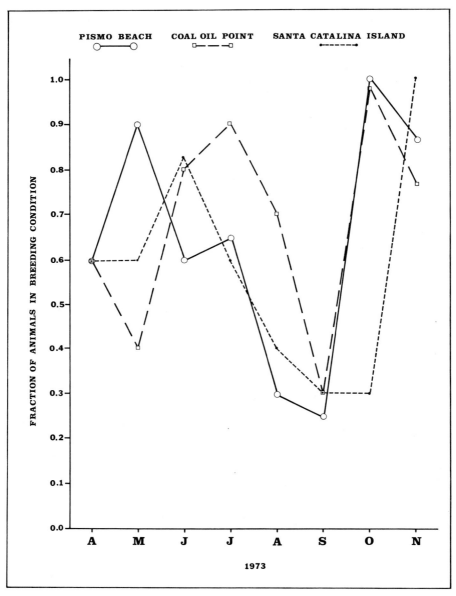

Fig. 2.   Fraction of animals in breeding condition on
          monthly census between April and November, 1973,
          inclusive.

between the two samples for a two-tailed test (=.001).
Figure 3 shows that there are more oocytes in the 35 to 60
μm diameter in animals from Santa Catalina Island than in
animals from Coal Oil Point. Since the animals from Coal Oil
Point have both larger and smaller oocytes than those
reported from Santa Catalina Island, the difference between
the two samples of oocytes is not due to either a stunting
or enlargement of oocytes at either locality.

DISCUSSION

Comparison of reproductive parameters indicate no
differences in reproduction between animals in natural oil
seep areas (Coal Oil Point) and animals not chronically
exposed to natural oil seepage (Pismo Beach, Santa Catalina
Island). In the period between April and November 1973
there were two breeding peaks at each site. The lowest
level of breeding at all sites was during summer when the
water temperatures were the highest. Jessee (1976) has
found an absence of reproductive synchrony between M.
californianus from different intertidal levels. Hence, the
fact that the peaks in breeding were not synchronized from
all three sites would not appear abnormal. Effectively, all
three populations have a spring and fall peak in breeding
activity. This is in agreement with studies at other
locations (Coe and Fox, 1942; Stohler, 1930; Bartlett, 1973).
Jessee (1976) recorded that the diameter of oocytes
still attached to the follicle wall in animals from Santa
Catalina Island ranged from 10.0 to 63.0 μm. This would
suggest that all but approximately 20% of the oocytes in the
present sample were still attached to the follicle wall. In
contrast, the sample from Coal Oil Point had both larger and
smaller oocytes than those recorded from Santa Catalina
Island. Their differences appear related to the differences
in breeding state in the two populations. Namely, the animals
from Coal Oil Point reached a peak in breeding in October
and were showing a decline in breeding condition in November
(Fig. 2). This coincides with the presence of large oocytes
which would indicate some residual breeding, and very small
oocytes which would indicate that oocytes were being
regenerated. The animals from Santa Catalina Island were in
a ripe phase just prior to breeding. Hence, the difference
in the oocytes appears related to the breeding condition of
the animals and not to the presence of petroleum.
The relationship between tissue weights and shell length
indicate similarity between animals from Coal Oil Point and
Pismo Beach. The animals from Santa Catalina Island were

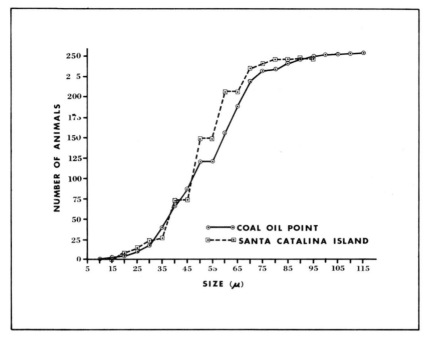

Fig. 3. Cumulative curve of oocyte diameter in M.
californianus from Coal Oil Point and Santa
Catalina Island, November 1973.

relatively lighter than those from the more northern sites.
There could be a food and/or water temperature effect
because the animals from Santa Catalina Island are closer to
the warmer extreme of the range of this species than those
from the two northern sites. Previous studies (Harger, 1967;
Harger and Straughan, 1972; Jessee, 1976) have indicated
that this relationship is consistent within any area, but
there are no similar comparisons between sites as widely
separated as in the present study.

Sections of gonadal tissue were relatively thick, were
only examined by light microscopy, and were examined to
determine gross details of gametogenesis. However, no
malformations of the type recorded by La Roche (1972) in Mya
arenaria gonads were recorded in Mytilus californianus either
from Coal Oil Point or Santa Catalina Island. He pointed out
that while the increase in these tumorous lesions appears
related to the distribution of No. 2 fuel oil, this does not
prove cause and effect. La Roche also speculated on possible

synergistic effects with other pollutants in the area where
M. arenaria were collected.

When comparing the data presented herein with that of
La Roche (1972), it must also be remembered that the data
were obtained from a different species of mollusc, in a
different ecological habitat, exposed to a different dose of
a different type of petroleum. Any one of these, but
particularly the difference between the chemical composition
of the crude petroleum and the refined product, could account
for a difference in the observations.

In conclusion, this comparative field study of size in
terms of shell length - dry body weight ratios, and of reproduction, did not reveal any differences between the
specimens from Coal Oil Point and specimens from areas not
chronically exposed to petroleum that could be attributed to
the effects of natural chronic exposure to petroleum.

ACKNOWLEDGMENTS

The research described in the paper is a portion of the
field study of the sublethal effects of natural chronic
exposure to oil seepage conducted at the Allan Hancock
Foundation and funded by the American Petroleum Institute,
Contract No. 093022.

The author is grateful to the following personnel for
assistance with this research: L. Masuoka, W. Jessee, D.
Hadley, S. B. De Lawter, and to B. Allen for typing the
manuscript. Figures are by T. Licari.

LITERATURE CITED

Bartlett, B. R. 1973. Reproductive ecology of the
        California sea mussel Mytilus californianus (Conrad).
        Unpublished Masters Thesis, University of the Pacific.
Coe, W. W. and D. L. Fox. 1942. Biology of the
        Californian sea mussel (Mytilus californianus) 1.
        Influence of temperature, food supply, sea and age, on
        the rate of growth. J. Exp. Zool. 90: 1-30.
Harger, J. R. E. 1967. Population studies on Mytilus
        communities. Ph.D. thesis, University of California,
        Santa Barbara.
_____ and D. Straughan. 1972. Biology of sea mussels
        Mytilus californianus Conrad and M. edulis (Linn.)
        before and after the Santa Barbara oil spill (1969).
        Water, Air and Soil Pollution 1: 380-388.

Jessee, W. N. 1976. The effects of water temperature, tidal cycles, and intertidal position on spawning in *Mytilus californianus* (Conrad). Unpublished Master of Arts Thesis, Humbolt State College, Arcata, Ca. 146 pp.

Kanter, R. 1974. Susceptibility to crude oil with respect to size, season, and geographic location in *Mytilus californianus* (Bivalvia). Pub. University of Southern California Sea Grant Program, Los Angeles, California 90007. USC-SG-4-74: 42 pp.

_____, D. Straughan, and W. N. Jessee. 1971. Effects of exposure to oil in *Mytilus californianus* from different localities. Proc. Joint. Conference on Prevention and Control of Oil Spills. API, EPI, USCG.,: 485-488.

La Roche, G. 1972. Biological effects of short-term exposures to hazardous materials. Proc. 1972 National Conference on Control of Hazardous Material Spills: 199-206.

Siegel, S. 1956. Nonparametric Statistics for the Behavioral Sciences. McGraw-Hill. 312 pp.

Soot-Ryen, T. 1955. A report on the family Mytilidae (Pelecypoda). Allan Hancock Pacific Exped. 20(1): 1-175.

Stohler, R. 1930. Beitrag zur kenntniss des geschtszyklus von *Mytilus californianus* (Conrad). Zool. Anz. 90: 263-268.

Straughan, D. 1971. Breeding and larval settlement of certain invertebrates in the Santa Barbara Channel following pollution by oil. In: Biological and Oceanographical Survey of the Santa Barbara Channel Oil Spill 1969-1970, 1: 245-254. Sea Grant Publ. No. 2. Pub. Allan Hancock Foundation, University of Southern California, Los Angeles, 90007.

_____ 1972. Factors causing environmental changes after an oil spill. J. Pet. Tech: 250-254.

_____ 1976. Sublethal effects of natural chronic exposure to petroleum in the marine environment. American Petroleum Institute Publ. No. : 120 pp.

Effects of Varying Concentrations of Petroleum Hydrocarbons in Sediments on Carbon Flux in *Mya arenaria*

EDWARD S. GILFILLAN[1], DANA W. MAYO[2], DAVID S. PAGE[2],
DANA DONOVAN[2], and SHERRY HANSON[1]

[1]Bigelow Laboratory for Ocean Sciences
West Boothbay Harbor, Maine 04575

[2]Department of Chemistry
Bowdoin College
Brunswick, Maine 04011

Gilfillan (1975) showed that stress from lowered salinity could reduce carbon flux (assimilated carbon - respired carbon) in <u>Mytilus</u> <u>edulis</u> and <u>Modiolus</u> <u>demissus</u>. Reduction in carbon flux represents a reduction in carbon available for reproduction and growth. Bayne (1975) has shown that stress imparted by high temperature and low food concentrations can reduce the scope for growth (=assimilated calories - respired calories) in <u>Mytilus</u> <u>edulis</u>. Other stressful conditions also have been shown to upset energy metabolism in <u>M</u>. <u>edulis</u> (Gabbott and Bayne, 1973; Widdows and Bayne, 1971; Bayne and Thompson, 1970).

Gilfillan (1975) also showed that carbon flux in both <u>M</u>. <u>edulis</u> and <u>M</u>. <u>demissus</u> could be severely reduced by exposure to seawater extracts of crude oils. Similarly, Stegeman and Teal (1973) have shown that exposure to #2 fuel oil can reduce the rate of biodeposition in <u>Crassostrea</u> <u>virginica</u>.

On July 22, 1972 the Norwegian tanker, TAMANO, struck Soldier Ledge in Portland Harbor, releasing 340 tons of #6 fuel oil into Casco Bay, Maine. An investigation was begun which compared carbon flux in <u>Mya</u> <u>arenaria</u>, on a monthly basis, from a heavily oiled site with <u>M</u>. <u>arenaria</u> from a

relatively clean site in Casco Bay. Results of this study
showed that carbon flux in M. arenaria from the heavily
oiled (1700 ppm in sediments) site was only 50% of that seen
in the unoiled population. These results posed the question
whether or not M. arenaria's growth rate could be
significantly reduced by low level petroleum hydrocarbon
contamination.

In August 1974 an investigation was begun in which
carbon flux was determined on a monthly basis for 8 populations of M. arenaria in Casco Bay. Total petroleum hydrocarbon contamination at these sites ranged from 9 to 228 ppm
in sediments. This paper is a report on data obtained from
the first 4 sampling periods of that study.

Sampling Locations

Figure 1 shows the location of the sampling sites in
Casco Bay. These sites were picked since they were contaminated to varying degrees at the time of the TAMANO oil spill.
Site selection was aided by a preliminary survey of sediment
hydrocarbon concentration done in May 1974. At Falmouth Town
Landing two sites about 25 yards apart were picked to
investigate the spatial distribution of contamination.
Falmouth Town Landing Site A is in the upper one-third of the
intertidal zone; Falmouth Town Landing Site B is about midway
down the beach. Other sampling sites are approximately at
mid-tide level.

MATERIALS AND METHODS

Samples of both M. arenaria and sediment were taken at
each of the stations in August, September and October, 1974,
and in February, 1975. Samples were collected from within a
6m x 6m square at each sampling site. Approximately 20 M.
arenaria 20-30mm in length were collected for physiological
experiment. These were transported to the laboratory in
insulated containers.

Several aliquots of sediment were taken with a pentane-
rinsed stainless steel ladle from the sediment dug to obtain
M. arenaria. These were thoroughly mixed in a pentane-
rinsed enamel pan. From this mixture a 150-200 gm aliquot
was placed in a 250ml glass jar with a foil lined cap.
Samples of M. arenaria for hydrocarbon analysis were wrapped
in sheets of aluminum foil, placed in plastic bags, and tied
shut. Both tissue and sediment samples were stored below
-18°C until analysis.

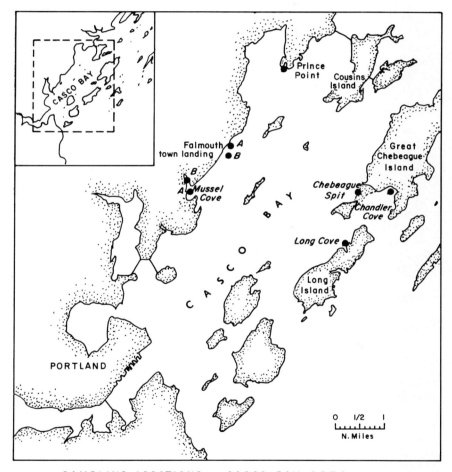

SAMPLING LOCATIONS — CASCO BAY AREA

Fig. 1. Map showing the locations of sampling sites in
         Casco Bay, Maine.

Sea water samples were collected at the low tide line at each location. The salinity was measured with a refractive index salinometer ($\pm$.1 o/oo); temperature was determined with a YSI electronic thermometer ($\pm$0.1°C). Chlorophyll a content of the water was also determined (Yentsch and Menzel, 1963).

Experimental animals were allowed to acclimate for 24 hours in flowing sea water in the laboratory. Following this, respiration rates were determined polarigraphically using a Gilson Medical Electronics Oxygraph in a water jacketed plexiglass cell. Respiration rates were calculated as μlO_2/mg dry wt. Carbon respired was calculated as respiration x 0.43 (assumes an RQ of 0.8). Respiration rates were determined for 5 animals from each station.

Filtration rates were determined by placing individual animals in 4l plastic beakers filled with glass-fiber filtered seawater to which sufficient ^{14}C labeled Dunaliella tertiolecta had been added to equal the chlorophyll a concentration observed at the collecting site. Filtration rates were determined for 5 animals from each station. The volume swept clear of particles, V, was calculated as: $V=v(\ln C_o - \ln C_t)/t$ where v is the volume of the beaker; C_o is the initial radioactivity in dpm/ml; C_t is the final radioactivity in dpm/ml; t is the duration of the experiment in hours. Experiments were run for 3 hours at the temperature observed at the sampling site.

After 3 hours beakers and animals were rinsed with raw seawater and refilled with raw seawater to which unlabeled D. tertiolecta had been added. They were then held 24-30 hours to allow them to clear their guts of radioactive algae. Following this, they were freeze dried, weighed and the radioactivity in their bodies determined by liquid scintillation counting techniques. Assimilation rates were calculated as dpm present in the animal's tissues/dpm consumed by the animal.

In order to calculate comparable carbon budgets for each of the populations, respiration rates and filtration rates were expressed in terms of 100 mg dry weight. This was done by taking advantage of the linear relation between these two rate functions and the common logarithm of the animals dry weight. In calculating carbon budgets the carbon concentration available to the animals was calculated as though all the chlorophyll a present at the sampling site was present as Dunaliella tertiolecta; a mean of the chlorophyll a concentrations at each of the stations was used to calculate carbon available according to the relation 1 μg chlorophyll a/l = 215 μg carbon/l.

Gross carbon intake was calculated as environmental carbon concentration ($\mu g/l$) x volume filtered (l/h). Assimilated carbon was calculated as gross carbon intake x assimilation ratio. Respired carbon was calculated as μl O_2 respired x 0.43. Assuming that the clams were metabolizing a mixture of protein and carbohydrate, a respiratory quotient of 0.8 was chosen arbitrarily to calculate carbon loss. Environmental stresses could change the R Q values; in all probability R Q values would increase and as a result, use of an R Q of 0.8 would underestimate respiratory carbon loss. Hence, use of an assumed R Q of 0.8 in calculating the results of a comparative study such as this should not affect the conclusions. Carbon flux was calculated as assimilated carbon - respired carbon.

Hydrocarbon Analysis of Sediments

Sediment samples were thawed at room temperature and air dried for 24 hours. Following this, 150 g of sediment was weighed and placed on a previously pentane extracted thimble. It was then soxhlet extracted for 24 hours with re-distilled pentane. After 24 hours the pentane was removed (Fraction I) and the sample extracted for 24 hours with another aliquot of pentane (Fraction I). Following this the sample was extracted with 50% (V/V) pentane/benzene (Fraction II). The two pentane portions (Fraction I) were combined; both Fraction I and Fraction II were reduced to ca. 25 ml and dried 12 hours over anhydrous Na_2SO_4. Both fractions were decanted onto activated copper columns, eluted with either pentane (I) or pentane/benzene (II) as appropriate, dried and weighed.

Fraction I was then taken up in pentane, put on an alumina/silica column and eluted with pentane, yielding Fraction III. The Fraction I flask was rinsed with 10% (V/V) benzene/pentane and Fraction II was taken up in the same mixture. They were put together on an alumina/silica gel column and eluted with 10% (V/V) benzene/pentane, yielding Fraction IV. The flask containing Fraction II was rinsed with 50% (V/V) pentane/benzene. This material was then put on the same alumina/silica column as Fraction II and eluted with 50% (V/V) pentane/benzene, yielding Fraction V. Fractions III, IV, and V were evaporated near dryness, taken up in the appropriate solvent and transferred to clean ½ dram vials. The vials were then dried to constant weight and weighed. Fraction III was then further analyzed by gas chromatography.

Hydrocarbon Analysis of Clam Tissue Samples

Samples of M. arenaria were thawed in their original
bags in warm water. Approximately 140-180 grams of clam
tissue were shucked into a clean glass beaker, saving the
water within the shell. Samples were then transferred to a
preextracted soxhlet thimble and extracted for 24 hours with
methanol (Fraction I). The tissue was then extracted for 24
hours with a 50% (V/V) mixture of methanol and benzene
(Fraction II). Both Fraction I and Fraction II were then
partitioned against pentane. Both pentane Fractions I and
II were then reduced to ca. 25 ml and dried over anhydrous
$Na_2 SO_4$ for at least 12 hours.
    Fraction I was then placed on an alumina/silica gel
column and eluted with pentane; after being reduced to dryness, the residue was weighed. Fraction II was then put on
the same column and eluted with 10% (V/V) pentane benzene.
It too was reduced to dryness and the residue weighed. The
column was eluted again with pentane and 10% (V/V) pentane/
benzene until each of the elutriates was constant in weight.
Both fractions were transferred to ½ dram vials, evaporated
to dryness and weighed. Fraction I was then further analyzed
by gas chromatography.

RESULTS

    The annual cycle of carbon flux in a population of M.
arenaria in Casco Bay (Cousins Island) is shown in Figure 2.
Data shown in Figure 2 were collected from November 1972 -
November 1973. The main trends seen in Figure 2 are that
carbon flux is low and negative during the winter, high and
negative during the spring and high and positive during the
summer. During the fall carbon flux values decline toward
winter values. It should be noted that no values are shown
in Figure 2 for September 1973.
    Carbon flux data for the eight populations of M.
arenaria investigated in this study are shown in Tables IA -
IH. Data for all eight populations of M. arenaria show
similar trends; carbon flux is high in August and declines
toward winter. A similar trend is seen in Figure 2. Values
for September 9, 1974 are probably abnormally low because of
the presence in Casco Bay of a bloom of the toxic dino-
flagellate Gonyaulax tamarensis. The toxin produced by this
algae causes paralytic shellfish poisoning in humans as well
as partial paralysis of M. arenaria (Gilfillan and Hanson,
1975).

CARBON FLUX

Mya arenaria

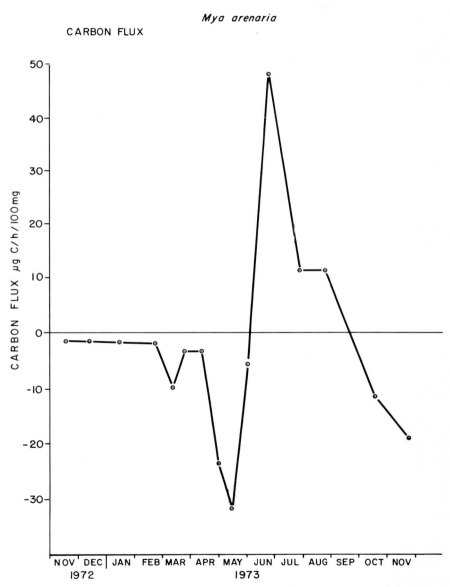

Fig. 2. Yearly cycle of carbon flux for <u>Mya</u> <u>arenaria</u> from Cousins Island, Casco Bay, Maine.

TABLE I-A   Elements of Monthly Carbon Budgets for M. arenaria
at Falmouth Town Landing--Site B.*

Date	Gross C	Assim.	Net C	Resp. C	C. Flux
8/9/74	321.43	.854	274.66	37.16	237.50
9/9/74**	135.33	.559	75.65	48.03	27.62
10/28/74	281.98	.425	119.84	48.31	71.81
2/3/75	12.92	.143	1.85	6.18	− 4.33

TABLE I-B   Elements of Monthly Carbon Budgets for M. arenaria
at Falmouth Town Landing--Site A.*

Date	Gross C	Assim.	Net C	Resp. C	C. Flux
8/9/74	185.80	.477	88.66	28.09	60.57
9/9/74**	135.41	.536	72.47	50.26	22.21
10/28/74	232.62	.650	151.20	42.86	108.34
2/3/75	17.75	.139	2.47	14.00	−11.53

TABLE I-C   Elements of Monthly Carbon Budgets for M. arenaria
at Mussel Cove--Site B.*

Date	Gross C	Assim.	Net C	Resp. C	C. Flux
8/9/74	223.39	.669	149.52	28.48	121.04
9/9/74**	149.21	.276	41.18	35.34	5.84
10/28/74	133.14	.275	36.61	35.44	0.77
2/3/75			FROZEN OVER		

*   All values are in terms of micrograms carbon/hr/100 mg
dry weight.

** Values for carbon flux in September are abnormally low as
a result of partial paralysis of the clams by paralytic
shellfish poisoning toxin (red tide).

TABLE I-D   Elements of Monthly Carbon Budgets for M. arenaria
            at Mussel Cove--Site A.*

Date	Gross C	Assim.	Net C	Resp. C	C. Flux
8/9/74	210.73	.306	64.43	24.33	40.10
9/9/74**	117.87	.411	48.44	31.71	16.73
10/28/74.	187.11	.119	22.27	31.01	-8.74
2/3/75	32.67	.297	9.70	6.30	3.40

TABLE I-E   Elements of Monthly Carbon Budgets for M. arenaria
            at Prince Point.*

Date	Gross C	Assim.	Net C	Resp. C	C. Flux
8/9/74	378.61	.831	314.75	39.28	275.48
9/9/74**	83.22	.658	54.72	46.60	8.12
10/28/74	263.64	.493	129.97	28.16	101.82
2/3/75	12.80	.198	2.53	25.08	-22.55

TABLE I-F   Elements of Monthly Carbon Budgets for M. arenaria
            at Chebeague Spit.*

Date	Gross C	Assim.	Net C	Resp. C	C. Flux
8/9/74	440.36	.462	203.62	82.50	121.12
9/9/74**	121.00	.287	34.76	35.78	-1.02
10/28/74	204.99	.120	24.60	38.149	-13.55
2/3/75	13.36	.344	4.60	8.78	-4.15

 * All values are in terms of micrograms carbon/hr/100 mg
    dry weight.

** Values for carbon flux in September are abnormally low as
    a result of partial paralysis of the clams by paralytic
    shellfish poisoning toxin (red tide).

TABLE I-G   Elements of Monthly Carbon Budgets for M. arenaria
at Chandler Cove--Chebeague Island.*

Date	Gross C	Assim.	Net C	Resp. C	C. Flux
8/9/74	730.23	.417	304.51	51.20	253.31
9/9/74**	104.31	.421	43.94	27.95	15.99
10/28/74	61.98	0	0	45.11	-45.11
2/3/75	10.42	.445	4.64	3.07	1.57

TABLE I-H   Elements of Monthly Carbon Budgets for M. arenaria
at Long Cove--Long Island.*

Date	Gross C	Assim.	Net C	Resp. C	C. Flux
8/9/74	784.77	0.304	238.57	49.80	+188.8
9/9/74**	89.44	.236	21.07	42.65	-21.58
10/28/74	0.00	0	0	42.05	-42.05
2/3/75	12.62	.296	3.74	₋7.38	-3.64

*   All values are in terms of micrograms carbon/hr/100 mg
dry weight.

**   Values for carbon flux in September are abnormally low as
a result of partial paralysis of the clams by paralytic
shellfish poisoning toxin (red tide).

In spite of the similarity of trends seen in the data shown in Tables 1A-1H, there are important differences between carbon budgets obtained for M. arenaria from the various locations. These differences are best seen in Table 2, where the stations are arranged in order of decreasing carbon flux. The values of carbon flux shown in Table 2 have been obtained by multiplying the hourly carbon flux data shown in Tables 1A-1H by 360 and summing to get a value for total carbon gained. It is important to note that values of carbon flux shown in Table 2 vary by a factor of 7.

It is important also to note that these differences in total carbon flux are not brought about by single very large or very low values. Comparing the data from the two locations having the highest values of carbon flux (Prince Point, Falmouth B) with the two having the lowest values (Chebeague Spit, Mussel Cove A) it is evident that the differences are largely brought about by lower assimilation ratios at the poorer stations. This situation may also be augmented by either lower gross carbon intake (Mussel Cove A) or higher respiration rates (Chebeague Spit).

Results of hydrocarbon analysis of both sediment and clam tissue samples are also shown in Table 2. Total sediment hydrocarbons are a result of the gravimetric analysis (sum of Fractions III, IV, and V expressed as ppm). Values of total tissue hydrocarbons are the sum of Fractions I and II as ppm. Values for tissue aliphatic fractions are Fraction I as ppm. Values for tissue aromatic fractions are Fraction II as ppm.

DISCUSSION

Trends seen in the carbon flux data for each of the populations of M. arenaria are the same. There are, however, important differences between the carbon flux values obtained for the .8 populations (Table 2). Some populations (Mussel Cove A) gained carbon at 1/7th the rate seen in some others (Prince Point, Falmouth B).

Decreases in both carbon flux (Gilfillan, 1975; Gilfillan, et al., in press) and 'scope for growth' (Bayne, 1975) have been shown to be responses to stresses in two different bivalve molluscs. Both of these quantities appear to give measures of either carbon or energy available to the animals for growth and reproduction. Gilfillan (in press) has shown that carbon flux as measured by the methods used in this study can account for 90-100% of the increase in dry weight expected for 100 mg M. arenaria in Casco Bay. As a result, it appears that relative differences in carbon

TABLE 2   Total Carbon Flux in ug Carbon Gained per 100 mg animal from August 1974 - February 1975. (Also shown are a number of measures of hydrocarbon contamination.)

Location.	Total Carbon Flux	Total* Sediment H-C	Total** Tissue H-C	Tissue** Aliphatic Fraction	Tissue** Aromatic Fraction	Peak*** Height Ratio
Prince Point	130178	228	9	8	1.1	0.9
Falmouth B	119736	92	8	6	1.8	1.6
Chandler Cove	81274	11	11	10	1.0	1.6
Falmouth A	64648	60	8.5	6	2.1	2.4
Mussel Cove B	45939	51	11	6	2.0	2.4
Long Cove	43427	111	11	7	2.7	2.7
Chebeague Spit	36864	9	8	5	2.6	3.1
Mussel Cove A	18537	41	11	8	2.6	4.1

*   Average of available values in ppm.

**   Values shown are means of August, September and October 1974 in ppm.

***   August 1974 (Peak height ratios taken from the gas chromatographs.)

flux between populations of M. arenaria reflect differences in the animals' relative growth rate.

Differences in carbon flux can be expected to reflect different levels of stress imparted to the animals at each location. Although Mussel Cove site A receives some domestic sewage and the salinity at Mussel Cove site B is frequently low, environmental variables such as temperature, salinity, and chlorophyll a content are not very different at each of the sites. The greatest changes between sites are in the hydrocarbon content of both sediments and tissues. These vary considerably from one site to another (Table 2), and changes in the level of stress imparted to the animals by hydrocarbons are the logical source of the variations in carbon flux.

One of the most surprising aspects of the results shown in Table 2 is that there is no apparent correlation between total hydrocarbon content of sediments and carbon flux in M. arenaria. Clams from the Prince Point site show the highest carbon flux and yet these animals are living in sediments with the highest total concentration of hydrocarbons. Clams at Mussel Cove Site A show the lowest carbon flux, yet they are living in sediments with the third lowest total concentration of hydrocarbons (Table 2).

Total concentration of hydrocarbons in an animal's tissues might be expected to be a realistic measure of stress imparted by hydrocarbons. However, the results shown in Table 2 show such is not the case. Values of total tissue hydrocarbons vary only slightly between the stations (9-11 ppm); there is no apparent correlation between carbon flux and total hydrocarbon content of the clam's tissues (Table 2).

Assuming that the activity influencing the carbon flux is contained in the "aromatic" low molecular weight portion of the unresolved envelope of the gas chromatogram, then the ratio of the maximum height of the low molecular weight side of the unresolved envelope to that of the unresolved envelope corresponding to the high molecular weight residual hydrocarbons should be related to the amount of low molecular weight aromatics present, and should therefore correlate directly with the carbon flux data ("Peak height ratio", Table 2). For samples collected during the four month period from July-October 1974, we now have twenty-two chromatograms of tissue samples. With a single exception all of these data (Mussel Cove B - September 1974) correlate directly with the metabolic values on a month by month basis. The one exception is Chandler Cove, where only one sample has been analyzed. The average values for this period also correlate with carbon flux.

Further substantiation of the above results is found in the gravimetric data. The second (methanol-benzene) extraction fraction should reflect more of the "aromatic" envelope material than the first fraction, which should be higher in the amount of aliphatic material. (There should be, therefore, a closer correlation of the data for the second fraction with the carbon flux data than the first fraction.) Twelve of the fifteen second fraction (aromatic) values obtained as of this date correlate closely with the animal metabolic rate data on a month by month basis, while there appears to be only a random distribution of data in the first fraction (aliphatic) weights.

As Table 2 demonstrates, there is an excellent correlation between large concentrations of light aromatic hydrocarbons (as indicated by the peak height ratio) and low carbon flux. Thus it appears that, when dealing with low concentrations of highly weathered oils, perhaps from several sources, total hydrocarbon content of either sediments or animal tissue is not an accurate measure of the stress imparted by the oils. Both the data on total aromatic content of tissues and gas chromatographic data on peak height ratios argue very strongly that the best measure of stress imparted to M. arenaria by hydrocarbons is the total concentration of aromatic hydrocarbons in the animals' tissues. This is certainly the expected result in view of the known toxicity of this class of hydrocarbons (Baker, 1971; Ottway, 1971). What is surprising is that such low concentrations of aromatic compounds can have such large effects. These results are especially significant in view of the fact that Stegeman (1974) and Anderson (1973) have shwon that the lighter aromatic fractions are more readily retained by bivalves than other oil fractions.

CONCLUSIONS

1. Carbon flux measurements made by the methods employed in this study are sensitive measures of stress imparted to M. arenaria by low levels of hydrocarbon contamination.
2. When dealing with low concentrations of highly weathered oils, perhaps from several sources, total hydrocarbon content of either sediments or animal tissues are not valid measures of stress imparted to M. arenaria.
3. The concentration of aromatic hydrocarbons present in the tissues of M. arenaria correlates very closely with reductions in carbon flux and hence stress imparted to the clams.

4. In the situation studied, very low concentrations of
    aromatic hydrocarbons in clam tissues (3-4 ppm) can
    cause large reductions in carbon flux.

LITERATURE CITED

Anderson, J. W. 1973. Uptake and depuration of specific
    hydrocarbons from fuel oil by the bivalves Rangia
    cuneata and Crassostrea virginica. In: Background
    papers for a workshop in Inputs, Fates, and Effects
    of Petroleum in the Marine Environment, Vol. 2, pp.
    690-780. Washington, D. C.: U. S. National Academy
    of Sciences.
Baker, J. 1971. Comparative toxicities of oils, oil
    fractions, and emulsifiers. In: The Ecological
    Effects of Oil Pollution on Littoral Communities,
    E. B. Cowell, ed. Applied Science, Barking, U. K.,
    1971. pp. 78-87.
Bayne, B. L. 1975. Reproduction in bivalve molluscs under
    stress. In: Physiological Ecology of Estuarine
    Organisms, F. J. Vernberg, ed. University of South
    Carolina Press, Columbia, 1975. pp. 259-277.
_____ and R. J. Thompson. 1970. Some physiological
    consequences of keeping Mytilus edulis in the
    laboratory. Helgolander Wiss. Meeresunters 20: 526-
    552.
Gabbott, P. A. and B. L. Bayne. 1973. Biochemical effects
    of temperature and nutritive stress on Mytilus edulis
    L. J. Mar. Biol. Ass. U.K., 53: 269-286.
Gilfillan, E. S. 1975. Decrease of net carbon flux in two
    species of mussels caused by extracts of crude oil.
    Mar. Biol. 29(1): 53-58.
_____ and S. A. Hanson. 1975. Effects of paralytic
    shellfish poisoning toxin on the behavior and
    physiology of marine invertebrates. In: Proc. First
    Int. Conf. on Toxic Dinoflagellate Blooms. V. R.
    LoCicero ed., Massachusetts Science and Technol.
    Found. Wakefield. pp. 367-376.
_____, D. Mayo, S. Hanson, D. Donovan, and L. C.
    Jiang. Reduction in carbon flux in Mya arenaria
    caused by a spill of #6 fuel oil. In press.
Ottway, S. 1971. The comparative toxicities of crude oils.
    In: The Ecological Effects of Oil Pollution on
    Littoral Communities, E. B. Cowell, ed. Applied
    Science, Barking, U. K. pp. 78-87.
Stegeman, J. J. 1974. Hydrocarbons in shellfish chroni-
    cally exposed to low levels of fuel oil. In:

Pollution and Physiology of Marine Organisms. F. J.
Vernberg and W. B. Vernberg, eds. Academic Press. New
York. pp. 329-347.
_____ and J. M. Teal. 1973. Accumulation, release,
and retention of petroleum hydrocarbons by the oyster
Crassostrea virginica. Mar. Biol. 22: 37-44.
Widdows, J. and B. L. Bayne. 1971. Temperature acclimation
of Mytilus edulis with reference to its energy budget.
J. Mar. Biol. Ass. U. K., 51: 827-843.
Yentsch, C. S. and D. W. Menzel. 1963. A method for the
determination of phytoplankton chlorophyll and
phaeophytin by fluorescence. Deep Sea Res., 10: 221-
231.

The Effect of Pure Oil Subcomponents on *Mytilus edulis,* the Blue Mussel

CHARLES W. MAJOR

Zoology Department
University of Maine
Orono, Maine 04473

The toxicity of crude oil and its refined fractions has
been repeatedly demonstrated (Moulder and Varley, 1971;
Vernberg and Vernberg, 1974), and a multiplicity of organisms
has been exposed to test solutions of such oil mixtures.
Prolonged exposure is usually highly toxic, even lethal, but
the mechanism of the effect is not yet clear. This is due to
two factors. The first is simply the problem of resolving
the specific toxicant from oil, a mixture of relatively
similar hydrocarbon components. In an effort to narrow this
down, a great deal of highly sophisticated research effort
has been done on the differential rates of uptake, retention,
and release of the major oil constituents (Anderson et al.,
1974; Lee, Sauerheber, and Benson, 1972). This is further
complicated by the differing relative susceptibilities of
various organisms. The principal toxicants, nevertheless,
seem to be the di- and tri-aromatic hydrocarbons.

The second factor contributing to the haziness of these
is the problem of resolving the level of response. Heitz et
al. (1974) did a comprehensive analysis of eleven enzymes in
mullet, a shrimp, and the American oyster after exposure to
oil. They used four tissues: gill, liver, brain, and muscle,
and found essentially no changes occurred. Dunning and Major
(1974) noted a decrease in dye clearance in the blue mussel
during oil exposure and presumed that ciliary activity was
being decreased. Many have observed increased respiration
on exposure of test organisms to oil. The whole organism may

respond by running or swimming away from the irritant, or it may show tissue responses such as cilia beat changes. These responses may be occurring at the tissue or cell level, in neural or hormonal control mechanisms, or in the regulation of cellular responses.

We have approached the problem using the method of Struhsaker et al. (1974) of exposing organisms to purified oil subcomponents. The organism we have used is Mytilus edulis, the common blue mussel, a rugged intertidal bivalve; this is one of the organisms that potentially would be most exposed to any inshore oil spill. One of the most useful characteristics of M. edulis is that it does not run away from irritants. The normal response of this bivalve is to close down and exclude the irritant, and this type of avoidance response has some experimental utility. While the animals can close down and avoid immediate exposure to fuel oil components, they do succumb if the level of volatile components is maintained for two days. One of the oils which has few volatile components, #6 oil, does not result in immediate toxicity, at least over a 12 day exposure period (Dunning and Major, 1974).

MATERIALS AND METHODS

Blue mussels, Mytilus edulis, 3.3 + .2 cm in overall length, were collected from a bed close to mean low tide at Lamoine Beach on Frenchman's Bay in the Gulf of Maine. They were returned to the laboratory where they were kept for up to one week in a continuously illuminated circulating sea water tank (32.5 gm/kg salinity) at 5°C. During this holding period there is no decline in respiration (Scott, 1971).

Animals were removed to a 10°C tray of saline, and a hole was drilled over the adductor muscle on the left side using a Dremel #280 motodrill with a number 118 burr. These animals were then kept in continuous light at 10°C, usually overnight, but in some early runs for up to three days before being used.

Three randomly selected mussels were placed in individual 135 ml Warburg flasks and covered with 20 ml of sterile seawater. The flask sidearm received 1 ml of 10% KOH solution with a 1.4 x 2.3 cm Whatman #1 filter paper wick to increase the surface area. These flasks served as the controls.

Experimental flasks received 20 ml or fraction thereof of this same seawater saturated with the test hydrocarbon; the volume was made up to 20 ml, as necessary, with seawater. The saturated solutions were prepared by agitation, with a motor-driven glass stirrer, of seawater in the presence of an excess of the hydrocarbon for thirty minutes at 10°C. The

saturated solution was then stored at 10°C under a layer of
the hydrocarbon or, in the case of solids, such as
naphthalene, in the presence of excess solid hydrocarbon for
20 hrs. The seawater phase was removed by pipetting. This
must not be done with serological pipettes calibrated to the
tip, as introduction of any trace of the pure hydrocarbon
phase completely changes the values measured. The hydrocarbon
saturated seawater was added at the last possible moment
through the already sealed flask's manometer attachment.

Control flasks were compensated with a thermobarometer
flask containing only seawater and KOH, while experimental
flasks were similarly corrected using a "volatile's" thermobarometer (vtb) containing the equivalent volume of test
solution and three heat-killed mussels, chopped fine, with
their shells. The shell volume was treated as a precluded
volume and the chopped tissue as a fluid phase volume. To
establish flask constants for those flasks containing live
mussels, the volume of 100 mussels, with adductors cut, was
determined by seawater displacement. The mussels were then
shucked, the volume of their shells determined, and this
value was then prorated to correct for that portion of the
total flask volume that was to be treated as precluded. The
live mussel volume was similarly prorated but added to the
fluid phase volume.

Before arriving at the vtb used herein, other possibilities were tried and several problems were encountered.
Whole heat-killed organisms do not give results comparable
to the finely chopped mussels. The volume is smaller than
the volume of live organisms and the mass more compact. It
is, indeed, so compact it must be considered as a precluded
volume which is not available for meaningful oxygen or hydrocarbon penetration. The difference between these two thermobarometer types is measurable, the chopped dead tissue
showing an uptake of hydrocarbons significantly greater than
the whole killed organism.

Flasks were placed in a modified Precision refrigerated
bath at 10 ± .02°C, agitated at 90 strokes/min, equilibrated
for 15 minutes, closed and read at intervals thereafter up to
4 hours. The flasks were then removed from the bath and the
flask contents examined for byssal thread formation. The
responsiveness of each organism was tested by pinching the
foot with a watchmaker's forceps. The organisms were then
removed from the shell into previously tared beakers, weighed,
and dried for 28 hours at 85°C.

RESULTS

It was immediately noted that time of holding after cutting the adductor changed the mean value of the controls and that, over six days post-surgery, this difference could be proven significant with a P value of less than 1%. The mean value at 15 hours was 197 μl/gm dry wt/hr and at six days 491 μl/gm dry wt/hr. To obtain meaningful statistics it was necessary to run paired comparisons; thus each experimental flask was compared to the adjacent control flask containing mussels with an identical history. Subsequent analysis was by t test of mean difference between the pairs.

Two types of volume error are found in the presence of the hydrocarbons. The first is the expected addition of hydrocarbons to the gas phase, increasing it as a function of the size of the gas phase and the vapor pressure of the hydrocarbon. The vapor pressure is determined by the temperature, and the total amount available from the saturated seawater phase is limited by the water solubility of the test compound. The second type error is due to the solubility of the available hydrocarbon in the lipids of the organism. It should be noted that Type I errors cause underestimation of respiration while Type II cause overestimation. Whether they are meaningful and significant is variable. In the case of benzene (Fig. 1) both are significant, while in the case of hexane the results are trifling. Type I errors are generally insignificant after the first 30 minutes, while Type II errors accumulate and are difficult to compensate for since the lipid phases of a heat-killed organism cannot be considered exactly comparable to the lipid of the normal mussel. When a droplet of a pure test hydrocarbon of high volatility is introduced into a flask by mistake, both Type I and Type II errors become so large as to make detection of respiration impossible.

Exposure to benzene caused a significant decrease in respiration only at the fully saturated dose level (Table 1). There was no byssal thread attachment, and test organisms were unresponsive to pinching when first examined. Subsequent exposure to air, leading presumably to loss of benzene, restored the irritability of the mussels in about 15 minutes. Naphthalene at the fully saturated level caused a similar decrease in respiration and byssal thread formation, but without the anesthetic effect on irritability. These two subcomponents, in combination, showed additive effects. The true value for the benzene inhibition is probably underestimated since substantial, i.e. visible, pigment extraction from the mussel tissue was noted during the four hours, occurring in the corresponding vtb as well. Such pronounced

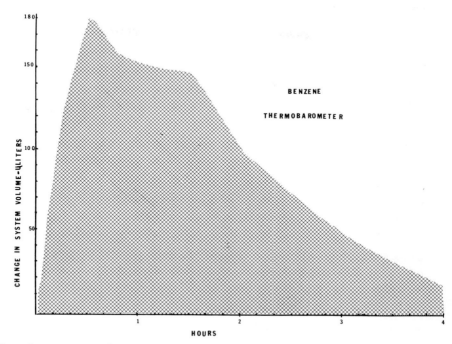

Fig. 1.   Cumulative gas changes in typical benzene thermobarometer. The changes in system volume are for a
"volatile's" thermobarometer corrected for ambient
changes by use of the regular thermobarometer.

extraction is presumably accompanied by extraction of some
tissue lipids, which in turn probably leads to a diminished
dry weight reading with magnification of respiration when
expressed on this basis. Hexane exposure caused increased
respiration in the mussel, decreased byssal thread formation,
and no loss of irritability. Data on the interactions of the
pure components are still very tentative. Most are negative
to date except for the benzene-naphthalene interaction noted
above. Unless a component has reached the saturated level,
it has no significant effect. Subsaturated combinations
have no significant effects. The hexane stimulation and the
benzene depression cancel each other in net effect when
mussels are exposed to seawater simultaneously saturated with
both components.

Table 1
Effect of oil subcomponents on Mytilus edulis.

Test Solution	Number of Mussels	Change in Respiration Over 4 Hrs.	P	Byssal Thread Formation
Benzene	72	-40%	<.01	1%
Hexane	24	+46%	<.01	0%
Naphthalene	24	-20%	<.08	0%
Benzene + Naphthalene	24	-72%	<.10	8%
Benzene + Naphthalene (3/4 concentration)	60	+13%	n.s.	0%
Controls	280	--	--	49%

DISCUSSION

The hexane response would seem anomalous. Little hexane actually goes into solution, and most of this probably ends up, at first, in the gas phase then shifts into the tissue lipids. The amount is small, and it may be that, at this level, it is merely an irritant leading to increased water siphoning in a purely reflex manner. Increased respiration can of course be toxic, even lethal, if it distorts metabolic patterns or exhausts substrates without concomitant energy storage. The failure of byssal formation proves that the hexane was detected by the organism and a withdrawal response initiated.

Short chain aliphatics generally cause an anesthetic effect in higher organisms, but they are not extremely effective when compared with ethers or the halogenated aliphatics. The normal sequence of response to an anesthetic agent is a transient excitation phase preceding the anesthetic depression. Apparently seawater saturation with hexane only

produces levels high enough for excitation in the blue
mussel. Since so little hexane is required to saturate seawater, the hexane effect will be produced immediately after a
spill as a complex function of water mixing and air blow off of
the volatiles. It will persist until the hexane in the oil
slick is inadequate to saturate a significant body of seawater.

The benzene effect, depression of respiration, is
impressive, but again it would not appear to be a long-term
factor in a real spill since the organism recovers its
irritability rapidly and readily after benzene exposure.
Benzene at high levels would be most likely on a drenched
beach or a shallow bay and may contribute to failure of shell
closure (Dunning and Major, 1974). While being open may
theoretically subject the mussels to predation, it is a moot
point whether any of its predators would escape a similar
anesthetic effect. The predators, furthermore, might not
recognize or even like benzene soaked mussels. Further work
is obviously required on the benzene toxicity.

The depression due to naphthalene is the most serious
since levels of this subcomponent could persist for
significant intervals. As long as the organism could remain
closed it could avoid the naphthalene effect, but once open
the effect would appear. Further, as the organism probably
concentrates naphthalene in its lipid components, the toxic
threshold might be reached internally. Should this occur,
the naphthalene toxicity would be persistent. Long-term
studies of naphthalene exposure are now in order.

LITERATURE CITED

Anderson, J. W., J. M. Neff, B. A. Cox, H. E. Tatem, and
        G. M. Hightower. 1974. Characteristics of dispersions
        and water-soluble extracts of crude and refined oils
        and their toxicity to estuarine crustaceans and fish.
        Mar. Biol. 27(1): 75-89.
Dunning, A. and C. W. Major. 1974. The effect of cold sea-
        water extracts of oil fractions upon the blue mussel,
        Mytilus edulis. In: Pollution and Physiology of
        Marine Organisms, pp. 349-366, ed. by F. J. Vernberg
        and W. B. Vernberg. New York: Academic Press.
Heitz, J. R., L. Lewis, J. Chambers, and J. D. Yarborough.
        1974. The acute effects of empire mix crude oil on
        enzymes in oysters, shrimp, and mullet. In: Pollution
        and Physiology of Marine Organisms, pp. 311-329, ed. by
        F. J. Vernberg and W. B. Vernberg. New York: Academic
        Press.

Lee, R. F., R. Sauerheber, and A. A. Benson. 1972.
    Petroleum hydrocarbons: uptake and discharge by the
    marine mussel Mytilus edulis. Sci., 177: 344-346.
Moulder, D. S. and A. Varley. 1971. A Bibliography on Marine
    and Estuarine Oil Pollution. Plymouth: The Marine
    Biological Association of the United Kingdom.
Scott, D. 1971. Thesis: The effect of copper (II) on
    survival, respiration, heart rate, and electrocardio-
    gram in the common blue mussel, Mytilus edulis. Univ.
    of Maine.
Struhsaker, J., M. B. Eldridge, and T. Echeverria. 1974.
    Effects of benzene (a water soluble component of crude
    oil) on eggs and larvae of Pacific herring and northern
    anchovy. In: Pollution and Physiology of Marine
    Organisms, pp. 253-284, ed. by F. J. Vernberg and W. B.
    Vernberg. New York: Academic Press.
Vernberg, F. J. and W. B. Vernberg. 1974. Pollution and
    Physiology of Marine Organisms. New York: Academic
    Press.

Controlled Ecosystems: Their Use in the Study of the Effects of Petroleum Hydrocarbons on Plankton

RICHARD F. LEE[1], M. TAKAHASKI[2], JOHN R. BEERS[3],
WILLIAM H. THOMAS[3], DON L. R. SEIBERT[3], P. KOELLER[2],
and D. R. GREEN[2]

[1]Skidaway Institute of Oceanography
Savannah, Georgia   31406

[2]Institute of Oceanography
University of British Columbia
Vancouver, British Columbia, Canada

[3]Institute of Marine Resources
University of California
San Diego, California   92093

The controlled ecosystem pollution experiments (CEPEX) located at Saanich Inlet in western Canada are an effort to determine the effects of pollutants on pelagic marine ecosystems. Natural and pollutant treated populations are maintained in plastic enclosures suspended in the inlet. The facility for CEPEX and the results of earlier replication experiments have been described by Parsons (1974) and Takahashi et al. (1975).

In our experiments the pollutant was fuel oil which was added to a quarter-scale enclosure (ca. 2 m diameter and 15 m deep - 60,000 liters). The effects on species composition and standing stock of the phytoplankton, microzooplankton, and zooplankton were determined during the experiment. In addition, chlorophyll, nutrients, and photosynthetic productivity were measured in the control and fuel oil treated enclosures. Earlier reports dealt with the effects on phytoplankton of adding fuel oil to a controlled ecosystem (Lee and Takahashi, 1976; Parsons, Li, and Waters, 1976). Fuel oil, a refined petroleum product with a water extract high in naphthalenes, was selected because of its known toxicity to various marine organisms (Anderson et al., 1974; Gordon and Prouse, 1973). Crude oils which have been studied to date show effects only at relatively high concentrations (Anderson et al., 1974; Gordon and Prouse, 1973; Hodson et al., 1976; Pulich et al., 1974; and Vaughan, 1973).

MATERIALS AND METHODS

Polyethylene enclosures (ca. 2 m diameter and 15 m deep) were filled with 60,000 liters of water from Saanich Inlet. The petroleum added to one of these was a water extract of fuel oil #2 (American Petroleum Institute standard) as described by Anderson et al. (1974). Carboys containing 17 liters of seawater from the CEPEX site were stirred for 24 hours with 35 ml of fuel oil #2. The phases were allowed to separate for 12 hours; the water extract from each carboy was pumped through a diffusion ring throughout the water column of the enclosure. The date of the addition was June 1, 1975 (day 6). Biological sampling was carried out at 2 or 3 day intervals both before and after petroleum addition. Chloro-phyll, nutrients (nitrate, phosphate and silicate), particle size spectra, and photosynthetic productivity were measured on integrated water samples taken with a peristaltic pump from three depth intervals (0-5 m, 5-10 m, and 10-13 m). Also studied were the species composition and standing stock of the phytoplankton, microzooplankton, and zooplankton. Microzooplankton population structure and abundance were monitored every 2-3 days. Samples were collected within 2-3 hours of mid-day using a diaphragm pumping system (pump: PAR Model 36970-0000, 12 V.D.C., Jabsco, Costa Mesa, Calif.) to obtain material over an integrated profile down the center of each enclosure. Further description of the methods used for the biological work has been described by Takahashi et al. (1975).

In bioassay experiments water collected from control enclosure F on day 18 was poured into 2 one liter glass containers, and fuel oil #2 was added to one container to give a final concentration of 90 ppb. Samples of the water were taken daily from both containers for measurement of the amount of chlorophyll and to enumerate the phytoplankton species. In a second experiment water from F was filtered through a 5 μm membrane filter (Millipore Co.) and addition of fuel oil was added as described above.

Microbial Degradation

Water from various depths in the enclosures was collected with a Niskin sampler. The bacterial degradation of hydrocarbons was measured by adding ^{14}C-labeled hydrocarbons to 100 ml water samples in 250 ml flasks capped with silicone stoppers. After incubation for 6 to 48 hours at the in situ temperature (12°C) in the dark, the respired [14]CO$_2$ was collected and counted in a liquid scintillation counter (Beckman LSC-100). The procedure for collecting [14]CO$_2$ in sodium hydroxide and transferring it to a second flask has been described by Hodson, Azam, and Lee (1976). Controls were water samples containing 2 ml of 2 N H$_2$SO$_4$. All samples were run in triplicate for each concentration and time interval. Turnover times were calculated by the equation: turnover time = t/f where t is the incubation time and f is the fraction of the labeled hydrocarbon degraded to [14]CO$_2$ during the incubation period. The radioactive hydrocarbons used were 2-methylnaphthalene-8-[14]C (7.98 mCi/mmol-California Bionuclear Corp.), 3,4-(benz-3,6-^{14}C) pyrene (21 mCi/mmol-Amersham-Searle), ^{14}C-1-naphthalene (5.1 mCi/mmol-Amersham-Searle), ^{14}C-9-fluorene (2.57 mCi/mmol-California Bionuclear Corp.), [14]C-(methyl)-benzene (17.5 mCi/mmol-Amersham-Searle), [14]C (U)-benzene (75 mCi/mmol-Amersham-Searle), ^{14}C-1-hexadecane (54.4 mCi/mmol-Amersham-Searle), ^{14}C-1-heptadecane (13.5 mCi/mmol-ICN), ^{14}C-1-octadecane (25 mCi/mmol-Amersham-Searle). All compounds, except benzene and toluene, were purified on silicic acid thin layer chromatograms before use.

Chemical Analysis

For total non-volatile hydrocarbons we used infrared procedures similar to those described by Brown et al. (1973) and Vaughan (1973). Analysis was at 2930 cm^{-1} using a Perkin-Elmer infrared spectrophotometer (Model 467). Four liters of seawater were extracted with spectrograde CCl$_4$. The extract was concentrated to 1 ml and then treated with activated Florisil to remove nonhydrocarbon material. The infrared method measured only aliphatic hydrocarbons, but was related back to a fuel oil standard to give total hydrocarbons. A second method of measuring total hydrocarbon was used based on the ultraviolet method of Levy (1972). Two liters of water were extracted with hexane and the absorption of the extract between 210 and 350 nm was measured after chromatography on Florisil. This method measures aromatic hydrocarbons; in the case of fuel oil most of the absorption is due to naphthalenes (between 210 and 240 nm). The area under the peak is integrated to give total hydrocarbon relative to

a fuel oil standard. For gas-liquid chromatography 6 liters
of water were extracted with 300 ml of hexane. Hexane
extracts were concentrated to a small volume on a rotary
evaporator and applied to activated silicic acid thin-layer
plates (Merck). A band corresponding to the area of aromatic
hydrocarbons, including all the naphthalenes, was scrapped
and eluted with benzene. This fraction was run on a Packard
gas chromatograph with an 8' column of 10% SP-2100 on
100/120 Suplecoport (Supelco Co.) and temperature programmed
from 70°C to 220°C. Hydrocarbon standards were used for
reference identifications. Naphthalene, methylnaphthalenes,
and dimethylnaphthalenes were quantified using an internal
standard. Identification of these compounds is tentative
since it is based only on comparisons of retention times
with standards. We selected naphthalenes because they are
the principal water soluble components of fuel oil (Anderson
et al., 1974). The ultraviolet method of quantitating
naphthalenes (Neff and Anderson, 1975) gave good agreement
with the GLC method.

For fluorescence spectroscopy 1.5 liters of seawater
were extracted with methylene chloride, followed by analysis
using the combined high speed liquid chromatography-
fluorescence spectroscopy method of Cretney and Wong (1974).
Excitation was at 308 nm and emission at 383 nm. The
readings were related to a chrysene equivalent.

For analysis of sedimented material approximately
200 grams of wet sediment collected from the bottom of the
enclosure were extracted with chloroform:methanol by the
method of Folch, Lees, and Sloane-Stanley (1957). The
organic phase was dried, redissolved in hexane, and the
hydrocarbon fraction collected from a silicic acid column.
A portion of this fraction was weighed and the rest was
applied to activated silicic acid thin-layer plates (Merck).
After running in benzene a band corresponding to the area of
aromatic hydrocarbons, including all the naphthalenes, was
scrapped and eluted with benzene. This fraction was run
with gas-liquid chromatography under the conditions described
above.

RESULTS

Hydrocarbon Concentrations in the Water

Hydrocarbons in the water extracts of fuel oil #2 have
been discussed by Anderson et al. (1974) and Boehm and Quinn
(1974). Our method of making the stock water extracts was
similar to that of Anderson et al. (1974); analysis of our

fuel oil extracts by gas-liquid chromatography indicated
that the concentrations of the major nonvolatile hydrocarbons, namely naphthalene (880 ppb), methylnaphthalenes
(800 ppb) and dimethylnaphthalenes (220 ppb), were similar
to those reported by Anderson et al. (1974). The major
volatile hydrocarbons were benzene (400 ppb), toluene
(900 ppb), and xylene (1000 ppb).

After pumping fuel oil into the enclosure, water
samples were collected at various time intervals for hydrocarbon analysis. The two enclosures used were labeled G
(petroleum treated) and F (control). The concentration of
non-volatile hydrocarbons in enclosure G after the addition
of a water extract of fuel oil #2 was 50 ppb by the infrared
method. The hydrocarbon concentration by the ultraviolet
method was 60 ppb. Enclosure F had a hydrocarbon concentration of approximately 10 ppb, although this varied from
5 to 20 ppb during the course of the experiment. Thus, the
concentration of petroleum hydrocarbons in enclosure G was
approximately 40 ppb. After 3 days the hydrocarbon
concentration in G had dropped to 30 ppb and was at baseline
levels on day 15.

The concentration of naphthalene, methylnaphthalenes,
and dimethylnaphthalenes in G on day 7 (one day after fuel
oil addition) by both gas-liquid chromatography and ultraviolet spectroscopy was 5 ppb, 5 ppb, and 2 ppb, respectively
(Fig. 1). Assuming uniform distribution, the total amount of
different naphthalenes on day 7 in the water would be
approximately 300 mg of naphthalene, 300 mg of methylnaphthalenes, and 120 mg of dimethylnaphthalenes. After 3
days their respective concentration had dropped to 2 ppb,
2 ppb, and 1 ppb, and by day 16 all naphthalenes were below
detectable levels in G (less than 0.5 ppb). No naphthalenes
were detected in F by gas-liquid chromatography, however,
some ultraviolet absorption was seen in the naphthalene
range (220-230 nm). Chrysene equivalents were 0.17 ppb on
day 7 and 0.12 ppb on day 11 compared with a background of
0.04-0.08 ppb. The chrysene equivalent of the stock fuel
oil water extract was 11 ppb. Because of instrument
difficulties, fluorescence analysis at later times was not
performed. Later, a second experiment was carried out by
adding fuel oil dissolved in ethanol to the enclosure. For
this second experiment, chrysene equivalents for water from
7 m were 1.7 ppb, 1.3 ppb, 1.2 ppb, 0.6 ppb, and 0.5 ppb
after 1, 2, 3, 5, and 7 days respectively. After 3 days
water from 13 m had a slightly higher concentration of
chrysene equivalents than water from 7 m.

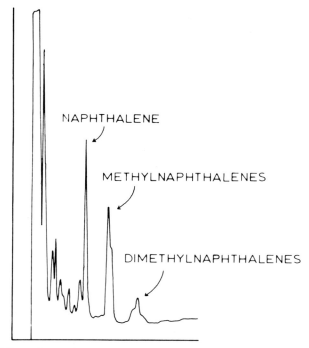

Fig. 1.   Chromatogram of hexane extract of water
from enclosure G on day 7. Concentrations
of naphthalene, methylnaphthalenes, and
dimethylnaphthalenes were 5 ppb, 5 ppb, and
2 ppb respectively.

Hydrocarbons Concentrations in Bottom Sediment

Sedimented material was collected from the bottom of
the enclosure every 4 days. This material was composed of
phytoplankton cells, zooplankton fecal material, and other
detrital particles. On day 10 (4 days after adding fuel oil
to G) the sediment from enclosure G had a total of 20 mg of
hydrocarbon. No naphthalene, methylnaphthalenes, or dimethylnaphthalenes were detected in sediment from enclosure F.
Associated with the sediment of enclosure G on day 10 was
2 mg of naphthalene, 5 mg of methylnaphthalenes, and 9 mg of
dimethylnaphthalenes. Analysis of naphthalenes was the only
method used to distinguish petroleum hydrocarbons from
biogenic hydrocarbons in the sediment of enclosure G. On
day 18 there were 25 mg and 33 mg of hydrocarbons in the

sediment of enclosures F and G respectively. Naphthalenes
were not detected in the sediment of either enclosure on this
date.

Microbial Degradation of Hydrocarbons

To measure the degradation rate of hydrocarbons various
^{14}C-hydrocarbons were added to water samples collected at
various periods from the enclosures. As a result of
petroleum addition to enclosure G there was a large increase
in the degradation rate of octadecane, hexadecane, hepta-
decane, naphthalene, and methylnaphthalene (Table 1). The
hydrocarbon degradation rate of different radiolabeled hydrocarbons before the addition of petroleum showed large
variations, resulting in a high standard error. However,
after the addition of petroleum there were only small
variations in degradation rates between different water
samples from the same depth interval. Both fluorene and
benzopyrene were not degraded in the water from enclosure F
or in the water of enclosure G before addition of oil.
After addition of oil,benzopyrene, but not fluorene,
degradation was measurable. The degradation rate of
naphthalene, a principal component of the water solubles
of fuel oil, increased from 0.1 to 2.5 µg/liter/day.
However, the degradation of methylnaphthalene, also a major
water soluble hydrocarbon, increased its degradation rate
only from 0.1 to 0.3 µg/liter/day. The turnover time and
degradation rates calculated indicate that microbial
degradation was responsible for much of the loss of
paraffinic and low molecular weight aromatics. A naphthalene
degradation rate of 2.5 µg/liter/day would account for the
mineralization in one day of 50% of the 300 mg of naphthalene
in G. On day 7 the water in G contained 300 mg of methylnaphthalenes, of which 18 mg would be mineralized with the
degradation rate of 0.3 µg/liter/day. Four days after
petroleum addition the 5-10 m water had 5 times the naphthalene degradation rate of 0-5 m water. Analyses indicated
that the 5-10 m water had a higher concentration of naphthalenes than 0-5 m water.
Processes such as adsorption to sinking particles or
metabolism by zooplankton are probably more important than
microbial degradation for losses of high molecular weight
aromatics from the water.

Phytoplankton

The standing stock of phytoplankton from day 7 to day
16 was between 280 and 700 µgC/liter in enclosure F while

TABLE 1

Microbial Degradation of ^{14}C-Hydrocarbons in Water Samples from Enclosure G

Hydrocarbon and Concentration	Collection Depth (meters)	Incubation Time (hours)	Time After Oil Addition (Days)	Degradation Rate (μg/liter/day) x 10²	Turnover Time (Days)
Benzpyrene (16ppb)	5-10	48	0	0	-
Benzpyrene (16ppb)	5-10	24	3	1±0.7	1400
Fluorene (30ppb)	5-10	48	0	0	-
Fluorene (30ppb)	5-10	48	3	0	-
Heptadecane (30ppb)	0-5	24	0	7±4	400
Heptadecane (30ppb)	0-5	16	3	50±3	60
Methylnaphthalene (50ppb)	0-5	24	0	10±6	500
Methylnaphthalene (50ppb)	0-5	24	3	26±4	200
Naphthalene (50ppb)	0-5	24	0	10±3	500
Naphthalene (50ppb)	0-5	16	3	250±7	22
Naphthalene (50ppb)	0-5	10	4	100±5	57
Naphthalene (50ppb)	5-10	10	4	500±11	10
Octadecane (30ppb)	0-5	24	0	16±7	200

*±Standard error

petroleum treated enclosure G had between 30 and 150 µgC/ liter during this same time period. The centrate diatom, Ceratualina bergonii, shown in Figure 2, was the dominant phytoplankter in enclosure F during the course of the experiment, accounting for between 50 and 95% of the total phytoplankton carbon (up to 600 µgC/liter on day 16). Between days 10 and 16 a brief microflagellate bloom (Fig. 3) in F accounted for 45% of the total phytoplankton carbon on day 13 (350 µgC/liter). In enclosure G, Ceratualina, although the dominant form at the beginning of the experiment, decreased in numbers after addition of petroleum, and the population of this diatom remained low during the remainder of the experiment (less than 10% of the phytoplankton carbon after day 8). After petroleum treatment, the phyto- plankton population in enclosure G was dominated by micro- flagellates, predominantly the Haptophyceae, Chrysochromulina kappa (3 to 7 µm) which accounted for between 60 and 80% of the total phytoplankton carbon after day 9 (up to 150 µgC/liter). The microflagellates in enclosure F were a mixture of several species, including Chrysochromulina kappa. Although C. kappa was the dominant flagellate in G immediately after petroleum addition, by day 19 Ochromonas sp. had replaced Chrysochromulina as the dominant flagellate, indicating a succession of dominant microflagellates. The cell counts on day 4, 10, and 19 are presented in Table 2; they indicate that pennate diatoms and dinoflagellates were not effected by the addition of fuel oil. Cell counts of the pennate diatom, Cylindrotheca closterium, were high, but because the organic carbon per cell was low it was a minor species. Standing stock of C. closterium increased to 10 µgC/liter on day 10 in enclosure G and up to 15 µgC/liter on day 13 in F.

Productivity in enclosure G reached a peak at day 10 (17 µgC/liter/hr) but decreased after that date (Fig. 4). Productivity measurements in F also showed a peak on day 10 (9 µgC/liter/hr), but the subsequent decrease was smaller than observed in enclosure G. After the addition of petroleum, G had a lower phytoplankton standing stock than F, but the primary productivity was much higher in G on day 10. This suggests a very high growth rate by the microflagellates of G.

Phytoplankton data beyond day 19 are not presented because of a storm on day 20 which resulted in the addition of outside water containing large numbers of Chaetoceros sp. to both enclosures.

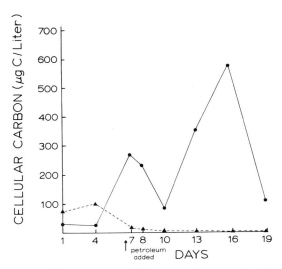

Fig. 2. Changes in the standing stock of the diatom,
Ceratualina bergonii, over 0-10 m.
▲ -G- contains 40 ppb of fuel oil added on
day 6
● -F- control enclosure

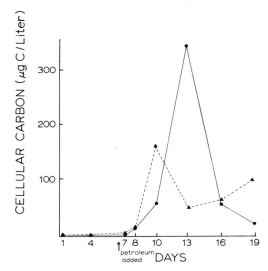

Fig. 3. Changes in the standing stock of microflagellates
over 0-10 m.
▲ -G- contains 40 ppb of fuel oil added on
day 6
● -F- control enclosure

TABLE 2

Phytoplankton in the 0-10m Water in Enclosures F&G

Fuel Oil added on day 6

Number per liter

Phytoplankton	Day 4		Day 10		Day 19	
Diatoms	Enclosure G	Enclosure F	Enclosure G	Enclosure F	Enclosure G	Enclosure F
Chaetoceros chains (5-15µm)	700	0	108,000	32,000	0	6,100
(16-25µm)	200	0	0	5,100	0	9,700
Rhizosolenia stolterforthii	0	0	0	1,800	0	800
Thalassiosira spp	0	0	0	2,500	800	2,300
Cerataulina bergonii	177,000	42,000	20,000	530,000	800	44,000
Cylindrotheca closterium	2,200	9,900	1,100,000	300,000	350,000	45,000
Nitzschia delicatissima	600	430	94,000	9,000	20,000	0
unidentified pennate (5-20µm)	0	150	0	1,200	4,600	9,000
diatoms (21-40µm)	0	0	0	7,500	5,500	20,000
(41-60µm)	0	0	1,900	3,100	1,800	0
(61-80µm)	0	0	0	2,500	1,800	0
Flagellates						
Amphidinium	200	200	4,900	1,200	4,600	4,600
Gymnodinium spp	2,600	500	17,000	15,000	18,000	2,200
Peridinium spp	200	0	1,000	3,100	2,500	900
unidentified dinoflagellates	0	0	0	11,000	7,000	1,400
Chrysochromulina kappa	20,000	18,000	6,400,000	740,000	1,000,000	600,000
Chrysochromulina ericina	8,700	4,000	2,600,000	290,000	270,000	40,000
Dicrateria inornata	33,000	36,000	5,000,000	1,700,000	5,700,000	840,000
Ochromonas sp.1	15,000	8,000	2,200,000	160,000	14,000,000	2,000,000
Ochromonas sp.2	100,000	84,000	4,000,000	800,000	4,600,000	680,000
unidentified flagellates (0-5µ)	19,000	16,000	88,000	200,000	540,000	80,000
(5-10µ)	13,000	10,000	1,100,000	120,000	810,000	120,000

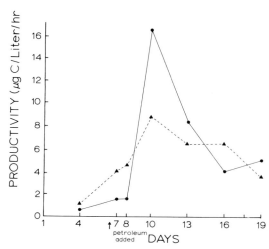

Fig. 4.   Changes in primary productivity over 0-10 m.
●-G- contains 40 ppb of fuel oil added on
day 6
▲ -F- control enclosure

Bioassay

A stock solution of fuel oil #2 (9 mg/liter of non-
volatile hydrocarbons) was added to water collected from
control enclosure F on day 18 to give a final concentration
of 90 μg/liter. The phytoplankton, predominantly the centrate
diatom, Ceratualina bergonni, and microflagellates grew
logarithmically in the control bottle, but in the treated
bottle a marked decrease in the phytoplankton population
occurred (Fig. 5). However, after day 4 logarithmic growth
began in the treated bottle, and the dominant forms were
microflagellates and pennate diatoms, including Cylindrotheca
closterium, Nitzchia delicatissima, and N. bilobata. After
day 4 growth rates were the same in the control (1.04
doublings/day) and treated (1.13 doubling/day) water.
A second experiment involved a preliminary filtration of
the water through a 5 μm filter before addition of the fuel
oil solution. The filtrate consisted of microflagellates
smaller than 5 μm. The control population grew logarithmically at a growth rate of 0.60 doublings/day. The treated
population decreased to one-third of the initial standing
stock, but then began logarithmic growth after 4 days at

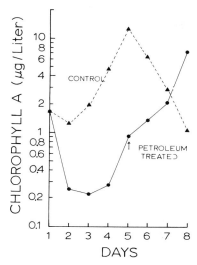

Fig. 5. Bioassay experiment. Effects of fuel oil #2 on
         phytoplankton from control enclosure F.   ● - fuel
         oil added to water (1 liter) from F to attain a
         final concentration of 90 ppb.   ▲ - one liter of
         water from F.

0.90 doublings/day. Thus, microflagellates initially in the
water were inhibited in their growth by fuel oil.

Microzooplankton

    Ciliates of the orders Oligotrichida and Tintinnida
were important protozoans in the microzooplankton populations.
There was an increase in both of these groups in enclosures
F and G although the increase was more pronounced in enclosure
G (Figs. 6 and 7). At its maximum the tintinnid population
in enclosure G (day 13) was strongly dominated by two species,
Helicostomella subulata (Fig. 6) and Eutintinnus pectinis.
The maximum for the microflagellate population in enclosure G
was at day 10 (Fig. 3). Over all sampling dates, sheathed
oligotrichs accounted for averages of 65% (+18%, one standard
deviation) and 57% (+25) of the total oligotrich numbers in
enclosures F and G respectively. The populations of sheathed
oligotrichs comprised five distinct species as well as
unidentified forms which were separated into three size
classes on the basis of length. Strombidium (Laboea)
delicatissima/vestita which had been a numerically important

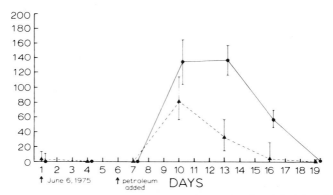

Fig. 6.   Changes in the concentration of the tintinnid,
Helicostomella subulata over 0-13 m.   The vertical
bars indicate the confidence interval (95% level).
●- G - contains 40 ppb of fuel oil added on
       day 6
▲ - F - control enclosure

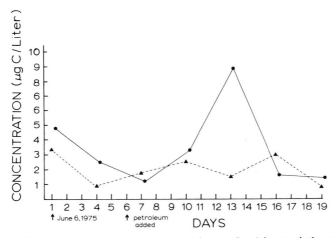

Fig. 7.   Changes in the concentration of oligotrich ciliates
(sheathed and non-sheathed).
● - G - contains 40 ppb of fuel oil added on
       day 6
▲ - F - control enclosure

component of the sheathed oligotrich population in both enclosures at the start of the experiment, but then declined, was virtually the only sheathed form in enclosure G after day 16 and was also a dominant sheathed form in enclosure F.

There was a large increase in both enclosures in the numbers of an unidentified species of rotifer between days 7 and 10. This species declined after day 10 in enclosure F but maintained its peak level for the next 7 days in enclosure G.

Standing stocks of naupliar copepods were quite similar in both F and G during the course of the experiment.

Zooplankton

The dominant zooplankton in both enclosures during the course of the experiment was the copepod, Pseudocalanus minutus, which accounted for between 44 and 91% of the total zooplankton population. Other important zooplankters were also copepods and included, in order of importance, Acartia longiremis, Epilabidocera amphitrites, and Centropages abdominalis. Minor species included larvaceans (Oikopleura sp.), ctenophores (Pleurobrachia pileus), and Medusae (mostly Phialidium sp.). No major differences were noted in the zooplankton standing stock of the two enclosures after the addition of petroleum. Bioassay experiments on Pleurobrachia pileus with fuel oil #2 showed a 24-hr LD_{50} of 590 µg/liter (Lee and Anderson, 1976). The copepod, Calanus plumchrus, also collected from the CEPEX area, had a 48-hr LD_{50} of 1350 µg/liter. Since the highest concentration of petroleum hydrocarbons in enclosure G was 40 µg/liter, a toxic effect on the zooplankton was not expected.

There was some indication that the growth rate of copepods in enclosure G was slower than in control enclosure F. Copepodids I-III comprised 42% of the P. minutus population in F, but only 29% in G on day 15. Also, the rate at which copepodid stages IV-V molted to adult appeared to be slower in enclosure G.

DISCUSSION

Effects

Several changes occurred in enclosure G after the addition of fuel oil on day 6 which were different from those observed in control enclosure F. First was a dramatic decrease in the phytoplankton carbon (30 µg/C/liter in G and compared with 280 µgC/liter in F on day 7) because of

the decline in the standing stock of the dominant phytoplankter, the centrate diatom, Cerataulina bergonii. A
subsequent increase in phytoplankton carbon was due to a
bloom of the microflagellate, Chrysochromulina kappa, with
later succession by a second microflagellate, Ochromonas sp.
The growth of a laboratory culture of C. kappa, isolated
from a fuel oil treated enclosure, has been shown to be
stimulated by fuel oil (Parsons, Li, and Waters, 1976). The
growth of the pennate diatoms, primarily Cylindrotheca
closterium and Nitzchia delicatissima, were not affected by
the fuel oil. Laboratory bioassays adding petroleum to water
from F showed a series of events similar to those earlier
observed in enclosure G, namely a decrease in phytoplankton
growth followed by an increase due to growth of microflagellates and pennate diatoms. Pulich et al. (1974) have
observed that fuel oil #2 at a concentration above 40 ppb
inhibits photosynthesis in the centrate diatom, Thalossiosira
pseudonana. A species of green and blue-green algae is
inhibited only at much higher concentrations of fuel oil.

   Addition of copper to CEPEX enclosures resulted in an
initial phytoplankton decrease, especially in the centrate
diatom population, and a selection for copper resistant algae,
which were certain species of microflagellates and pennate
diatoms (Thomas and Seibert, 1976). Menzel, Anderson, and
Randke (1970) have shown that the photosynthesis of green
flagellates is less inhibited by DDT than offshore diatoms.
Thus, high concentrations of DDT and PCB's in the enclosure
would probably also result in resistant species replacing
sensitive species, as shown with mixed cultures of phytoplankton (Mosser, Fisher, and Wurster, 1972).

   The large increase in the concentration of certain
microzooplankton, specifically ciliates of the order
Oligotrichida and Tintinnida and a species of rotifer, was
correlated with the increase in the population of the microflagellate, C. kappa. Presumably, the large diatom cells
dominant before day 6 do not suffice as food for those microzooplankton. Grazing pressure on C. kappa by microzooplankton
would account for the decrease of microflagellates by day 13.
The microflagellate bloom in control enclosure F on day 13
was not correlated with an increase in the microzooplankton
species as observed in G. Possibly the herbivores feeding on
the microflagellates in F could also feed on the large
centrate diatom, Certaulina bergonii, which remained the
dominant phytoplankter in F.

   Addition of petroleum did not appear to affect the
standing stock of zooplankton, although there was some
indication of a slower growth rate for Pseudocalanus minutus.
C. kappa, the dominant phytoplankter of G, may have been too

small (less than 10 µm) to be efficiently captured by the
filtering mouth parts of P. minutus. A lowered copepod
feeding rate because of petroleum in the water, as observed
by Reeve et al. (1976) with copper, could also result in a
slower growth rate.

We speculate that the only direct effects of petroleum
were on the phytoplankton, with centrate diatoms sensitive
and certain microflagellates and pennate diatoms either
insensitive or even stimulated in growth by petroleum. The
increase in the population of certain microzooplankton
species was probably due to the dominance of the microflagellates. The sequence of events from large diatoms to
small flagellates following nutrient exhaustion and eventual
return to diatoms is a normal cycle in Saanich Inlet and in
the control enclosures (Takahashi et al., 1975). Because of
the short duration of the petroleum experiments, it is
difficult to predict if and when the ecosystem would return
to a centrate diatom dominated system. In the present
experiment a return of these diatoms would be expected since
hydrocarbon concentrations returned to near baseline 9 days
after the addition of petroleum.

Fate

The approximate concentration of nonvolatile petroleum
hydrocarbons in the water of enclosure G on day 7 (one day
after addition of an extract of fuel oil #2) was 40 ppb.
Naphthalene, methylnaphalenes, and dimethylnaphthalenes were
important hydrocarbons on day 7 at concentrations of 5 ppb,
5 ppb, and 2 ppb respectively. The naphthalenes decreased to
near baseline levels by day 13. However, fluorescence
measurements indicated the continued presence of compounds
with a structure similar to chrysene (4-ring aromatic hydrocarbons).

Microbial degradation and adsorption to sinking
particles caused a decrease in the hydrocarbon concentration
of the water. The large increase in the microbial
degradation rate of hydrocarbons after addition of fuel oil
suggests changes in the microbial population, probably by
an increase in the number of microorganisms able to use
hydrocarbons as a carbon source. The small standard error in
the hydrocarbon degradation rate after addition of petroleum
may be the result of one species dominating hydrocarbon
degradation; whereas before addition, a mixture of microbial
groups was probably responsible for the degradation.
Bacteria and fungi were assumed to be the major degraders of
lower weight non-volatile hydrocarbons. The studies
suggested that most of the paraffinic and lower molecular

weight aromatics were broken down by microbes in the water.
The increase in the rate of microbial degradation of
naphthalenes after fuel oil addition resulted in
approximately 50% of the naphthalene in enclosure G being
broken down to CO_2 in one day. Large aromatic hydrocarbons,
such as benzopyrene and fluorene, had very low rates of
microbial degradation. This would explain the persistence
of chrysene-like compounds in the water after disappearance
of naphthalenes. A low rate of microbial degradation of
high molecular weight aromatics has also been noted for
estuarine waters (Lee and Ryan, 1976) and we therefore
hypothesize that microbial breakdown of high molecular
weight aromatics is not an important process in ocean
waters. Microbes in marine sediments may play a role in
breaking down these hydrocarbons and large organisms, such
as zooplankton (Lee, 1975), may play a role in degrading
these compounds in the water column.

ACKNOWLEDGMENTS

Our special thanks to the various CEPEX staff for their
assistance in the collection of data for this paper.
Kathryn D. Hoskins was responsible for much of the microzooplankton analysis. The identification of microflagellates
was by R. Waters at the Institute of Oceanography, University
of British Columbia. We thank W. Cretney and C. S. Wong for
facilities and equipment at the Department of the
Environment (Victoria, British Columbia). We acknowledge
the help of D. Menzel and T. Parsons in the planning of the
experiments. Support was provided by NSF/IDOE, grants GX-
39149 (CEPEX), IDO73-09761-AO1 (J. Beers) and GX-42582 (R.
Lee).

LITERATURE CITED

Anderson, J. W., J. M. Neff, B. A. Cox, H. E. Tatem, and
     G. M. Hightower. 1974. Characteristics of the
     dispersions and water-soluble extracts of crude and
     refined oils and their toxicity to estuarine crusta-
     ceans and fish. Mar. Biol. 27: 75-88.
Boehm, P. D. and J. G. Quinn. 1974. The solubility
     behavior of no. 2 fuel oil in seawater. Mar. Pollut.
     Bull. 5: 101-105.
Brown, R. A., T. D. Searle, J. J. Elliott, B. G. Phillips,
     D. E. Brandon, and P. H. Monaghan. 1973. Distribution
     of heavy hydrocarbons in some Atlantic Ocean waters.

Proc. 1973 Jt. Conf. on Prevention and Control of Oil
Spills, Washington, D. C., American Petroleum
Institute. pp. 505-519.

Cretney, W. P. and C. S. Wong. 1974. Fluorescence
monitoring study at ocean weather station "p".
Proceedings, Marine Pollution Monitoring (Petroleum)
Symposium and Workshop. May 13-17, 1974. Gaithers-
burg, Md.: National Bureau of Standards, pp. 175-177.

Folch, J., M. Lees, and G. H. Sloane-Stanley. 1957. A
simple method for the isolation and purification of
total lipids from animal tissues. J. Biol. Chem. 26:
497-509.

Gordon, D. C. and N. J. Prouse. 1973. The effects of three
oils on marine phytoplankton photosynthesis. Mar.
Biol. 22: 329-333.

Hodson, R. E., F. Azam, and R. F. Lee. 1976. Effects of
four oils on coastal marine microbial populations: a
controlled ecosystem pollution experiment. Bull. Mar.
Sci. (in press).

Lee, R. F. 1975. Fate of petroleum hydrocarbons in marine
zooplankton. Proceedings of the 1975 Conference of
Prevention and Control of Oil Pollution, March 25-27,
1975. American Petroleum Institute, Washington, D. C.,
pp. 549-553.

_____ and J. W. Anderson. 1976. Fate and effect of
naphthalenes in controlled ecosystem enclosures. Bull.
Mar. Sci. (in press).

_____ and C. Ryan. 1976. Biodegradation of petroleum
hydrocarbons by marine microbes. In: Biodeterioration
of Materials, Vol. 3. Applied Science Publishers,
Essex, England (in press).

_____ and M. Takahashi. 1976. The fate and effect of
petroleum in controlled ecosystem enclosure. Rapp. P.
-V. Reun. Cons. Int. Explor. Mer. (in press).

Levy, E. M. 1972. The identification of petroleum products
in the marine environment by adsorption spectrophotometry. Water Res. 6: 57-69.

Menzel, D. W., J. Anderson, and A. Randke. 1970. Marine
phytoplankton vary in their response to chlorinated
hydrocarbons. Sci. 167: 1724-1726.

Mosser, J. L., N. S. Fisher, and C. F. Wurster. 1972.
Polychlorinated biphenyls and DDT alter species
composition in mixed cultures of algae. Sci. 176:
533-535.

Neff, J. M. and J. W. Anderson. 1975. An ultraviolet and
spectrophotometric method for the determination of
naphthalene and alkylnaphthalenes in the tissues of
oil-contaminated marine animals. Bull. Environ.

Contamin. Toxicol. 14: 122-128.

Parsons, T. R. 1974. Controlled ecosystem pollution experiment (CEPEX). Environ. Conser. 1: 224.

_____, W. K. Li, and R. Waters. 1976. Some preliminary observations on the enhancement of phytoplankton growth by low levels of mineral hydrocarbons. Hydrobiologia (in press).

Pulich, W. M., K. Winters, and C. Van Baalen. 1974. The effects of a No. 2 fuel oil and two crude oils on the growth and photosynthesis of microalgae. Mar. Biol. 28: 87-94.

Reeve, M. R., J. C. Gamble, and M. A. Walter. 1976. The behavior of copepods and other zooplankton in copper contaminated enclosed water columns. Bull. Mar. Sci. (in press).

Takahashi, M., W. H. Thomas, D. L. R. Seibert, J. Beers, P. Koeller, and T. R. Parsons. 1975. The replication of biological events in enclosed water columns. Archiv. Hydrobiol. 76: 5-23.

Thomas, W. H. and D. L. R. Seibert. 1976. Effects of copper on enclosed marine phytoplankton communities: Taxonomic composition of the phytoplankton. Bull. Mar. Sci. (in press).

Vaughan, B. E. 1973. Effects of oil and chemically dispersed oil on selected marine biota - a laboratory. American Petroleum Institute Publication No. 4191. Battelle Pacific Northwest Laboratories, Richland, Washington.

Part IV.
Factor Interaction

Synergistic Effects of Exposure to Temperature and Chlorine on Survival of Young-of-the-Year Estuarine Fishes

DONALD E. HOSS, LINDA COSTON CLEMENTS, and DAVID R. COLBY

National Marine Fisheries Service
Atlantic Estuarine Fisheries Center
Beaufort, North Carolina 28516

Increasing demand for electric power has increased the demand for estuarine and marine water to be used in cooling steam powered electric plants. In recent years, as fresh water sources for cooling water have been exhausted, many power plants have been located in coastal areas where there is access to estuarine and ocean waters. One of the main advantages of proposed offshore floating nuclear power plants is the availability of cooling water (U. S. Atomic Energy Commission, 1974). Operation of power plants using marine waters for once-through cooling purposes increases water temperatures and requires that biocides be added to the water to prevent accumulation of marine fouling organisms on condenser tubes.

Chlorine, the most frequently used biocide in once-through cooling systems, is introduced in several forms and at various concentrations. Chlorine may be added to water as a gas (Cl_2), as aqueous solutions of hypochlorites (NaOCl or $Ca(OCl)_2$), or it may be generated on site by direct electrolysis of raw sea water (White, 1972). In general, the amount of fouling increases with increased salinity, and, therefore, biocide levels must be increased as more saline waters are used for cooling. Chlorination may be continuous (Nash,

[1]Research supported jointly by NMFS and ERDA Agreement No. AT(49-7)-5.

1974) or intermittent (Becker and Thatcher, 1973) with the
concentration at the point of injection ranging from 1 to
12 ppm free chlorine residual (USAEC, 1974; Becker and That
Thatcher, 1973).

The effects of heat and chlorine alone on fish have been
well documented, especially in fresh water (see reviews of
Brungs, 1973, Coutant et al., 1975, and Tsai, 1975). Little
information has been published on the combined effects of
heat and chlorine that would be encountered by marine fish
during power plant entrainment. In our previous research we
have developed laboratory techniques for simulating thermal
shock (Hoss, Hettler, and Coston, 1974) and exposure to
chlorine (Hoss et al., 1975). In this paper we simulate in
the laboratory several combinations of thermal shock,
chlorine, and exposure duration that entrained fish would be
exposed to at estuarine and offshore power plant sites. Also,
the interaction of thermal shock and chlorine as it affects
the survival of fish under simulated entrainment is reported.

MATERIALS AND METHODS

The experimental fishes, southern flounder
(Paralichthys lethostigma), striped mullet (Mugil cephalus),
Atlantic silverside (Menidia menidia), and mojarra
(Eucinostomus argenteus) are important components of the
estuarine ecosystem along the southeastern Atlantic coast of
the United States. Flounder, mullet, and mojarra spawn in
the ocean and the larvae are transported to the estuary
where they develop for the first year; they are therefore
subject to man-imposed stresses, such as entrainment in power
plant cooling systems in estuarine and offshore sites. The
silverside spawns in the estuary and the larvae are subject
to entrainment only in estuarine power plant situations.

Experimental fish were collected from wild stock in the
estuary near Beaufort, North Carolina, except for silversides
which were spawned in the laboratory. The fish were acclimated in the laboratory for 7 days at ambient temperatures
($15^\circ C$ for flounder and $25-26^\circ C$ for the other species) before
they were used in experiments. All fish used in these
experiments were fed brine shrimp nauplii (Artemia salina).
Relatively natural photo-periods were maintained throughout
the experiment with artificial light.

In tests using chlorine, water was pumped directly from
the estuary, filtered through crushed oyster-shells, and
passed into a 750 liter recirculating water tank. An

Advance chlorinator[2] was used for mixing chlorine gas and water in the recirculating tank. Chlorine was added to the water until a predetermined level of free available residual chlorine was reached. Free available chlorine levels were measured using a Wallace and Tiernan amperometric titrator and the experimental doses were based on proposed levels of chlorination for offshore power plants (USAEC, 1974). The pH of our experimental water was 7.5 to 7.8.

We used experimental concentrations of free available chlorine in these experiments based on information that proposed offshore power plants will probably chlorinate intake cooling water to an inlet concentration of about 1 ppm so that the free available chlorine concentration at the point of discharge will be approximately 0.1 ppm or less (USAEC, 1974). The levels of chlorine used in our experiments fell within the range of 0.1 to 1 ppm.

Our experimental design was defined in terms of the initial concentration of free available chlorine in the water at the beginning of the exposure period. We know, however, that the level of free available chlorine in the water decreased during the 10 min. maximum exposure period. Chlorine decay experiments indicate that a 30 to 35% reduction in free available chlorine may occur in 10 min. Similar reductions in free available chlorine would probably occur during actual entrainment.

We based our analysis on percent survival after 24 hours following timed exposures to various temperature, chlorine, or temperature-chlorine combinations, although we followed percent survival for seven days. In nearly all cases one day and seven day survival was similar, indicating that mortality due to the treatment occurred within 24 hours.

We have described our procedure for temperature shock previously (Hoss et al., 1974). In these experiments fish are exposed to nearly instantaneous changes in temperature (5.6 to 18°C, Auerbach et al., 1971) and/or chlorine over ambient temperature which simulate changes that occur during entrainment. The individual effects of chlorine alone on silverside, mojarra, and mullet were tested at three chlorine concentrations (0, 0.3, and 0.5 ppm) and at five exposure times (1, 3, 5, 7, and 10 min.). The combined effects of chlorine and temperature on flounder acclimated to 15°C and mullet acclimated to 26°C were tested at the same chlorine concentrations, exposure times, and at two temperatures (ambient and a 15°C shock for flounder and ambient and a 10°C shock for mullet). Our choice of shock temperatures

---

[2]Use of trade names does not imply endorsement by the National Marine Fisheries Service.

was based on previous experiments which showed that flounder acclimated to $15^{o}C$ can withstand a $15^{o}C$ thermal shock for up to 40 min. with no mortality (Hoss et al., 1974) and that mullet acclimated to $26^{o}C$ can withstand a $10^{o}C$ thermal shock for up to 10 min. with no appreciable mortality.

For those experiments in which both chlorine and thermal shock were experimental factors, multiple regression analysis was used to derive predictive models for percent survival. The containing model was:

$$Y = b_o + b_1(T) + b_{11}(T^2) + b_2(C) + b_{22}(C^2) + b_{12}(TXC)$$

Where: Y = percent survival
b_i and b_{ii} = constants
T = linear effect of exposure duration
T^2 = quadratic effect of exposure duration
C = linear effect of chlorine
C^2 = quadratic effect of chlorine
TXC = interaction between exposure duration and chlorine

Separate regression models were fitted to data for fish subjected to thermal shock and chlorine and to data for fish subjected only to chlorine. A reduced form of the containing model was derived for each data set by stepwise application of partial F-tests (Draper and Smith, 1966). The results of the analyses are presented as response surfaces for the reduced models.

RESULTS

Chlorine Alone

The effect of chlorine alone on percent survival of mullet, mojarra, and silverside 24 hours following chlorine exposure for various time periods is shown in Table 1. Each species responded differently to exposure to chlorine. Silverside appeared to be the most sensitive to chlorine and exhibited a graded response with increased exposure time. Exposure to as little as 0.13 ppm chlorine for 3 min. caused approximately 50% mortality. Mojarra, on the other hand, were able to withstand a concentration of 0.28 ppm for 7 min. with no mortality. A 10 min. exposure abruptly reduced survival to 25%. When the concentration was increased to 0.46 ppm, survival was reduced to 78% after 3 min. exposure and 0% survival after 5 min. exposure.

Survival of mullet was not appreciably reduced at any exposure time at either 0.28 or 0.50 ppm chlorine.

Table 1
Percent survival of young-of-the-year fish 24 hours following exposure to chlorine for various time periods. Numbers in parentheses are initial number of fish. Temperature, 25-26°C, salinity, 33-35 0/00.

Species ($\overline{X}$ Wet Wt. mg)	Chlorine ($\overline{X}$ ppm)	Exposure Time (min.)				
		1	3	5	7	10
64-day-old Silverside (63)	0.13	96(23)	54(23)	27(22)	20(22)	0(10)
Mojarra (60)	0.28	--	100(11)	100(11)	100(9)	25(12)
	0.46	100(12)	78(9)	0(10)	0(10)	0(10)
Mullet (343)	0.28	100(9)	100(10)	100(9)	90(10)	100(10)
	0.50	100(11)	100(10)	100(10)	100(10)	90(10)

Interaction of Chlorine and Temperature

The combined effects of chlorine and exposure time in the presence or absence of thermal shock on 24 hour percent survival of postlarval flounder and juvenile mullet were evaluated. For these species, response surfaces were fitted to observed survival under various combinations of chlorine, exposure time, and presence or absence of shock temperature. The results for flounder exposed to chlorine with no thermal shock are given in Figure 1A for chlorine alone and Figure 1B for chlorine plus a 15°C thermal shock. A 15°C thermal shock, coinciding with exposure to chlorine, reduced the tolerance of flounder to chlorine. A comparison of Figures 1A and 1B shows that the tolerance zone is greatly reduced when fish are exposed to both temperature and chlorine. More importantly, however, comparison of the isopleths of the two response surfaces reveals that the interaction between chlorine and thermal shock is

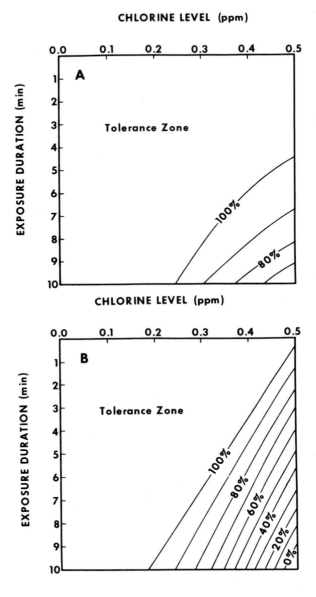

Fig. 1. Estimation of percentage survival of post-
         larval flounder based on response surfaces
         fitted to observed survival under combinations
         of chlorine and exposure time -- A without and
         B with a 15°C thermal shock.

synergistic. That is, the effect on survival of a simultaneous 15°C thermal shock and increased chlorine is more than additive. If the effects of thermal shock and chlorine were only additive, percent mortality due to chlorine in combination with thermal shock would be equal to the sum of mortality due to chlorine alone plus mortality due to thermal shock alone. Since thermal shock alone caused no mortality, it is obvious that the effect of thermal shock plus chlorine is more than additive and this is indicated by the isopleths in Figures 1A and 1B. In a strictly additive situation corresponding isopleths in Figures 1A and 1B would be parallel on the two surfaces, but the isopleths for the combined thermal shock and chlorine group are displaced to the left. We see that the effect of a given increment in the concentration of chlorine (at some fixed exposure duration) is much greater if the fish are simultaneously exposed to thermal shock. For example in Figure 1A, increasing the chlorine level alone from 0.4 ppm to 0.5 ppm for a 5 min. duration reduced percent survival from 100% to approximately 97%, a 3% reduction. On the other hand, in Fig. 1B, a like increase in the chlorine level for the same duration in the presence of a 15°C thermal shock reduced percent survival from 85% to 50%, a 35% reduction.

Mullet responded differently than flounder to the separate and combined effects of thermal shock and chlorine. When mullet were subjected to 0.5 ppm chlorine alone, there was no appreciable mortality (2 of 100 fish died). However, when fish were subjected to both temperature shock and chlorine simultaneously, drastic effects on survival were observed. As shown in Figure 2, for the higher chlorine and longer exposure times all the fish died. Again, synergism is indicated since the combined effect is much greater than the sum of the individual effects of chlorine and temperature alone.

DISCUSSION

Our research to determine the effects of entrainment on fish indicates that the magnitude of their responses to stress is dependent on the combination of several factors. Our previous reports (Hoss et al., 1974; Hoss, Coston, and Hettler, 1971) on the effects of temperature alone on fish show that response to thermal shock depends upon acclimation temperature and exposure time. This report on the combined effects of chlorine and temperature indicates that thermal shock increases the toxicity of chlorine to the species of fish that we have tested.

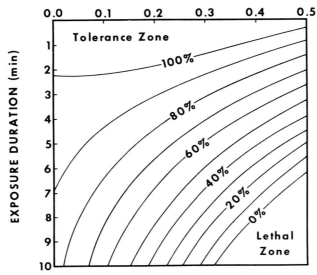

CHLORINE LEVEL (ppm)

Fig. 2. Estimation of percentage survival of
juvenile mullet based on response
surfaces fitted to observed survival
under combinations of chlorine and
exposure time with a 10°C thermal
shock.

Our previous data, on the effects of temperature alone,
indicate that the two shock temperatures alone would not
cause appreciable mortality for up to 10 min. exposure. Our
present data on the effects of chlorine alone indicate that
small changes in chlorine levels and exposure duration may
cause large changes in percent survival of silverside and
mojarra. Stober and Hanson (1974) have demonstrated similar
results working with pink and chinook salmon. We also found
what appears to be a difference in sensitivity between
species to chlorine alone. This difference, however, may be
in part due to differences in developmental stages. While
all of the fish we used were in their first year, they were
not all at the same stage of development. The only fish of a
known age were silversides, which were hatched in the
laboratory. Mullet, the most resistant species, were
approximately 5 times larger by weight than the other two
species (Table 1). Thus, on a per weight basis they would be
receiving a smaller chlorine dose.

The combination of thermal shock and chlorine had synergistic effects on the survival of flounder and mullet as shown in Figures 1 and 2. This reduced survival due to a thermal shock-chlorine interaction agrees with the results obtained for other species by other investigators (Wolf, Schneider, and Thatcher, in press; Meldrim, Gift, and Petrosky, 1974) and with our previous work on mullet, flounder, and menhaden (Hoss et al., 1975) although actual percent survival for both flounder and mullet was greater than that obtained in our previous experiments. This difference in absolute values could be due to differences in environmental conditions between years which may have affected the physiological condition of the fish or to differences in genetic stock. This points to a need for fish larvae of a known age and condition for use in laboratory experiments which attempt to measure complex interactions between various stress factors.

An important ramification of our finding of synergism is that mortality of estuarine larval fishes cannot be satisfactorily predicted from estimates of the separate effects of chlorine and thermal shock. Therefore, bioassays must simultaneously explore combinations of both factors.

Predictive equations of the type generated in this paper, although limited to a few levels of chlorine and temperature, are one part of the information needed to evaluate the impact of entrainment in once-through cooling systems on larval fish populations. They deal, however, only with short-term and direct consequences of exposing fish to entrainment. Equal attention must be devoted to consideration and elucidation of the long-term and indirect effects. In other words, we must ultimately relate larval fish mortality from entrainment to a broad ecological context. Therefore, we must as Marcy (1975) points out, determine natural survival rates and the carrying capacity of the system if we are to predict the cumulative effects of a number of power plants on the environment.

LITERATURE CITED

Auerbach, S. I., D. J. Nelson, S. V. Kaye, D. E. Reichler, and C. C. Coutant. 1971. Ecological consideration in reactor power plant siting. In: Environmental Aspects of Nuclear Power Stations, pp. 803-820, Vienna: IAEA.
Becker, C. D. and T. O. Thatcher. 1973. Toxicity of power plant chemicals to aquatic life. USAEC WASH-1249.
Brungs, W. A. 1973. Effects of residual chlorine on aquatic life. J. Wat. Pollut. Control Fed. 45: 2180-2193.

Coutant, C. C., S. S. Talmage, R. F. Carrier, and B. N.
     Collier. 1975. Thermal effects on aquatic organisms -
     annotated bibliography of the 1974 literature. ORNL-
     EIS-75-28.
Draper, N. R. and H. Smith. 1966. Applied Regression
     Analysis. New York: John Wiley and Sons, Inc.
Hoss, D. E., L. C. Coston, J. P. Baptist, and D. W. Engel.
     1975. Effects of temperature, copper and chlorine on
     fish during simulated entrainment in power-plant
     condenser cooling systems. In: Environmental Effects
     of Cooling Systems at Nuclear Power Plants, pp. 519-527.
     Vienna: IAEA.
Hoss, D. E., L. C. Coston, and W. F. Hettler, Jr. 1971.
     Effects of increased temperature on postlarval and
     juvenile estuarine fish. Proc. Conf. SEast. Ass. Game
     Fish Comm. 25: 635-642.
Hoss, D. E., W. F. Hettler, Jr., and L. C. Coston. 1974.
     Effects of thermal shock on larval fish - Ecological
     implications with respect to entrainment in power plant
     cooling systems. In: The Early Life History of Fish,
     pp. 357-371, ed. by J. H. S. Blaxter. Berlin: Springer-
     Verlag.
Marcy, B. C., Jr. 1975. Entrainment of organisms at power
     plants, with emphasis on fishes - an overview. In:
     Fisheries and Energy Production: A Symposium, pp. 87-106,
     ed. by Saul B. Saila. Lexington: D. C. Heath and Co.
Meldrim, J. W., J. J. Gift, and B. R. Petrosky. 1974. The
     effect of temperature and chemical pollutants on the
     behavior of several estuarine organisms. Ichthyological
     Associates, Inc., Bulletin No. 11.
Nash, C. E. 1974. Residual chlorine retention and power
     plant fish farms. Prog. Fish Cult. 36: 92-95.
Stober, Q. J. and C. H. Hanson. 1974. Toxicity of chlorine
     and heat to pink (Oncorhynchus gorbrischa) and chinook
     salmon (O. tshawytscha). Trans. Am. Fish. Soc. 103:
     569-576.
Tsai, Chu-fa. 1975. Effects of sewage treatment plant
     effluents on fish: a review of the literature.
     Chesapeake Research Consortium, Inc. CRC Publication
     No. 36.
U. S. Atomic Energy Commission, Directorate of Licensing.
     1974. A survey of unique technical features of the
     floating nuclear power plant concept.
White, G. C. 1972. Handbook of Chlorination. New York:
     Van Nostrand Reinhold Company.
Wolf, E. G., M. J. Schneider, and T. O. Thatcher. Bioassays
     on the combined effects of chlorine, heavy metals, and
     temperature on fish and fish good organisms. In:

Second Thermal Ecology Symposium.   Augusta, Ga.,
April 2-5, 1975.   In press.

The Effects of DDT and Mirex Alone and in Combination on the Reproduction of a Salt Marsh Cyprinodont Fish, *Adinia xenica*

CHRISTOPHER C. KOENIG

Department of Biology
College of Charleston
Charleston, South Carolina 29401

Although a great deal of attention has been directed toward responses of adult and juvenile saltwater fishes to various pesticides, relatively few toxicity studies have dealt with the sublethal effects of pesticides on their reproduction. A population may be adversely affected through a chronic impairment of its reproductive potential even though no direct reduction of the adult spawning population occurs (Macek, 1968). Certain non-polar contaminants which have low water and high lipid solubility, such as many organochlorine pesticides (including DDT, its metabolites DDD and DDE, and mirex) may accumulate in fish eggs, apparently in association with lipids in the yolk. Upon absorption of the yolk the embryo or larva is exposed to the pesticides and, depending upon the concentration, may display symptoms of acute toxicity. On the other hand, parental adults with similar concentrations may appear entirely normal (Holden, 1973).

Numerous recent pesticide monitoring studies confirm widespread DDT contamination of fishes as well as other aquatic organisms (Frank et al., 1974; Linko et al., 1974; MacGregor, 1974; Markin et al., 1974). Significantly, DDT residues are rarely present alone. Other persistent organochlorine chemicals such as polychlorinated biphenyls (PCB's), dieldrin, aldrin, endrin, lindane, heptachlor, chlordane, toxaphene, and mirex also contaminate fishes (Henderson et al., 1969; Henderson et al., 1971; Markin et al., 1974). How

these chemicals interact is virtually unknown. There are
very few studies of pesticide interactions in non-target
aquatic organisms. Existing studies serve to illustrate the
largely unpredictable nature of pesticide interactions.
Reviews of pesticide interactions mainly involve insect and
mammalian studies (Frawley, 1965; Durham, 1967; O'Brien, 1967;
Murphy, 1969; Casida, 1970; Kay, 1973). However, a recent
review (Livingston et al., 1974b) attempts to relate these
studies to fishes and other non-target organisms.

Recently mirex has been detected in fish samples in
some estuaries in the southeastern United States (Naqvi and de
la Cruz, 1973; Borthwick et al., 1974; Markin et al., 1974)
where applications for the control of fire ants (Solenopsis
saevissima) have been made. It has also been detected in fish
samples in Lake Ontario, far from any known source of
contamination (Kaiser, 1974), suggesting atmospheric transport
such as is known for DDT residues (Woodwell et al., 1971;
Cramer, 1973). Although fishes are relatively insensitive to
acute mirex toxicity (Jenkins, 1963; Van Valid et al., 1968;
Lowe et al., 1971) the possibility of effects on reproduction
and of synergism with DDT, another widespread environmental
contaminant, has not been considered.

The purpose of this study is to determine the relative
toxicity of DDT and mirex alone and in combination to the
early life stages of Adinia xenica, a salt marsh cyprinodont
fish. Adinia was chosen for this study mainly on the basis
of its well described life history, ecology, and development
(Hastings and Yerger, 1972; Koenig and Livingston, in press).

MATERIALS AND METHODS

Males and females of Adinia were collected during
natural spawning season by seining or trapping from tidal
marshes in the St. Mark's National Wildlife Refuge, Wakulla
County, Florida. The virtual lack of pesticide contamination
in Adinia from this area was confirmed by gas chromatographic
analysis. Of the pesticides looked for (p,p'-DDT, p,p'-DDD,
p,p'-DDE, dieldrin, and mirex) only DDE was detected
(<0.05 ppm). For comparative purposes two separate
experiments were conducted, one in July and the other in
August. In both, the conditions were virtually the same.
Koenig (1975) and Koenig and Livingston (in press) describe
the maintenance and artificial spawning of adults as well as
embryo and larval culture techniques.

Exposure of Parental Females to DDT and Mirex

Adult female Adinia (30-35 mm SL) were placed in separate 21.1 ℓ aquaria (9-13 fish per tank) in a stratified random arrangement within four hours after collection. Each aquarium was an open system with fresh conditioned seawater supplied at a rate of 250-300 ml/min (seawater conditioning system described by Livingston et al., 1974a). The water was monitored periodically for dissolved oxygen (Winkler method), ammonia (Solorzano, 1969), pH (electronic probe), salinity (refractometer), and temperature. Dissolved oxygen remained above 6.0 ppm, ammonia below 0.1 ppm, pH 8.0 $\pm$ 0.2, and salinity varied from 28-31 ppt. To simulate natural conditions during peak spawning, day length was set at 14 hours and the temperature maintained at 26-27°C. The fish were held at a density less than 15.0 g per aquarium. During the 36-48 hours after collection the fish were given two feedings (ad libitum) with Tetramin commercial fish food and the dying or injured were culled. The fish in each aquarium were then counted and weighed to the nearest 0.5 g in a tared beaker containing 300 ml of seawater.

The pesticides p,p'-DDT and mirex (99.9% purity) were delivered alone and in combination via the diet. Fish were fed Tetramin (spiked or control) at a rate of 3% of their body weight daily for 9 days. The average weight of dead fish was subtracted from the original weight and the daily ration recalculated. Each group of fish was fed slowly so that all of the food was consumed. On the tenth day after commencement of controlled feeding the females were stripped and the eggs were fertilized and incubated.

The diet formulations were made by dissolving the appropriate weight ($\pm$ 0.1 mg) of the pesticides in acetone, then mixing this solution with a quantity of corn oil (5% of final mixture). The acetone was then evaporated from the corn oil at room temperature in a slight draft under a hood. Tetramin was then mixed with the corn oil-pesticide solution with a stainless steel spatula for several minutes and tumbled on a revolving mixer (about 30 rpm) for 5-6 hours. The diet concentrations ($\pm$ 10%) were confirmed with a gas chromatograph. The control diet was prepared identically but without the addition of the pesticides. A check of the control food with a gas chromatograph confirmed no detectable DDT or mirex. The pesticide spiked and control diets were randomly matched (table of random numbers) with the respective aquaria.

Effects of DDT and Mirex on Embryos and Larvae

In all experiments, parental females were stripped the day after the last dose was fed. Males with ripe testes were chosen for fertilization from groups in separate aquaria fed clean Tetramin daily ad libitum. Eggs were designated as fertilized or unfertilized between 30-48 hours after mixing eggs and sperm. Prior to 30 hours (blastula and gastrula stages) truly fertilized eggs are difficult to distinguish from those merely activated. In the neurula stages or during early organogenesis, sorting may be done rapidly. A group of 5-10 fertilized embryos from each of several females was then randomly selected (taken singly from a culture dish without use of a microscope) and stored at 5°C in moist, air-tight vials for pesticide analysis.

The eggs were incubated according to the method of Koenig (1975). Briefly, this method involves raising and lowering cylindrical egg baskets (60 mm x 30 mm dia.) at the air-water interface of a temperature controlled incubation aquarium (21.1 ℓ) by means of a 6 rpm electric motor. Fresh conditioned seawater was supplied to the incubation aquarium at 50-100 ml/min. The incubation water was maintained at 27 $\pm$ 0.2°C and salinity varied from 28-31 ppt. Day length was set at 14 hours and light intensities were kept well below levels determined detrimental to fish embryos (Eisler, 1957). Periodic dissolved oxygen and ammonia determinations confirmed air saturation of the water and no detectable ammonia.

When the first larvae hatched out, the motor controlling movement of the egg baskets was plugged to an hourly timer. The timer was adjusted so that the baskets cycled a single 2 min interval per hour. The intermittent water exchange in the baskets reduced stress on the larvae, for continuous cycling of the baskets forced continuous swimming. Dissolved oxygen levels in the baskets apparently were not affected significantly by this procedure since there were no overt indications of respiratory stress in the larvae. Larval density per basket was kept below 30 individuals.

Development of embryos and larvae was observed daily noting both mortalities and stage of development until 27 days after fertilization. Larvae were fed brine shrimp (Artemia salina) nauplii once daily ad libitum for one hour.

The dose-mortality curves were calculated by "maximum likelihood" estimation of probit analysis (Finney, 1971). All calculations (LD_{50}'s, regression equations, etc.) are based on "total" mortality (i.e., embryo plus larval mortality). The relative potency of DDT alone to DDT-mirex combinations was determined according to the procedure of

Finney (1971). The regression lines of Figure 3 were
assumed parallel since x^2 tests of parallelism and hetero-
geneity appropriate for probit analysis (Finney, 1971) did
not controvert the assumption. Of course, the null hypothesis is that the lines are not parallel. Rejection of the
null hypothesis does not mean the lines are parallel, but
only that the possibility of parallelism exists. The
assumption of parallelism is made so that the relative
toxicity of the two treatment series may be compared over
the range of effective dosages. Adjustments for natural
mortality were made on the probit-regression curves for
embryo and larval mortality using "Abbott's formula"
(Finney, 1971). Standard t-tests or general one-factor
analyses of variance were used to compare sample means.

Gas Chromatographic Analysis

The pesticide extraction process was slightly modified
from that of A. J. Wilson, Jr. (Environmental Protection
Agency, Gulf Breeze, Florida 32561). The details of this
method are given by Koenig (1975).
Mean percentage recoveries for egg samples spiked with
DDD, DDE, DDT, and mirex varied from 86.1 to 89.9% for the
individual pesticides. The data were not corrected for
percentage recovery estimates. Of the total DDT residues in
the eggs, DDD plus DDE comprised less than 7%. DDT burdens
are expressed in terms of p,p'-DDT only.

RESULTS

Accumulation of DDT and Mirex in Eggs

The accumulation of DDT and mirex in the eggs of
female Adenia xenica given nine equal daily doses of the two
pesticides followed closely the non-linear regression equation
log Y = a + b log x (Figs. 1 and 2). The regression
equations were calculated from the combined values of the
July and August experiments (Table 1) since there were no
significant differences (p > 0.1) at specific dosage levels.
Egg burden data resulting from dosages of DDT and mirex fed
in combination were excluded from the calculations. The
correlation coefficients for the mirex (r = 0.983) and DDT
(r = 0.951) regression curves indicate close agreement of the
data with the above equation.
Comparisons of the mean egg-DDT burdens among specific
dosage levels (alone or in combination with mirex) were made
using one factor analysis of variance. At specific dosage

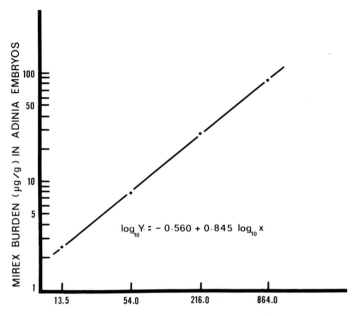

$$\log_{10} Y = -0.560 + 0.845 \log_{10} x$$

MIREX (µg/g) FED TO PARENTAL FEMALES

Fig. 1.   Accumulation of mirex in the eggs of female
          Adinia xenica fed 1.5, 6.0, 24.0, and
          96.0 µg/g/day under controlled conditions for
          9 days. The values along the abscissa
          represent the total dose (i.e., daily dose
          x 9).

levels there were no significant differences (p > .25) in the
DDT burdens of eggs from July and August experiments or from
doses combined with various levels of mirex.

Toxicity to Embryos and Larvae

A synergistic interaction between DDT and mirex on
larval mortality was evident from both July and August
experiments (Table 2). Figure 3 illustrates the influence of
small doses of mirex (about 1-3% of the LD_{50} of mirex alone)
on the toxicity of DDT to embryos and larvae combined. The
embryo and larval LD_{50} of DDT administered alone to parental
females was 40.3 µg/g/9 days (95% c.i. = 44.7 - 36.2) or
about 4.47 µg/g/day. This is equivalent to about 10.8 ppm
DDT in the eggs. The LD_{50} when 1.5 µg/g/day mirex was added

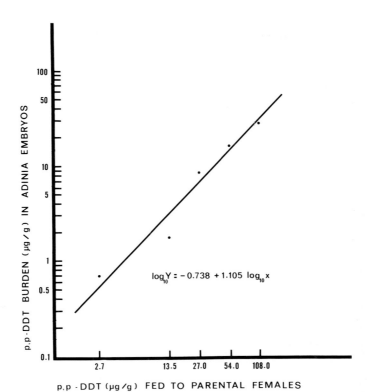

Fig. 2.   Accumulation of p,p'-DDT in the eggs of
female <u>Adinia</u> <u>xenica</u> fed 0.3, 1.5, 3.0,
6.0, and 12.0 µg/g/day under controlled
conditions for 9 days. The values along
the abscissa represent the total dose
(i.e., daily dose x 9).

to the DDT doses was reduced to 26.7 µg/g/9 days (95% c.i. =
29.9 - 23.9) or about 2.97 µg/g/day. This is equivalent to
about 7.56 ppm DDT and about 2.48 ppm mirex in the eggs (egg
burdens calculated from equations given in Figs. 1 and 2).
The relative toxicity of the two series of effective doses
(Fig. 3) is 1.5 with a 95% c.i. of 1.7 - 1.3. That is, the
addition of mirex to the DDT dosages significantly increased
the toxicity (p < .05) by a factor of about 1.5.

Comparison of the July and August data indicate close
agreement in the synergistic phenomenon. As a measure of
comparison, mean probit difference is used. The mean probit

TABLE 1

DDT AND MIREX BURDENS IN THE EGGS OF FEMALE ADINIA XENICA
FED THESE PESTICIDES ALONE AND IN COMBINATION

Pesticide	daily dose (μg/g/day)	total dose (μg/g/9 days)	July experiment DDT ($\bar{x}$ ± SD)	Mirex ($\bar{x}$ ± SD)	N[d]	August experiment DDT ($\bar{x}$ ± SD)	Mirex ($\bar{x}$ ± SD)	N
Control	0.00	0.00	ND[b]	ND	3	0.717 ± 0.44	ND	3
DDT	0.3	2.7	—[c]	—	—	—	—	—
"	1.5	13.5	1.73 ± 0.51	ND	3	9.44 ± 1.5	ND	3
"	3.0	27.0	7.80 ± 1.2	ND	3	15.5 ± 5.3	ND	5
"	6.0	54.0	18.0 ± 2.7	ND	3	28.8 ± 5.2	ND	5
"	12.0	108.0	26.1	ND	1	—	—	—
Mirex	1.5	13.5	ND	2.56 ± 0.09	2	—	—	—
"	6.0	54.0	ND	7.88 ± 1.3	3	—	—	—
"	24.0	216.0	ND	27.9 ± 13.5	3	ND	85.2 ± 14.2	2
"	96.0	864.0	—	—	—	—	—	—
Combinations DDT:Mirex	1.5:1.5	13.5:13.5	2.14 ± 1.9	1.30 ± 1.0	2	—	—	—
"	1.5:6.0	13.5:54.0	1.89 ± 0.41	5.01 ± 1.8	4	—	—	—
"	1.5:24.0	13.5:216.0	2.33 ± 0.60	23.4 ± 7.1	3	—	—	—
"	3.0:1.5	27.0:13.5	7.85 ± 0.73	2.99 ± 0.65	2	8.94 ± 1.3	2.65 ± 0.53	5
"	6.0:1.5	54.0:13.5	18.4 ± 0.97	2.46 ± 0.27	3	—	—	—
"	6.0:6.0	54.0:54.0	14.9 ± 2.6	12.8 ± 1.8	2	—	—	—
"	6.0:24.0	54.0:216.0	—	—	—	—	—	—
"	12.0:1.5	108.0:13.5	—	—	—	28.8 ± 3.9	ND	5

Pesticide concentration in eggs (μg/g)[a]

[a] wet weight
[b] ND = none detected
[c] — = no data
[d] N = number of egg samples from separate fish

TABLE 2

PERCENT MORTALITY OF EMBRYOS AND LARVAE OF FEMALES OF
ADINIA XENICA FED VARIOUS DOSES OF DDT AND MIREX
ALONE AND IN COMBINATION

Percent Mortality[a]

Pesticide	daily dose (μg/g/day)	July Experiment				August Experiment			
		Embryos	Larvae	Total	N[b]	Embryos	Larvae	Total	N
Control	0.00	3.5	1.2	4.7	171	4.5	0	4.5	202
DDT	1.5	5.1	0	5.1	78	—[c]	—	—	—
"	3.0	29.5	0	29.5	95	12.4	1.8	14.2	170
"	6.0	12.7	57.0	69.7	142	5.6	52.5	58.1	177
"	12.0	0	95.5	95.5	67	13.7	84.3	98.0	102
Mirex	1.5	5.4	2.3	7.7	91	—	—	—	—
"	6.0	2.8	4.0	6.8	109	—	—	—	—
"	24.0	6.9	0.9	7.8	102	—	—	—	—
"	96.0	—	—	—	—	2.5	77.2	79.7	202
Combination DDT:Mirex	1.5:1.5	8.8	0	8.8	57	—	—	—	—
"	1.5:6.0	12.8	0	12.8	74	—	—	—	—
"	1.5:24.0	16.0	6.2	22.2	162	10.4	23.3	33.7	193
"	3.0:1.5	4.0	56.5	60.5	124	—	—	—	—
"	6.0:1.5	5.1	83.0	88.1	118	—	—	—	—
"	6.0:6.0	0	87.9	87.9	165	—	—	—	—
"	6.0:24.0	0.7	87.4	88.1	151	—	—	—	—
"	12.0:1.5	3.1	94.6	97.7	130	1.0	93.7	94.7	207

[a] Percent mortality through 27 days after fertilization under controlled conditions.

[b] N = number of fertilized eggs used in the determination of percent mortality of embryos and larvae

[c] — = no data

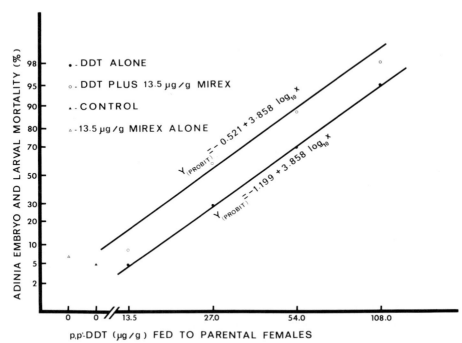

Fig. 3.   Log-probit regression of dosage-mortality data.
Percent mortality is of the offspring (embryos and
larvae) of the dosed females. Mature females were
fed 1.5, 3.0, 6.0, 12.0 µg/g/day p, p'-DDT alone
and in combination with 1.5 µg/g/day mirex for a 9
day period. The values along the abscissa as well
as the mirex dose represent the total dose (i.e.,
daily dose x 9). The mean egg-DDT burdens
corresponding to doses fed to parental females are
3.24, 6.95, 14.9, and 32.0, respectively (calculated
from regression equation, Fig. 2). The mean egg-
mirex burden corresponding to the 1.5 µg/g/day dose
is 2.48 ppm (calculated from regression equation,
Fig. 1).

difference between two parallel probit lines is defined as
the constant vertical difference between the lines (Finney,
1971). For the July experiment the mean probit difference is
0.678 (95% c.i. = 0.923 - 0.433), whereas the probit
difference between DDT dosage 3.0 µg/g/day and combination
dosage 3.0:1.5 µg/g/day from the August experiment is 0.650.

These August data therefore corroborate the finding that mirex synergizes DDT toxicity.

There was no apparent synergistic effect of mirex on DDT toxicity to embryos alone. However, mirex clearly increased DDT toxicity to larvae and this is the basis for the synergistic interaction resulting in an overall reduction in the reproductive potential (Figs. 3 and 4).

Larval mortality, as a result of DDT intoxication, occurred soon after hatching (Fig. 4). As the DDT dose increased, mortalities occurred sooner after hatching. Larval mortality due to mirex intoxication, in contrast, occurred over a longer period of time after hatching. An estimated mirex dose sufficient to produce 50% mortality was 15-20 times greater than a comparable DDT dose (Table 2).

No changes in embryo developmental rate or time to hatching were noted at any of the dosages. Overall, there were no embryo developmental stages more likely to be affected than others (Fig. 4).

Symptoms of DDT intoxication in larvae were completely different from those of mirex intoxication. Larvae affected by DDT were hyperactive and appeared uncoordinated with jerky, spastic movements. Soon after the onset of the symptoms the larvae ceased feeding (or never began) and developed a darkened appearance. The larvae became emaciated and soon died. Symptoms of mirex intoxication, however, were much more subtle. The larvae fed well and appeared normal. Death was preceded by a loss of equilibrium causing the larvae to drift disoriented in midwater, as if narcotized. These symptoms were brought on prematurely if the larvae were forced into physical exertion (such as when the culture basket was lifted from the water for several seconds). When this occurred, however, most of the affected larvae recovered within an hour and again appeared normal. Different mechanisms of toxicity apparently exist for the two pesticides.

DDT and mirex fed alone or in combination to parental females had no direct effect on egg production or percentage fertilization. This was determined by an analysis of variance among all dosage levels.

DDT was toxic to parental females at the highest doses. Mortality of parental females up to 80% occurred in the 6.0 µg/g/day dose. No parental female mortalities occurred in the controls, nor at any other dosage levels of DDT and mirex alone or in combination.

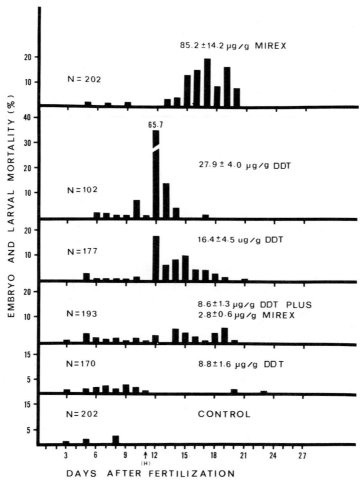

Fig. 4. Distribution of mortalities of embryos and larvae from female _Adinia zenica_ fed DDT alone, DDT:mirex combination, and mirex alone in doses of 3.0, 3.0:1.5, 6.0, 12.0, and 96.0 (mirex) µg/g/day (lower to upper graphs, respectively) for 9 days. Corresponding egg burdens ($\bar{x} \pm$ SD) and the number of fertilized eggs (N) are indicated on graphs. Mortalities occurring after day 11 (H = hatch) are of larvae. All data are from the August experiment.

DISCUSSION

Toxicity to Embryos and Larvae

Various studies on the effects of DDT on fish
reproduction have been reported. Burdick et al. (1964) found
a strong relationship between concentration of DDT in the
eggs of lake trout, Salvelinus namaycush, and mortality of
the fry just after absorption of the yolk sac. No
significant embryo mortalities occurred. Smith (1957) stated
that the bulk of the yolk of trout eggs is composed
essentially of phospholipid-protein complexes and triglyceride droplets. Apparently the phospholipid-protein complex
is utilized by the embryo and the triglycerides are
metabolized by the larvae. Burdick et al. (1964) suggested
that the glycerides probably contain most of the DDT
residues. When these oils are metabolized, the larvae are
exposed to acute concentrations of DDT. This suggestion may
explain the relatively high larval mortality in my study
(Fig. 4).

All groups of eggs in the study of Burdick et al. (1964)
containing 2.95 ppm DDT and above showed some degree of fry
mortality. Fry mortality was negligible in eggs containing
2.67 ppm or less. Therefore, the incipient lethal level of
DDT in the eggs was about 2.95 ppm. Data on Salvelinus
fontinalis (Macek, 1968), Salmo gairdneri (Dacre and Scott,
1971), Salmo salar (Locke and Harvey, 1972), and
Pseudopleuronectes americanus (Smith and Cole, 1973), also
suggest the incipient lethal level pf p,p'-DDT in eggs is
about 1-4 ppm.

The levels of DDT in the eggs of Adinia necessary to
induce mortality are quite similar to those indicated for
other fish species. If one assumes that the LD_1 (dose at
which 1% mortality occurs) is the lowest dose necessary to
induce mortality in Adinia embryos and larvae, the LD_1 under
the experimental conditions would be 10.0 µg/g/9 days with an
egg burden of 2.3 ppm as calculated from the equations of
Figure 2. The egg burden at LD_{10} (10% mortality) would be
4.7 ppm.

No clear dose-dependent effects of DDT on embryo
mortality were observed, although mortalities were generally
higher than controls (Fig. 4). No effect of DDT on fish
embryo mortality is reported in the literature. For example,
Burdick et al. (1964) and Macek (1968) reported no evidence
of increased embryo mortality in salmonid eggs. However,
salmonid embryo and larval mortality due to DDT residues was
reported by Locke and Havey (1972) and Dacre and Scott (1971).
Hopkins et al. (1969) reported increased trout embryo

mortality and no significant larval mortality. Smith and
Cole (1973) showed in winter flounder eggs dose-dependent
embryo mortalities as well as deformities in the larvae due
to DDT concentrations in the eggs. I observed variable
embryo mortalities in the DDT contaminated eggs of Adinia
(Table 2). The mortalities generally occurred in the latter
stages of embryonic development during which time there is a
considerable reduction in the volume of the yolk. It is
possible that a certain amount of the DDT residues are
carried into the developing embryos during this period. The
mechanism may be related to the finding that DDT binds to
lecithin (Tinsley et al., 1971).

The toxicity of mirex to larvae of Adinia showed a somewhat delayed effect occurring over a relatively long period
of time (Fig. 4). This same phenomenon was noted in various
estuarine crustaceans (Lowe et al., 1971), and the freshwater
crayfish, Procambarus blandingi (Ludke et al., 1971).

Although the delayed toxicity of mirex appears to be a
general phenomenon, the symptoms of mirex toxicity to
crustaceans are quite different from the symptoms in larvae
of Adinia. Crustaceans exhibit initial hyperactivity or
irritability, followed by sluggishness, loss of equilibrium
and appetite, and final paralysis (Ludke et al., 1971; Lowe
et al., 1971). In contrast, Adinia larvae feed well and
appear normal. Death is immediately preceded by loss of
equilibrium and paralysis. No hyperactivity, irritability,
or loss of appetite was noted. This suggests that the
mechanism of mirex toxicity is different for crustaceans than
for fish.

The symptoms of DDT and mirex intoxication in larvae of
Adinia indicate different mechanisms of toxicity. The
toxicity of DDT is generally attributed to its effect on the
nervous system where it produces an excitatory effect on
axons (O'Brien and Matsumura, 1964; Holan, 1969; Metcalf,
1973). The general symptoms of DDT poisoning, hyperactivity
and tremoring suggest a primary action on the nerves. Some
recent studies have attempted to explain the action of DDT in
terms of active transport (Janicki and Kinter, 1971; Phillips
and Wells, 1974). The evidence suggests that Na^+, K^+, Mg^{++}-
dependent ATPase, involved in active transport of sodium
across cell membranes, is inhibited by DDT. The primary
action of mirex is also unknown but may be related to the
inhibition of respiratory enzyme systems (Cruz and Naqvi,
1973; McCorkle and Yarbrough, 1974).

Figure 5 illustrates the influence of increasing sublethal doses of mirex on the toxicity of two constant DDT
doses. At the 13.5 µg/g/9 day DDT dose (near threshold of
toxicity), an increase in the mirex dose caused an increase

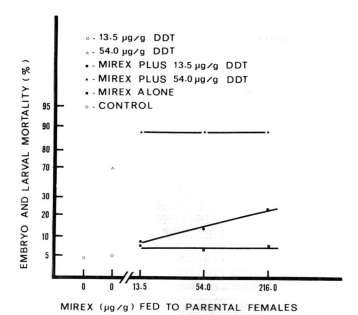

Fig. 5.   Log-probit regression of dosage-mortality
data.

in the toxicity to embryos and larvae. The bottom horizontal
line (mirex alone) confirms mirex as sublethal over the range
of doses. The DDT doses 13.5 and 54.0 µg/g/9 days correspond
to DDT-egg burdens of 3.24 and 14.9 ppm, respectively. The
mirex doses 13.5, 54.0, and 216.0 µg/g/9 days correspond to
mirex-egg burdens of 2.48, 7.98, and 25.7 ppm (DDT and mirex
egg burdens calculated from regression equations, Figs. 4 and
5). At a high lethal level of DDT (54 µg/g/9 day), there was
no increase in toxicity with increase in sublethal mirex
dose. However, the addition of the lowest dose of mirex to
the high lethal DDT dose caused a significant increase in
toxicity over that of the same dose of DDT alone. At the
threshold DDT level the lowest dose of mirex caused only a
small increase in toxicity over that of the same level of DDT
alone. This suggests that the synergistic effect reaches a
plateau (at higher DDT levels) and that increase in the sublethal mirex dose will not cause an increase in the DDT
toxicity. Apparently, the maximum synergistic effect is
reached at doses of mirex lower than the experimental level
when the DDT level is high; and, conversely, higher levels
of mirex are necessary to produce the maximum synergistic

effect at low levels of DDT. Based on these conclusions, the regression curve of the series of combination dosages of Figure 3 is probably not linear at the extreme lower end. Therefore, extrapolation of lower lethal combination dosages would not be valid.

Environmental Implications

At present the use of DDT in the U.S.A. has been curtailed considerably and applications of mirex to coastal areas have been prohibited. Nevertheless, pesticide monitoring studies indicate that both compounds are widespread environmental contaminants. Although widespread contamination of mirex and DDT residues exist, it is difficult to estimate the concentrations in the eggs of contaminated fish. Modin (1969) found concentrations of 0.59 ppm total DDT (0.076 ppm p,p'-DDT) in California halibut eggs (Paralichthys californicus). In a survey of the organo-chlorine contamination of fishes of the lower Colorado River Basin, Johnson and Lew (1970) found concentrations of p,p'-DDT in the ovaries of green sunfish (Lepomis cyanellus) as high as 2.46 ppm (10.46 ppm total DDT). Extrapolation of adult fish pesticide burdens to their eggs is probably not valid; however, Burdick et al. (1964) showed that a rough correlation existed. Assuming that the effect of DDT on embryo and larval mortality is similar for most species of teleost fish, the present level of contamination has probably affected the reproductive potential of fishes in certain coastal and freshwater areas of the United States. The same cannot be said of mirex when considered alone. Mirex concentrations in fishes are generally low, probably due in large part to the low rate of application. The high concentrations in eggs necessary to induce larval mortality probably would occur only in situations of accidental spills where fish could feed directly on the mirex bait for a period of days. Synergistic interaction of DDT and mirex may occur in areas where the two compounds occur jointly, and especially where DDT contamination is high.

ACKNOWLEDGMENTS

This paper is based on portions of a Ph.D. thesis submitted to the Department of Biological Science, Florida State University. I am indebted to Drs. R. J. Livingston, R. C. Harriss, R. W. Yerger, W. F. Herrnkind, and J. E. Byram for their criticism and advice and to T. C. Lewis, C. R. Cripe, and J. S. Koenig for their help and suggestions.

Partial funding was provided by NOAA, Office of Sea Grant, U. S. Department of Commerce under grant number 04-3-158-43.
LITERATURE CITED

Baetcke, K. P., J. D. Cain, and W. E. Poe. 1972. Mirex and DDT residues in wildlife and miscellaneous samples in Mississippi-1970. Pest. Monit. J. 6: 14-22.

Borthwick, P. W., G. H. Cook, and J. M. Patrick, Jr. 1974. Mirex residues in selected estuaries in South Carolina-June 1972. Pest. Monit. J. 7: 144-145.

Burdick, G. E., E. J. Harris, H. J. Dean, T. M. Walker, J. Skea, and D. Colby. 1964. The accumulation of DDT in lake trout and the effect on reproduction. Trans. Amer. Fish. Soc. 93: 127-136.

Casida, J. E. 1970. Mixed-function oxidase involvement in the biochemistry of insecticide synergists. J. Agr. Food Chem. 18: 753-772.

Cramer, J. 1973. Model of the circulation of DDT on earth. Atmospheric Environ. 7: 241-256.

Cruz, de la, A. A. and S. M. Naqvi. 1973. Mirex incorporation in the environment: uptake in aquatic organisms and effects on the rates of photosynthesis and respiration. Arch. Environ. Contam. Toxicol. 1: 255-264.

Dacre, J. C. and D. Scott. 1971. Possible DDT mortality in young rainbow trout. N. Z. J. Mar. Freshwater Res. 5: 58-65.

Durham, W. F. 1967. The interaction of pesticides with other factors. Residue Reviews 18: 21-103.

Eisler, R. 1957. Some effects of artificial light on salmon eggs and larvae. Trans. Am. Fish. Soc. 87: 151-162.

Finney, D. J. 1971. Probit Analysis (3rd ed.). Cambridge Univ. Press, Cambridge.

Frank, R., A. E. Armstrong, R. G. Boelens, H. E. Braun, and C. W. Douglas. 1974. Organochlorine insecticide residues in sediment and fish tissues, Ontario, Canada. Pest. Monit. J. 7: 165-180.

Frawley, J. P. 1965. Synergism and Antagonism. pp. 69-83. In: C. O. Chichester, ed. Research in Pesticides. Academic Press, New York.

Hastings, R. W. and R. W. Yerger. 1971. Ecology and life history of the diamond killifish, Adinia xenica. Amer. Midl. Nat. 86: 276-291.

Henderson, C., A. Inglis, and W. L. Johnson. 1971. Organochlorine insecticide residues in fish-Fall 1968. Pest. Monit. J. 5: 1-11.

_____, W. L. Johnson, and A. Inglis. 1969. Organochlorine insecticide residues in fish. Pest. Monit. J. 3: 145-171.

Holan, G. 1969. New halocyclopropane insecticides and the mode of action of DDT. Nature 221: 1025-1029.

Holden, A. V. 1973. Effects of Pesticides on Fish. pp. 213-253. In: C. A. Edwards, ed. Environmental Pollution by Pesticides. Plenum Press, New York.

Hopkins, C. L., S. R. B. Solly, and A. R. Ritchie. 1969. DDT in trout and its possible effect on reproduction potential. N. Z. J. Mar. Freshwater Res. 3: 220-229.

Janicki, R. H. and W. B. Kinter. 1971. DDT: disrupted osmoregulatory events in the intestine of the eel (Anguilla rostrata). Sci. 173: 1146-1148.

Jenkins, J. H. 1963. A review of five years research on the effects of the fire ant control program on selected wildlife populations. (Abstract) Bull. Georgia Acad. Sci. 21: 3.

Johnson, D. W. and S. Lew. 1970. Chlorinated hydrocarbon pesticides in representative fishes of southern Arizona. Pest. Monit. J. 4: 57-61.

Kaiser, K. L. E. 1974. Mirex: an unrecognized contaminant of fishes from Lake Ontario. Sci. 185: 523-525.

Kay, K. 1973. Toxicology of pesticides: recent advances. Environmental Res. 6: 202-243.

Koenig, C. C. 1975. The effects of DDT and mirex alone and in combination on the reproduction of a salt marsh cyprinodont fish, Adinix xenica. Doctoral Dissertation, Florida State University, Tallahassee, Florida.

_____ and R. J. Livingston. 1976. Embryological development of the diamond killifish, Adinia xenica. Copeia (In press).

Linko, R. R., J. Kaitaranta, P. Rantamaki, and L. Eronen. 1974. Occurrence of DDT and PCB compounds in Baltic herring and pike from the Turku Archipelago. Environ. Pollut. 7: 193-207.

Livingston, R. J., C. R. Cripe, C. C. Koenig, F. G. Lewis, III, and B. D. DeGrove. 1974a. A system for the determination of chronic effects of pollutants on the physiology and behavior of marine organisms. St. Univ. Syst., Florida Sea Grant Program. Rep. 4. 15 pp.

_____, C. C. Koenig, J. L. Lincer, A. Michael, C. McAuliffe, R. J. Nadeau, R. E. sparks, N. Thompson, and B. E. Vaughn. 1974b. Synergism and modifying effects: interacting factors in bioassay and field research. pp. 225-304. In: Marine Bioassay Workshop Proceedings, 1974. Marine Technological Society, Washington, D. C.

Locke, D. O. and K. Harvey. 1972. Effects of DDT upon salmon from Schoodic Lake, Maine. Trans. Amer. Fish. Soc. 101: 638-643.

Lowe, J. I., P. R. Parrish, A. J. Wilson, Jr., P. D. Wilson, and T. W. Duke. 1971. Effects of m rex on selected estuarine organisms. Trans. 36th N. Amer. Wild. and Nat. Res. Conf., March 7-10, 1971. Portland, Oregon, pp. 171-186.

Ludke, J. L., M. T. Finely, and C. Lusk. 1971. Toxicity of mirex to crayfish, Procambarus blandingi. Bull. Environ. Contam. Toxicol. 6: 89-96.

Macek, K. J. 1968. Reproduction in brook trout, Salvelinus fontinals, fed sublethal concentrations of DDT. J. Fish. Res. Bd. Canada 25: 1787-1796.

MacGregor, J. S. 1974. Changes in the amount and proportions of DDT and its metabolites, DDE and DDD, in the marine environment off Southern California. Fish. Bull. 72: 275-294.

Mahood, R. K., M. D. McKenzie, D. P. Middaugh, S. J. Bollar, J. R. Davis, and D. Spitzbergen. 1970. A report on the cooperative blue crab study-South Atlantic states. U. S. Dept. Interior, Washington, D. C. 32 pp.

Markin, G. P., J. C. Hawthorne, H. L. Collins, and J. H. Ford. 1974. Levels of mirex and some other organo-chlorine residues in seafood from Atlantic and Gulf coastal states. Pest. Monit. J. 7: 139-143.

Metcalf, R. L. 1973. A century of DDT. J. Agr. Food Chem. 21: 511-518.

McCorkle, F. M. and J. D. Yarbrough. 1974. The in vitro effects of mirex on succinic dehydrogenase activity in Gambusia affinis and Lepomis cyanellus. Bull. Environ. Contam. Toxicol. 11: 364-370.

Modin, J. C. 1969. Chlorinated hydrocarbon pesticides in California bays and estuaries. Pest. Monit. J. 3: 1-7.

Murphy, S. D. 1969. Mechanisms of pesticide interactions in vertebrates. Residue Reviews 25: 201-221.

Naqvi, S. M. and A. A. de la Cruz. 1973. Mirex incorporation in the environment: residues in nontarget organisms-1972. Pest. Monit. J. 7: 104-111.

O'Brien, R. D. 1967. Insecticides: Action and Metabolism. Academic Press, New York. 332 pp.

_____ and F. Matsumura. 1964. DDT: a new hypothesis of its mode of action. Sci. 146: 657-658.

Phillips, J. B. and M. R. Wells. 1974. Adenosine triphos-phatase activity in liver, intestinal mucosa, cloacal bladder, and kidney tissue of five turtle species following in vitro treatment with DDT. J. Agr. Food Chem. 22: 404-407.

Reinert, R. E. and H. L. Bergman. 1974. Residues of DDT in lake trout (Salvelinus namaycush) and coho salmon (Oncorhynchus kisutch) from the Great Lakes. J. Fish.

Res. Bd. Canada 31: 191-199.
Smith, R. M. and C. F. Cole. 1973. Effects of egg
concentrations of DDT and dieldrin on development in
winter flounder (Pseudopleuronectes americanus). J.
Fish. Res. Bd. Canada 30: 1894-1898.
Smith, S. 1957. Early development and hatching. pp. 323-
359. In: M. E. Brown, ed. The Physiology of Fishes.
Academic Press, New York.
Solorzano, L. 1969. Determination of ammonia in natural
waters by the phenolhypochlorite method. Limnol.
Oceanogr. 14: 799-801.
Tinsley, I. J., R. Haque, and D. Schmedding. 1971. Binding
of DDT to licithin. Sci. 174: 145-147.
Van Valin, C. C., A. K. Andrews, and L. L. Eller. 1968.
Some effects of mirex on two warm-water fishes. Trans.
Amer. Fish. Soc. 97: 185-196.
Woodwell, G. M., P. P. Craig, and H. A. Johnson. 1971. DDT
in the biosphere: where does it go? Sci. 174: 1101-
1107.

Part V.
General

Some Temperature Relationships in the Physiology of Two Ecologically Distinct Bivalve Populations

B. L. BAYNE, J. WIDDOWS, and C. WORRALL

Institute for Marine Environmental Research
67 Citadel Road
Plymouth PL1 3DH
England

In attempts to understand the multiple effects of
temperature on animal populations, we make a distinction
between the limits of the individual's capacity to compensate
physiologically for a change in temperature and the limits to
its thermal tolerance. But these limits are not absolute;
they change with differences in the thermal histories of
individuals, with changes in animal size and condition, and
with geographical and micro-geographical distribution. The
ambient temperature itself may vary hourly, daily, or over
longer periods, so that the time-course, as well as the
amplitude and the rate of change of temperature must also be
considered. These various aspects of the relationships
between temperature and physiological response have been the
subject of a great deal of study (see reviews by Gunter, 1957;
Kinne, 1963, 1970; Naylor, 1965; Newell, 1970). Nevertheless,
when we are required to try to predict the effects of a
change in temperature on the physiology of an individual, and
to define optimal and sub-optimal thermal environments, our
understanding is often found to be inadequate.

In this paper, therefore, we describe some of the
relationships between temperature and the physiology of
Mytilus edulis from two populations in different thermal
environments, in an attempt to define more clearly the kind
of information necessary for the assessment of the effects of
thermal discharges on sessile invertebrates.

MATERIAL

Mussels, <u>Mytilus</u> <u>edulis</u> L., were sampled from two populations near Plymouth on the south coast of England. One of
these (the 'Lynher population') is situated at the junction
of the Lynher and Tamar estuaries. This is a typical
estuarine mud-flat mussel population, with a high density of
individuals and long-term stability of numbers. The monthly
mean water temperature varies from 6° to 18°C, and the
diurnal range experienced by sublittoral individuals is no
greater than ± 1°C of the mean. The second population (the
'power-station population') is situated on wharf pilings
adjacent to the outflow from the cooling-water system of an
electricity-generating station at Cattedown, in the Plym
estuary. These mussels are subjected, on the average, to
temperatures 2°-6°C higher than the mussels in the Lynher.
Furthermore, water temperatures at the power station are
extremely variable over short periods of the day (Fig. 1);
diurnal variations may be as great as 6°C either side of the
mean. This population is considerably more unstable numerically than the Lynher population, and shows signs of being
under a greater degree of general physiological stress
(unpublished data).

METHODS

Rates of oxygen consumption and filtration (= feeding)
were determined as described by Bayne (1971, 1973) and
Widdows and Bayne (1971). Experiments with fluctuating
temperatures were carried out as described by Widdows (1976).
Mussels of mean length 5.5 cm (± S.D. 0.38) were collected in
September when ambient temperatures were 13°-15°C, and
gradually exposed to higher temperatures at the rate of 2°C
per day. Groups of these mussels were then maintained at
21°, 25°, or 29°C. Others were held in a fluctuating
temperature regime in which water temperature was cycled
from 21° to 29°C over six hours, and then returned to 21°C
over the following six hours. All mussels were continuously
fed with the diatom <u>Phaeodactylum</u> <u>tricornutum</u> at a concentration of 12,000 cells per ml. From regular determinations
of the rates of filtration and the rates of oxygen consumption and from measurement of the efficiency with which
ingested food was assimilated the "scope for growth" (Warren
and Davis, 1967) for individual mussels was calculated as
described previously (Widdows and Bayne, 1971; Bayne,
Thompson, and Widdows, 1973). Mussels from the two populations were also collected in August and subjected to sudden

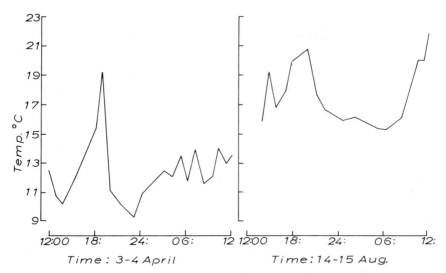

Fig. 1.   Water temperatures recorded continuously over 24
          hours at the 'power-station' population in April
          and in August, 1974.

changes in temperature between $10°$ and $28°C$. After one hour
at each temperature, their rates of filtration were measured.
      In the first experimental series our aim was to deter-
mine whether there were population differences in the
responses of <u>Mytilus</u> to sublethal temperatures. In another
series of experiments we determined the median mortality
time (or median resistance time; Sprague, 1969) for individuals of different size, at different times of the year, from
the two populations. Mussels were acclimated in the laboratory at $10°$ or $20°C$ (salinity 32-34 o/oo) with continuous
feeding, for 14-16 days. The temperature was then raised
rapidly to $27.5°$ or to $30°C$ in small tanks of running sea
water. The animals were fed throughout each experiment with
<u>Phaeodactylum</u>. Frequent inspection of the tanks established
when a mussel was no longer responsive to touching of the
mantle edge or to squeezing the shell valves shut after
gaping; this was recorded as a time of death. The data
were analyzed graphically by plotting cumulative mortality
against the logarithm of time on probability paper (Litchfield, 1949) and by determining the median mortality time as
the time at which 50% of the population died.
      We also record here, for purposes of population com-
parison, selected data from a larger study (Bayne and
Widdows, unpublished) on seasonal changes in the rates of

various physiological processes in mussels from the Lynher and power-station populations. Regular monthly visits were made to each site, during which rates of oxygen consumption by individuals of various sizes were measured under ambient conditions of temperature, salinity, and the concentrations of particulate matter. The data from each visit were analyzed by linear regression (least squares) of $\log_{10}$ oxygen consumption (ml O_2 per hour) against $\log_{10}$ dry body weight (grams). These regressions were compared by covariance analysis, which established that a common slope (= regression coefficient) of 0.7 could be used to describe the relationship between the rate of oxygen consumption and flesh weight. Consequently, in order to reduce the oxygen consumption data to a common value for animal size, they were recalculated to include the 0.7 regression coefficient, i.e., the rate of oxygen consumption as ml O_2 per hour was divided by $W^{0.7}$ ("metabolic body size"), where W is the dry-flesh weight in grams. In this way we excluded much of the variability in the rate measurements that is due to differences in body size.

RESULTS

Lethal Temperatures

1. Effects of season. Figure 2 is a cumulative mortality plot for mussels of similar size from the power station in May and September 1975, when subjected to 30°C after acclimation to 20°C. Thermal tolerance was greater in September (Table 1). In the Lynher populations there was also a marked seasonal change in tolerance; the median mortality time in February was 2.8 times greater than in June (Table 1). These seasonal changes can be related, at least in part, to the reproductive condition of the animals. The "condition index" in Table 1 is the ratio of dry flesh weight to shell volume. When the gonads are fully developed, just prior to spawning, the condition index is high (June in the Lynher) and thermal tolerance is at a minimum. Earlier in the gametogenic cycle (February in the Lynher), the condition index is low and thermal tolerance is high. The mussels in the power-station population do not show the marked seasonal changes in condition index that are typical of the Lynher and many other mussel populations. This is reflected in less seasonal variability in the median mortality times for these animals. In general, however,

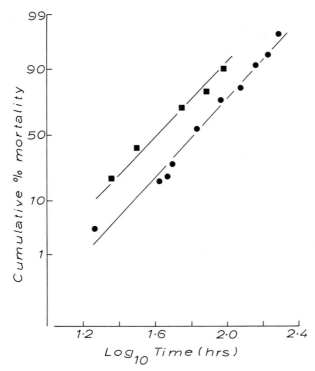

Fig. 2.   Seasonal differences in cumulative percentage mortality in <u>Mytilus</u> <u>edulis</u> from the power-station population plotted on probability paper against the logarithm of time. Mussels between 3.0 and 3.5 cm length were acclimated at 20°C and tested at 30°C; ■ , May; ● , September.

the presence of ripe gametes within the gonad appears to lower the thermal tolerance of the mussels.
2.   Effects of size. In Figure 3 are plotted the results of one experiment in which median mortality time was estimated for animals of different size. In both the power-station and the Lynher populations (Table 2), small individuals (< 2 cm shell length) were always more tolerant of high temperatures than were large mussels (> 5 cm shell length). Mussels in the medium-sized category (3-4 cm shell length) were usually, but not always, more tolerant of high temperatures than the larger mussels.

Table 1
Median Mortality Times (MMT) for *Mytilus edulis* from Two Populations to Show the Effects of
Season. n, Number of Animals Tested; TA, Acclimation Temperature; TE, Experimental Temperature.
The Condition Index is the Ratio of Dry Flesh Weight to Shell Volume.

Population	Mean Size $\pm$ SD (cm)	n	Month	TA ^{o}C	TE ^{o}C	MMT (hrs)	95% Confidence Limit MMT	Condition Index $\pm$ SD
Power Station	3.0 $\pm$ 0.19	10	May	20	30	40.8	26.7 – 62.5	68.8 $\pm$ 14.0
	3.5 $\pm$ 0.29	30	Sept	20	30	66.2	53.6 – 81.7	79.5 $\pm$ 23.2
Lynher	3.3 $\pm$ 0.39	10	Feb	20	30	109.1	59.9 –198.6	37.1 $\pm$ 15.9
	3.6 $\pm$ 0.31	30	June	20	30	39.4	38.5 – 40.3	63.2 $\pm$ 14.1

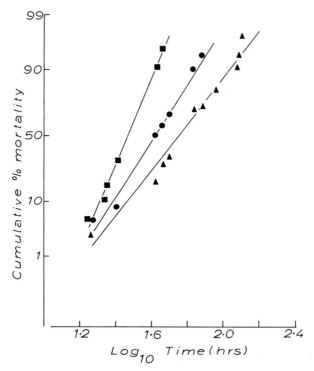

Fig. 3. Size differences in cumulative percentage
mortality in <u>Mytilus</u> <u>edulis</u> from the power
station population plotted on probability
paper against the logarithm of time.
Mussels were acclimated at 10°C and tested
at 30°C. ■ , shell length > 5 cm; ● , 3-
4 cm; ▲ , 1-2 cm.

3.   Effects of acclimation temperature (TA). With one excep-
     tion, mussels acclimated at 20°C had an enhanced thermal
     tolerance over mussels acclimated at 10°C (Table 3).
     For example, animals from the Lynher population had a
     median mortality time at 30°C three times greater after
     acclimation at 20°C than following acclimation at 10°C.
     The one exception to this, which we cannot explain, is
     the Lynher mussels (3.6 cm length) in June, when animals
     at TA20/TE30 showed a median mortality time of 39.4
     hours, compared with 45.5 hours for TA10/TE30.
4.   Effects of metabolic rate. The rate of oxygen consump-
     tion by <u>M</u>. <u>edulis</u> is known to vary seasonally (Bayne,

Table 2
Median Mortality Times (MMT) for Mytilus edulis from Two Populations, to Show the Effects of Animal Size. n, Number of Animals Tested; TA, Acclimation Temperature; TE, Experimental Temperature. The Condition Index is the Ratio of Dry Flesh Weight to Shell Volume.

Population	Mean Size ± SD (cm)	n	Month	TA°C	TE°C	MMT (hrs)	95% Confidence Limit MMT	Condition Index ± SD
Power station	1.8 ± 0.32	29	Sept	10	27.5	>1000	-	41.3 ± 8.0
	3.6 ± 0.25	27	"	10	27.5	344.7	244.5 - 486.0	81.2 ± 31.0
	5.4 ± 0.22	20	"	10	27.5	118.2	87.6 - 159.6	86.0 ± 22.7
	1.8 ± 0.18	30	"	10	30	56.8	47.5 - 67.8	70.3 ± 17.0
	3.6 ± 0.33	27	"	10	30	41.9	36.0 - 48.7	84.6 ± 21.3
	5.4 ± 0.51	20	"	10	30	29.6	26.4 - 33.3	76.7 ± 12.5
Lynher	1.8 ± 0.17	30	June	10	27.5	>500	-	65.0 ± 14.2
	3.6 ± 0.29	30	"	10	27.5	283.6	177.3 - 453.8	69.7 ± 14.1
	5.6 ± 0.46	20	"	10	27.5	42.0	26.8 - 65.9	48.2 ± 15.3
	1.8 ± 0.12	30	June	10	30	59.3	52.1 - 67.4	64.7 ± 21.0
	3.6 ± 0.30	30	"	10	30	45.5	40.8 - 50.8	74.0 ± 19.5
	5.6 ± 0.35	20	"	10	30	35.5	29.8 - 42.2	56.1 ± 14.8

Table 3
Median Mortality Times (MMT) for Mytilus edulis from Two Populations to Show the Effects of Acclimation Temperature. n, Number of Animals Tested; TA, Acclimation Temperature; TE, Experimental Temperature. The Condition Index is the Ratio of Dry Flesh Weight to Shell Volume.

Population	Mean Size ± SD (cm)	n	Month	TA°C	TE°C	MMT (hrs)	95% Confidence Limit MMT	Condition Index ± SD
Power station	5.4 ± 0.22	20	Sept	10	27.5	118.2	87.6 - 159.6	86.0 ± 22.7
	5.3 ± 0.32	15	"	20	27.5	162.7	117.3 - 225.7	72.5 ± 20.4
	5.4 ± 0.51	20	Sept	10	30	29.6	26.4 - 33.3	76.7 ± 22.5
	5.3 ± 0.34	15	"	20	30	42.3	31.5 - 56.6	92.8 ± 22.4
Lynher	5.6 ± 0.46	20	June	10	27.5	42.0	26.8 - 65.9	48.2 ± 15.5
	5.7 ± 0.53	20	"	20	27.5	349.3	228.3 - 534.4	46.5 ± 17.0
	5.6 ± 0.35	20	June	10	30	35.5	29.8 - 42.2	56.1 ± 14.8
	5.7 ± 0.31	20	"	20	30	105.4	87.1 - 127.2	56.4 ± 17.9

1973); hence, we wondered to what extent seasonal changes in thermal tolerance might be related to variation in metabolic rate. The rates of oxygen consumption by mussels in the power station (May and September) and Lynher (February and June) populations, each reduced to unit 'metabolic body size' ($W^{0.7}$), are listed in Table 4. We found no simple relationship between metabolic rate and thermal tolerance.

Table 4
The Rates of Oxygen Consumption (Mean Values Based on 10-12 Determinations) and Median Mortality Times for Two Populations of <u>Mytilus</u> <u>edulis</u> at Different Seasons. Size Range: 3.0 - 3.6 cm. Temperatures Reported Were Ambient at Time of Measurement of Oxygen Consumption. Median Mortality Times Were Determined for Mussels Acclimated to $20^{\circ}C$ and Subjected to $30^{\circ}C$.

Population	Month	Temp ($^{\circ}C$)	(ml O_2 h^{-1} per $W^{0.7}$)	Median Mortality Time (hrs)
Power station	May	15.0	0.812	40.8
	Sept	18.5	0.489	66.2
Lynher	Feb	11.5	0.655	109.1
	June	14.5	0.474	39.4

5. Population differences. Our experiments with thermal tolerance demonstrate that in comparing populations one must consider animals of similar size and reproductive condition. These criteria are met, in general, for mussels from the Lynher in June and from the power station in September. However, the considerable difference in condition index among some of the large-size groups renders comparisons between these large mussels in the two populations impossible in some cases (e.g., large Lynher animals in June, TA20/TE30, mean condition 56.7; large power-station animals in September, TA20/TE30, mean condition 92.8). With these constraints in mind, we have selected from the data the results shown in Table 5

Table 5
Median Mortality Times (MMT) for Mytilus edulis from Two Populations to Illustrate Population Differences. n, Number of Animals Tested, TA, Acclimation Temperature, TE, Experimental Temperature. The Condition Index is the Ratio of Dry Flesh Weight to Shell Volume.

Population	Mean Size ± SD (cm)	n	Month	TA°C	TE°C	MMT (hrs)	95% Confidence Interval MMT	Condition Index ± SD
Lynher	3.6 ± 0.29	30	June	10	27.5	283.6	177.3 – 433.8	69.7 ± 14.2
Power station	3.6 ± 0.25	27	Sept	10	27.5	344.7	244.5 – 486.0	81.2 ± 31.0
Lynher	5.6 ± 0.46	20	June	10	27.5	42.0	26.8 – 65.9	48.2 ± 15.5
Power station	5.4 ± 0.22	20	Sept	10	27.5	118.2	87.6 – 159.6	86.0 ± 22.7
Lynher	1.8 ± 0.12	30	June	10	30	59.3	52.1 – 67.9	64.7 ± 21.0
Power station	1.8 ± 0.18	30	Sept	10	30	56.8	47.5 – 67.8	70.3 ± 17.0
Lynher	3.6 ± 0.30	30	June	10	30	45.5	40.8 – 50.8	74.0 ± 19.5
Power station	3.6 ± 0.33	27	Sept	10	30	41.9	36.0 – 48.7	84.6 ± 21.3
Lynher	3.6 ± 0.31	30	June	20	30	39.4	38.5 – 40.3	63.2 ± 14.1
Power station	3.5 ± 0.29	30	Sept	20	30	66.2	53.6 – 81.7	79.5 ± 23.3

in order to compare thermal tolerance in the two populations. Following acclimation to 10°C mussels from the power station were more tolerant of 27.5°C than mussels from the Lynher. This difference was not apparent in TA10/TE30 conditions though medium-sized animals from the power station at TA20/ TE30 had greater thermal tolerance than mussels from the Lynher.

Acclimation of the Rates of Oxygen Consumption and Filtration to Changes in Temperature

Our previous studies have demonstrated that M. edulis can fully acclimate its oxygen consumption and filtration rates to changes in temperature between 5 and 20°C within 14 days (Widdows and Bayne, 1971; Bayne et al., 1973). More recently Widdows (1976) has shown that mussels can also acclimate fully under conditions of fluctuating temperatures between 6 and 14°C and between 11 and 19°C. Under a constant-temperature regime mussels are unable to acclimate their metabolic and feeding rates to temperatures greater than 20-22°C. Fig. 4a shows the rates of oxygen consumption by mussels that have been maintained at 10, 15, 20, and 25°C for 24 days. Acclimation is complete up to 20°C, but the increased rate between 20 and 25°C signifies the failure to acclimate metabolic rate at the higher temperature. Figure 4b shows the filtration rates of mussels maintained at 10, 15, 20, 25, and 30°C. Here the failure to acclimate is marked by a reduction in filtration rate at temperatures greater than 20°C. The observed increase in routine oxygen consumption rate and decline in filtration rate at 25°C results in a depression of the scope for growth which can eventually force the animal to utilize its body reserves for energy metabolism (Bayne et al., 1973). In these experiments, therefore, a temperature of 25°C imposed a significant sublethal stress on M. edulis.

Response to 'Acute' Change in Temperature

The results of sudden exposures to temperatures between 10 and 28°C (Fig. 5) showed that mussels from the Lynher population were much more susceptible to these acute temperature changes than animals from the power station. Filtration rates by the power-station animals were virtually independent of temperature change over a range from 10 to 25°C. In both populations filtration rates declined between 25 and 28°C, and in animals from the Lynher there was a marked decline also between 20 and 25°C.

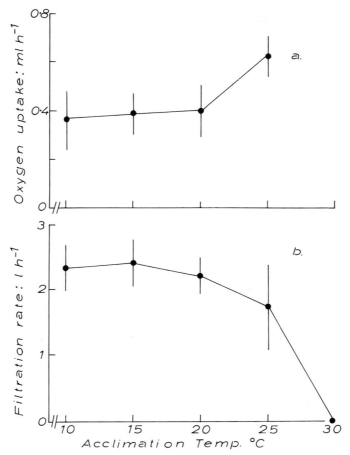

Fig. 4a. The rates of oxygen consumption by
          <u>Mytilus</u> <u>edulis</u> held for 24 days at 10,
          15, 20, and 25°C. All values are mean
          + S.D. for 5-8 determinations. Mussels
          from the Lynher population.

     b.   The rates of filtration by <u>Mytilus</u> <u>edulis</u>
          held for 14-21 days at 10, 15, 20, and
          25°C, then for 2-4 days at 30°C. All
          values are means + S.D. for 5-8
          determinations. Mussels from the Lynher
          population.

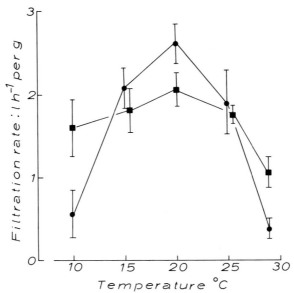

Fig. 5.   The effects of 'acute' temperature
changes on the rates of filtration by
Mytilus edulis from the power station
(■) and from the Lynher (●) populations.
Values are means ± 1 standard deviation.

Response to Fluctuating Temperatures

     Drawing from the two populations under discussion,
Widdows (1976) recently examined the responses of Mytilus
to temperature fluctuations between 21 and 29°C, as
compared with constant temperature conditions of 21, 25, and
29°C. Results for filtration rates of mussels from the
Lynher are shown in Figure 6. When held at constant
temperatures, there was a high mortality of animals at 29°C,
as was expected from our tolerance experiments. Survivors
at this temperature had zero rates of filtration. However,
in mussels that experienced 29°C twice daily within a
fluctuating temperature regime, filtration rate was
suppressed, but significantly above zero at this high
temperature (Fig. 6a). When mussels which had been maintained under the fluctuating temperature conditions for 21
days were placed at a constant 29°C for three days, their
rates of filtration declined (Fig. 6b) and there was an
increasing mortality with time. Within the laboratory,

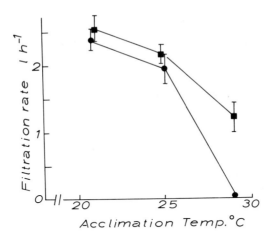

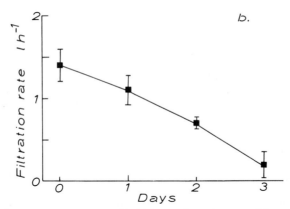

Fig. 6a.　Rates of filtration by <u>Mytilus</u> <u>edulis</u>
(Lynher population) at 21, 25, and 29°C;
■ , mussels maintained at fluctuating
temperatures, 21-29°C;　● , mussels at
constant temperatures.　Values are means
± 1 standard deviation.

b.　Rates of filtration by <u>Mytilus</u> <u>edulis</u>
(Lynher population) which had previously
been maintained at fluctuating
temperatures, and then held at a constant
29°C.　Values are means ± 1 standard
deviation.

therefore, acclimation to fluctuating temperatures enhanced the mussels' capacity to function at high temperature.

The Scope for Growth

We have shown earlier that the scope for growth (i.e., the calories available for growth and for the production of gametes after the demands of maintenance and activity metabolism have been met from the assimilated ration) is reduced in mussels maintained at temperatures greater than 20°C (Bayne et al., 1973). In our earlier experiments with animals from a North Sea population the scope for growth at 25°C was invariably negative. In mussels from the Lynher population, however, at similar levels of ration, the scope for at 25°C is positive, although depressed below the scope at 20°C (Fig. 7), possibly reflecting their more southerly distribution. The scope for growth at 29°C is negative. Mussels held at fluctuating temperatures (21 to 29°C) show a "rotation" (anti-clockwise) in the curve relating scope to acclimation temperature and the degree of rotation is more marked for individuals from the power station than from the Lynher (Fig. 7).

DISCUSSION

Mytilus edulis has been the subject of much research to establish maximal limits of thermal tolerance. Read and Cumming (1967) determined an upper lethal temperature of 27°C for M. edulis by raising the temperature one degree C every 3.5 days. This appeared to provide an ecologically meaningful estimate since Wells and Gray (1960) had stated earlier that the southern distribution limit for this species coincided with a mean summer water temperature of 26.7°C. Wallis (1975) determined an upper incipient lethal temperature of 28.2°C for M. edulis in eastern Australia. Our experiments have illustrated how the thermal tolerance of mussels (estimated as the median mortality time) may vary according to season, size, and previous thermal history. Nevertheless, a lethal temperature in the range 27 to 29°C is an acceptable estimate for the species.

However, there is disagreement in the literature on the influence of size upon thermal tolerance in bivalves. Waugh and Garside (1971) could find no such influence in Modiolus demissus over a range in shell length from 30-70 mm. For Mya arenaria and Macoma balthica, Kennedy and Mihursky (1971) found that young individuals of both species were more

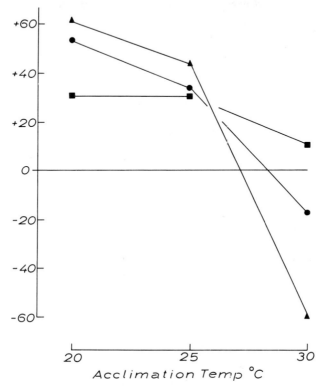

Fig. 7.   The scope for growth (see text) in <u>Mytilus</u>
          <u>edulis</u> at 21, 25, and 29°C; ▲ , mussels
          from the Lynher population at constant
          temperatures; ● , mussels from the Lynher
          population held at fluctuating temperatures,
          21-29°C; ■ , mussels from the power-station
          population held at fluctuating temperatures,
          21-29°C.

tolerant of high temperature than older individuals, and they
related this difference to changes in distribution within the
sediments as the bivalves aged. Wallis (1975), however,
found an increase in temperature tolerance with increase in
size for <u>M. edulis</u> over a size range <3 to >5 cm shell
length. Our experiments indicated the opposite, that smaller
mussels were more tolerant of high temperatures than larger
mussels. We can offer no reasonable explanation for these

differences at the present time; they must be resolved by future research.

Our experiments have also shown a marked seasonal change in thermal tolerance, in both populations, that can be related to the condition index. Individuals with a high condition index and with a large proportion of fully grown gametes within the gonad, are less tolerant of high temperatures than individuals after spawning, or much earlier in the gametogenic cycle. These changes in thermal tolerance do not appear to be related in a simple manner to the rate of oxygen consumption exhibited by the whole animal. They may be a function of increased thermal sensitivity by the gametes.

The mussels from the power station, which experience higher ambient temperatures than do those from the Lynher, generally have a higher thermal tolerance when the comparison is made for individuals of similar size and reproductive condition. In Figure 8 we have brought together some data from the literature on thermal influence on filtration rate for mussels from the Baltic, North Sea, English Channel, and the Mediterranean. In all these areas filtration rates are at a maximum at 15°C, but a decline in filtration rate occurs at higher temperatures in mussels further south. The observation that there is a geographical component to differences in thermal tolerance is not new (Kinne, 1970); the point is made here to emphasize that previous thermal history must be considered when discussing thermal optima.

Another aspect of thermal history that must be borne in mind is the extent of temperature fluctuation within the habitat. Laboratory experiments suggest that the higher thermal tolerance shown by mussels from the power station might be due not only to their having experienced large fluctuations in temperature, but also to the higher mean ambient temperature at this site as compared with the Lynher. Enhancement of thermal tolerance following acclimation to fluctuating temperatures has been demonstrated by Costlow and Bookhout (1971), Otto (1974), and Feldmeth, Stone, and Brown (1974).

Our experiments recorded here and in earlier publications (Widdows and Bayne, 1971; Bayne, Thompson, and Widdows, 1973; Thompson and Bayne, 1974; and Bayne, 1975) suggest that calculation of the scope for growth from data on ingested ration, assimilation efficiency and metabolic rate, provides an efficient means of representing the extent of a stress experienced by a population. A negative scope for growth signifies a marked degree of stress under conditions approaching the lethal limit. This is particularly important when we try to assess the consequences of a thermal discharge on a

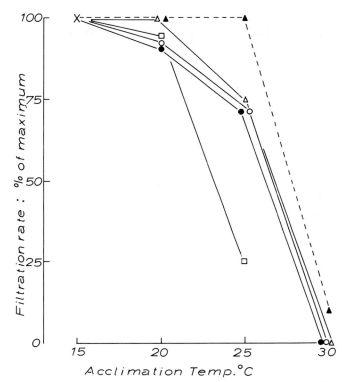

Fig. 8.   Rates of filtration by <u>Mytilus</u> <u>edulis</u> from
          different areas, plotted as % of maximum,
          against acclimation temperature. ■ ,
          mussels from the Baltic (Theede, 1963); ● ,
          North Sea (Widdows, 1973); ● , Lynher
          population; ▲ , Brittany, France (Widdows
          and Bayne, unpublished data); ▲ , Gulf of
          La Spezia, Italy (Schulte, 1975).

sessile invertebrate population. The measurement of thermal
tolerance of itself may be irrelevant in such circumstances,
and often has little biological meaning. Measurement of the
scope for growth, however, produces greater understanding of
the biological processes that are operating and of the
probable longer-term (and often rather subtle) effects with
which we should be concerned, and which are more effectively
measured by the scope for growth than by the thermal limit to
life.

ACKNOWLEDGMENTS

We are grateful for the skilled help of Christine
Scullard. This work forms part of the estuarine and near-
shore program of the Institute for Marine Environmental
Research, a component of the Natural Environment Research
Council, U. K. It was commissioned in part by the
Department of the Environment.

LITERATURE CITED

Bayne, B. L. 1971. Oxygen consumption by three species of
    lamellibranch mollusc in declining ambient oxygen
    tension. Comp. Biochem. Physiol. 40a: 955-970.
    _____. 1973. Physiological changes in Mytilus edulis L.,
    induced by temperature and nutritive stress. J. Mar.
    Biol. Ass. U.K. 53: 39-58.
    _____. 1975. Aspects of physiological condition in
    Mytilus edulis L., with special reference to the
    effects of oxygen tension and salinity. In: Proc. 9th
    Europ. Mar. Biol. Symp., pp. 213-238, ed. by H. Barnes.
    Aberdeen, Aberdeen University Press.
    _____, R. J. Thompson, and J. Widdows. 1973. Some
    effects of temperature and food on the rate of oxygen
    consumption by Mytilus edulis L. In: Effects of
    Temperature on Ectothermic Organisms, pp. 181-193,
    ed. by W. Wieser. Berlin, Springer-Verlag.
Costlow, J. D. and C. G. Bookhout. 1971. The effect of
    cyclic temperature on larval development in the mud-
    crab Rhithropanopeus harrisii. In: Fourth Europ. Mar.
    Biol. Symp., pp. 211-220, ed. by D. J. Crisp.
    Cambridge, Cambridge University Press.
Feldmeth, C. R., E. A. Stone, and J. H. Brown. 1974. An
    increased scope for thermal tolerance upon acclimating
    pupfish (Cyprinodon) to cycling temperatures. J. Comp.
    Physiol.89: 39-44.
Gunter, G. 1957. Temperature. In: Treatise on Marine
    Ecology and Palaeoecology, Vol. 1, Geol. Soc. Amer.,
    Memoir 67, pp. 159-184, ed. by J. W. Hedgpeth.
Kennedy, V. S. and J. A. Mihursky. 1971. Upper temperature
    tolerances of some estuarine bivalves. Chesapeake
    Sci., 12: 193-204.
Kinne, O. 1963. The effect of temperature and salinity on
    marine and brackish water animals. I. Temperature.
    Oceanogr. Mar. Biol. Ann. Rev. 1: 301-340.

_____. 1970. Temperature-Invertebrates. In: Marine Ecology, Vol. 1., Part 1, pp. 407-514, ed. by O. Kinne. London, Wiley-Interscience.

Litchfield, J. T. 1949. A method for rapid graphic solution of time-percent effect curves. J. Pharmac. Exp. Ther. 97: 399-408.

Naylor, E. 1965. Effects of heated effluents upon marine and estuarine organisms. Adv. Mar. Biol. 3: 63-103.

Newell, R. C. 1970. Biology of Intertidal Animals. London, Logos Press, 555 pp.

Otto, R. G. 1974. The effects of acclimation to cyclic thermal regimes on heat tolerance of the Western mosqui-tofish. Trans. Amer. Fish. Soc. 103: 331-335.

Read, K. R. H. and K. B. Cumming. 1967. Thermal tolerance of the bivalve molluscs Modiolus modiolus L., Mytilus edulis L., and Brachidontes demissus Dillwyn. Comp. Biochem. Physiol. 22: 149-155.

Schulte, E. H. 1975. Influence of algal concentration and temperature on the filtration rate of Mytilus edulis. Mar. Biol. 30: 331-341.

Sprague, J. B. 1969. Measurement of pollutant toxicity to fish. I. Bioassay methods for acute toxicity. Water Res. 3: 793-821.

Theede, H. 1963. Experimentelle Untersuchunger über die Filtrations-leistung der Miesmuschel Mytilus edulis. Kieler Meeresforsch. 19: 20-41.

Thompson, R. J. and B. L. Bayne. 1974. Some relationships between growth, metabolism and food in the mussel Mytilus edulis. Mar. Biol. 27: 317-326.

Wallis, R. L. 1975. Thermal tolerance of Mytilus edulis of eastern Australia. Mar. Biol. 30: 183-191.

Warren, C. E. and G. E. Davis. 1967. Laboratory studies on the feeding, bioenergetics and growth of fish. In: The Biological Basis of Freshwater Fish Production, pp. 175-214, ed. by S. D. Gerking. Oxford, Blackwell.

Waugh, D. L. and E. T. Garside. 1971. Upper lethal temperatures in relation to osmotic stress in the ribbed mussel Modiolus demissus. J. Fish. Res. Bd. Canada, 29: 527-532.

Wells, H. W. and I. E. Gray. 1960. The seasonal occurrence of Mytilus edulis on the Carolina coast as a result of transport around Cape Hatteras. Biol. Bull. Mar. Biol. Lab., Woods Hole, 119: 550-559.

Widdows, J. 1973. Effect of temperature and food on the heart beat, ventilation rate and oxygen uptake of Mytilus edulis. Mar. Biol. 20: 269-276.

_____. 1976. Physiological adaptation of Mytilus edulis to cyclic temperatures. J. Comp. Physiol. 105: 115-128.

_____ and B. L. Bayne. 1971. Temperature acclimation of *Mytilus edulis* with reference to its energy budget. *J. Mar. Biol. Ass. U.K.* 51: 823-843.

Variations in the Physiological Responses of Crustacean Larvae to Temperature

A. N. SASTRY[1] and SANDRA L. VARGO[1, 2]

[1]Graduate School of Oceanography
University of Rhode Island
Kingston, Rhode Island   02881

[2]Chesapeake Biological Laboratory
University of Maryland
Solomons Island, Maryland   20688

Many benthic crustaceans from coastal and estuarine waters possess pelagic larvae. These larvae are a dispersal mechanism (Scheltema, 1971), and play an important role in the pelagic food web. A vulnerable link in the life cycle, their survival through complete development is important for the successful recruitment of young to the adult populations (Sandifer, 1975). Generally the larvae of a species are released into the environment when conditions are optimal for their development and growth. During their pelagic existence larvae are exposed to varying combinations of temperature, salinity, light, food, and other factors. Of these factors, temperature, acting singly or in combination with others, often is a major parameter influencing the time of development, survival, and distribution of larvae. The effects of man-made alterations of the environment on the distribution and survival of pelagic larvae are beginning to attract attention (Mileikovsky, 1970).

In recent years, environmental requirements of larvae and their capacities for physiological adaptation to environmental changes have been investigated. An understanding of these larval responses is important to their basic biology and their evaluation as bioassay organisms for assessing the effects of pollutants (Kalber and Costlow, 1966; Vernberg and Costlow, 1966; DeCoursey and Vernberg, 1972; Sastry and McCarthy, 1973; Forward and Costlow, 1974; Vernberg, DeCoursey, and O'Hara, 1974; Bookhout and Costlow, 1975). In this paper some experimental data on survival and physiology of larvae of several species of crustaceans are compared to illustrate the variable responses of larvae to alterations in temperature, singly and in combination with a variety of other factors.

MATERIALS AND METHODS

Larvae of seven species of crustaceans, Cancer borealis, Cancer irroratus, and Homarus americanus from the sublittoral zone, and Pagurus longicarpus, Palaemonetes pugio, Panopeus herbstii, and Rhithropanopeus harrisii from estuarine waters in Narragansett Bay and vicinity have been cultured in the laboratory according to the methods described previously by Sastry (1970; 1976a,b,c). The general methods for incubation of eggs and larval culture are the same except for minor modifications with two species. Eggs removed from ovigerous animals brought to the laboratory were incubated until hatching in 30 o/oo sea water at a temperature suitable for each species (Tables 1 and 2). In the case of H. americanus and P. pugio, the eggs were not removed and the ovigerous animals were maintained at the incubation temperature until hatching. To determine temperature and salinity requirements for complete development and effects on development rate, larvae were then cultured individually (n $\geq$ 54) in compartmented plastic boxes, or, in the case of H. americanus, in crystallizing dishes (n = 24) containing 100 ml of sea water. Larvae of all the species were cultured in standing sea water and exposed to a 14:10 L:D cycle. The sea water was changed daily and the larvae were fed on a diet of freshly hatched Artemia nauplii. Molting and deaths were recorded daily to reveal patterns of survival and the complete development time of each species.

To assess the effects of varying temperatures, C. irroratus larvae hatched from eggs incubated at 15°C were reared individually (n = 108) under 10°-20°, 15°-25°, and 12.5°-17.5°C daily cyclic temperature regimes. Larvae hatched from eggs incubated at the above temperature cycles were also reared individually (n = 108) under their respective cyclic temperature regimes. The larvae were exposed to a square wave daily temperature cycle with a 14:10 L:D cycle, warm temperatures coinciding with the light period and colder temperatures with the dark period.

Table 1
Distribution, ecology and breeding period of seven species of crustaceans used in the study.

Species	Geographical Distribution	Habitat*	Narragansett Bay and Vicinity		
			Location	Breeding Period	Temperature °C and Salinity (o/oo)
Pagurus longicarpus	Nova Scotia to Northern Florida and from Sanibel Island, Florida to Texas	Common on harbor beaches, and in shallow littoral on a variety of bottoms.	Boat basin	Late April to mid-June	7.5-17.0 C 31 o/oo
Palaemonetes pugio	Massachusetts to Texas	Estuarine waters especially in submerged vegetation	Bissell's Cove	Mid-June to early September	20-25 C 10-30 o/oo
Rhithropanopeus harrisii	Canada to Mexico, northeast Brazil, introduced to West Coast of U.S. and in parts of Europe	Estuarine, found in places always providing shelter	Narrow River	Mid-June to late August	22-27 C 13-26 o/oo
Cancer irroratus	Labrador to South Carolina	Low water mark to 600 m, shallow bay in north and deep waters in south	Narragansett Bay	April to early July	6-19 C 31 o/oo
Cancer borealis	Nova Scotia to Tortugas, Florida and Bermuda	Between tides in rocks to 870 m; shallow bays in north and deep waters in south	Narragansett Bay	July	18-23 C 31 o/oo
Homarus americanus	Nova Scotia to New Jersey	Sub-littoral rocky bottom	Narragansett Bay and vicinity	July	18-23 C 31 o/oo
Panopeus herbstii	Massachusetts to Brazil; Bermuda	Estuarine, bottom composed of soft mud and oyster shells	Pawcatuck River	July	18-23 C 31 o/oo

*Source - Williams (1965)

403

Table 2

Egg incubation and larval culture conditions for six species of crustacea.

Species	Egg Incubation	Temperature (°C)	Larval Culture Conditions Salinity (o/oo)	Number of Combinations
Cancer irroratus	15°C, 30 o/oo	10,15,20,25	10,15,20,25,30,35	24
Cancer borealis	15°C, 30 o/oo	10,15,20,25	10,15,20,25,35	24
Homarus americanus	20°C, 30 o/oo	10,15,20,25	15,20,25,30,35	20
Pagurus longicarpus	15°C, 30 o/oo	10,15,20,25	15,20,25,30,35	20
Rhithropanopeus harrisii	20°C, 25 o/oo	20,25,30	10,15,20,25,30,35	18
Palaemonetes pugio	20°C, 30 o/oo	10,15,20,25,30	5,10,15,20,25,30,35,40	40

To determine temperature effects on metabolism, the oxygen consumption rates for the larvae stages of C. borealis, C. irroratus, and P. pugio were measured over a graded series of test temperatures. All larvae used were mass cultured in the temperature-salinity combination giving their highest survival rate. Cancer irroratus larvae were also cultured under a cyclic temperature regime (10°-20°C) and their oxygen consumption rates compared to those of larvae reared under a constant temperature regime (15°C). The oxygen consumption rates were measured with all glass differential microrespirometers (Grunbaum et al., 1955). The oxygen consumption rates were measured for 10-12 individuals of first, second, or third stage zoeae, 4-5 fourth or fifth stage zoeae, and one megalops of C. irroratus and C. borealis introduced into each respiration flask. One to four larvae of P. pugio were introduced into each respiration flask depending on the stage of development. A number of replicate runs (3-12) were made for each life cycle stage of the three species at the respective test temperatures. After measuring the oxygen uptake, the larvae were dried for 48 hours at 60°C and then weighed to the nearest microgram on a Cahn Electro-balance. The mean oxygen consumption rate and the standard deviation of the life cycle stages at each test temperature were computed. The regression analysis of oxygen consumption as a function of weight of each life cycle stage was performed on a computer and tested for significance by F test. The mean weight-specific oxygen consumption rates or those values determined from the regressions when those were significant were plotted against temperature to represent the metabolic temperature response patterns. These values were used to compute the Q_{10} for 5°C test temperature intervals to reveal changes in metabolic response to temperature (Sastry and McCarthy, 1973).

Acute temperature tolerance limits for the larvae of C. irroratus, H. americanus, and P. pugio were determined under saturated oxygen conditions using larvae reared in the temperature-salinity combination yielding their maximum percent survival. All the tests were made in glass 250 ml tissue culture flasks filled with sea water of the same salinity as the mass cultures. The flasks were bubbled with compressed air throughout the test to maintain oxygen at saturation levels. They were held at the test temperatures in water baths set at 2^{O}C intervals over the survival range. Ten animals were used at each test temperature. Cessation of heartbeat was the criterion of death. After exposure the larvae were returned to their culture temperature and observed for 24 hours. No change in mortality was noted.

The LD$_{50}$ values for temperature were determined using the graphical method (Goldstein, 1964).

RESULTS AND DISCUSSION

Temperature and Salinity Requirements

A number of studies have previously determined the effects of temperature and salinity on crustacean larval development (Costlow and Bookhout, 1964 and 1965). In order to establish the culture requirements for complete development with optimal survival, the temperature and salinity requirements have been determined for the larvae of six species discussed in this study (Figs. 1 and 2). Larvae of C. borealis had very specific requirements, completing development only in 30 o/oo at 20°C with a survival rate of 8.9%. Larvae of the congeneric species, C. irroratus, completed development to crab stage in salinities from 25-30 o/oo at 10° and 20°C and 20-35 o/oo at 15°C. Larval survival was maximum in 30 o/oo at 15°C (Fig. 2). The survival rate decreased as conditions deviated from this optimum.

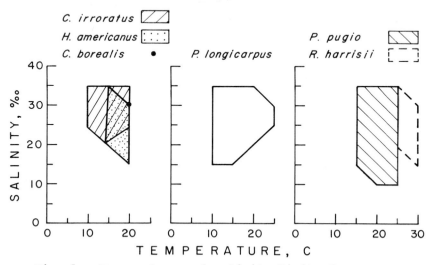

Fig. 1. Temperature and salinity limits for complete larval development of six species of crustaceans from Narragansett Bay and vicinity..

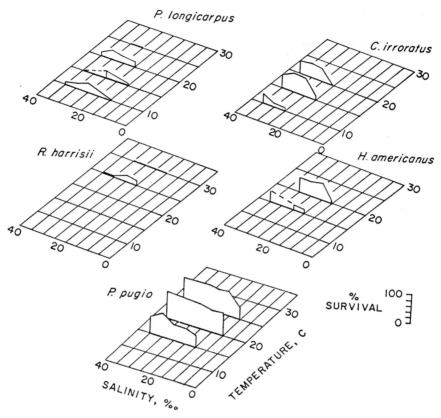

Fig. 2.   Percentage survival of larvae to the
post-larval stages for five species
of crustaceans reared in different
combinations of salinity and
temperature.

Larvae of H. americanus completed development to the
post-larval stage in salinities between 20-35 o/oo at 15°C
and 15-30 o/oo at 20°C (Fig. 1).   Survival of larvae to the
post-larval stage was higher at 35 o/oo at 15°C and in
salinities of 20-30 o/oo at 20°C (Fig. 2).   In this species
temperature appears to restrict development more than
salinity.
Pagurus longicarpus larvae developed over a wider
temperature and salinity range than larvae of any other
species, reaching post-larval stage in salinities between
15-35 o/oo at 10° and 15°C, 20-30 o/oo at 20°C, and 25-
30 o/oo at 25°C (Fig. 1).   This range agrees with limits of
18-30.5 o/oo reported by Roberts (1971).   Larval survival was

highest in 25 o/oo at 10°, 15°, and 20°C (Fig. 2). In this species salinity restricts development more than temperature.

The larvae of R. harrisii completed development to the crab stage in 20-35 o/oo at 25°C and in 15-30 o/oo at 30°C (Fig. 1). The larval survival was highest in 25 o/oo at 25°C (Fig. 2). These larvae required warmer temperatures for complete development than any of the other species. Costlow, Bookhout, and Monroe (1966) reported that R. harrisii larvae from North Carolina completed development in salinities from 2.5-40 o/oo at 20°, 25°, and 30°C. The survival of larvae from this southern population was highest in salinities between 5-33 o/oo at 20°, 25°, and 30°C. There appears to be population differences in the temperature and salinity limits for complete development as well as the optimal conditions for survival.

Complete development of P. pugio larvae occurred over a wide range of salinities from 10-35 o/oo at 20° and 25°C, and in 15-35 o/oo at 15°C (Fig. 1). Larval survival was uniformly high in all salinities from 15-35 o/oo at 20° and 25°C (Fig. 2). There does not appear to be an optimal combination of temperature and salinity for survival as observed in the other species.

Seasonal Variation

The temperature and salinity limits for complete development and maximum larval survival may vary seasonally within a species. Ovigerous C. irroratus are found in Narragansett Bay from November to early July (Jones, 1973), but larvae are only released from April to early July. Eggs collected at any season are hatched when incubated in 30 o/oo sea water at 15°C. The extensive data available on larval development of this species indicate that temperature and salinity limits for complete development, as well as survival under comparable culture conditions, vary for larvae hatched at different seasons (Fig. 3). Larvae resulting from winter hatches completed development in 30 o/oo at 10°C and in 25-35 o/oo at 15° and 20°C. In comparison, larvae from spring hatches completed development in salinities from 25-35 o/oo at 10° and 20°C and between 20-35 o/oo at 15°C. The larvae from summer hatches failed to complete development in any salinity at 10°C, although the larvae completed development in 20-35 o/oo at 15°C and 25-35 o/oo at 20°C.

Under comparable temperature and salinity conditions, larvae resulting from the spring hatches survived better than those from either winter or summer. The minimum temperature required for complete development for summer hatches was

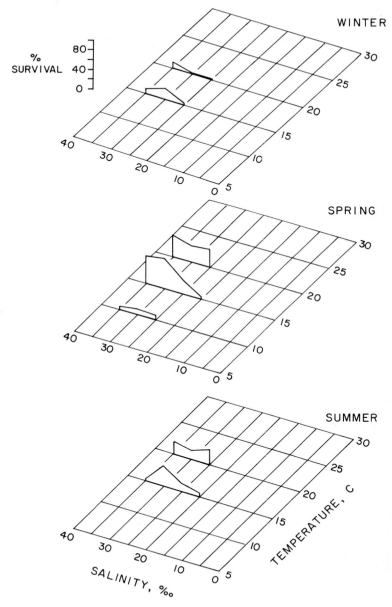

Fig. 3. Survival of C. irroratus larvae to the
post-larval stage in different combinations
of salinity and temperature. The larvae
resulted from eggs hatched in 30 o/oo
salinity at 15°C in different seasons.

15°C, compared to 10°C for winter and spring hatches (Fig. 3).
Since eggs collected in each season were hatched under the
same conditions, it would appear that the previous thermal
history and stage of embryonic development prior to
incubation influence the temperature and salinity limits for
development and maximum survival.

Effects of Constant and Cyclic Temperatures

    Under natural conditions, the larvae are exposed to
varying environmental temperatures as contrasted to constant
conditions generally used for laboratory studies. Although
it is recognized that a varying environment may affect
survival and physiology of organisms differently compared to
constant conditions (Precht et al., 1973) it is only recently
that effects of cyclic temperatures on crustacean larval
development have been examined. Costlow and Bookhout (1971)
found that larvae of R. harrisii survive better at a 30°-
35°C cycle compared to those at constant 30° or 35°C. Within
the temperature range of 10°-30°C, the culture of larvae at
10° or 5°C cycles did not significantly alter the survival
rate compared to those at constant temperatures. Larvae of
C. irroratus in Narragansett Bay are exposed to diurnal
variation in surface temperature of ± 2°C between April and
June, a seasonal variation from 5°-6°C in early April to 16°-
19°C in June (Hillman, 1964) and a vertical temperature
gradient of 5°-6°C in May (Hicks, 1958). In addition, as
indicated by the seasonal variation in survival (Fig. 3), the
previous thermal history of the eggs affects the survival of
the larvae. To examine whether a temperature variation
affects larval development, C. irroratus larvae were hatched
at constant and cyclic temperatures and reared under various
daily cyclic temperature regimes.
    Cancer irroratus larvae cultured under constant and at
comparable cyclic temperatures in 30 o/oo sea water exhibited
different survival and developmental rates (Fig. 4A). Larvae
resulting from eggs hatched at constant temperatures and
reared under cyclic temperature regimes survived much better
than those cultured at constant temperatures. Larval
survival increased 23% for the 10°-20°C cycle, and 20% for
the 15°-25°C cycle compared to that at constant 15°C and 20°C,
respectively. However, larval survival for the 12.5°-17.5°C
cycle decreased by 28% compared to constant 15°C and by 41%
compared to the 10°-20°C cycle.
    The differential response of larvae to constant and
cyclic temperatures is also evident in the time required for
complete development to crab stage. For constant temperature

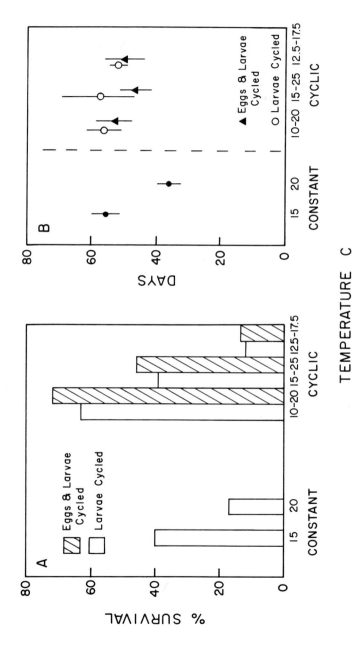

Fig. 4. Survival of C. irroratus larvae to the post-larval stage under constant and comparable cyclic temperature regimes (A). The duration of larval development under constant and comparable cyclic temperature regimes (B).

conditions, development time decreased with increasing
temperature (Fig. 4B). The time for complete development for
the 10°-20°C and the 15°-25°C cycle was about the same.
Slightly faster development was observed at the 12.5°-17.5°C
cycle compared to constant 15°C or the 10°-20°C cycle.
While development time at constant 15°C or the 10°-20°C cycle
did not differ significantly, larval development is
considerably delayed for the 15°-25°C cycle compared to that
at constant 20°C.

Larval survival from eggs hatched under cyclic
temperatures and reared under the same temperature cycles
showed a further increase in the survival to crab stage,
compared to larvae hatched at constant temperatures and
reared under cyclic temperatures. Survival increased by 9%
for the 10°-20°C cycle, 7% for the 15°-25°C cycle, and 1.5%
for the 12.5°-17.5°C cycle. Larval development time, whether
hatched under cyclic or constant temperatures, is not
significantly different for the 10°-20° and 12.5°-17.5°C
cycles. However for the 15°-25°C cycle, the development time
of larvae hatched under cyclic temperatures is less than that
for larvae hatched under constant conditions.

Variations of Geographically Separated Populations

In addition the larvae of populations of a species may
exhibit differences in survival and physiological response
relative to the geographic variations in the environment
(Vernberg and Costlow, 1966). Comparable data on larval
development of geographically separated populations are
available for some of the species examined in this study.
The development time and survival of P. herbstii and R.
harrisii larvae from North Carolina populations (Costlow,
Bookhout, and Monroe, 1962 and 1966) have been compared with
the results for Rhode Island populations (Table 3). The
survival of P. herbstii larvae from Rhode Island is higher
and development time is shorter than that for the North
Carolina population. This general tendency of climatic
adaptation to a faster rate of development for cold adapted
populations as compared to those from warmer environments is
not shown in the larval development of R. harrisii. Under
comparable temperature and salinity conditions, the larvae
of Rhode Island populations take longer to complete development to the crab stage and their survival is much lower as
compared to those from North Carolina. Schneider (1968)
compared the metabolic adaptation of R. harrisii populations
from Maine and North Carolina. The population differences
were most pronounced in the adults from the field, but the

Table 3

Variation in the survival and duration for complete development of geographically separated populations of three species of crustaceans.

Species	Rhode Island				North Carolina*			
	T (C), S (o/oo)	% Survival	Duration Days		T (C), S (o/oo)	% Survival	Duration Days	
Panopeus herbstii	20 C 32 o/oo	52	30-38		20 C 31.1 o/oo	0	-	
	25 C 32 o/oo	48	20-32		25 C 31.1 o/oo	3	33-35	
	25 C 25 o/oo	20.7	20-30		25 C 25 o/oo	83	15-23	
	25 C 35 o/oo	2.0	22		25 C 35 o/oo	23	16-21	
Rhithropanopeus harrisii	30 C 15 o/oo	1.89	23-24		30 C 15 o/oo	14	15-22	
	30 C 25 o/oo	4.08	23-30		30 C 25 o/oo	60	11-15	
Palaemonetes pugio	25 C 30 o/oo	91.8	18-31		25-27 C	65	17-21	

* Data from Costlow, Bookhout and Monroe (1962, 1966); Broad (1957)

adults of populations from Maine and North Carolina reared under identical conditions showed no differences between their acute or acclimated M-T curves. Apparently, irreversible phenotypic changes take place in R. harrisii under field conditions. Although differential responses of geographically separated larval populations occur, it is not clear in many cases whether these differences are genotypic or irreversible nongenetic adaptation (Kinne, 1962; Vernberg and Vernberg, 1974). Such adaptation to climatic differences over the geographical range of a species may produce physiological differences in rates of development, optimal conditions for survival, and limits for complete development.

Metabolic Adaptation

    Within their tolerance range, larval stages of each species exhibit variation in their metabolic response to temperature. The metabolic-temperature response curves of larvae of C. irroratus and C. borealis, congeneric species from Narragansett Bay, cultured at a temperature and salinity producing maximum survival to the crab stage demonstrate this variation (Fig. 5). There is both intra- and inter-specific variation in the metabolic response. The first stage larvae of C. irroratus are metabolically active over a temperature range of 5°-20^OC with a Q_{10} close to 2. Their metabolic rate is slightly depressed from 20°-25^OC. The temperature ranges for metabolic activity and depression remained the same, but differences in the temperature ranges for compensation and sensitivity were observed in successive stages. There are major changes in the metabolic-temperature response patterns between the first and second stages, third and fourth stages, and fifth and megalops stages. The temperature range for sensitivity in the second and third stage shifts from 5°-15^OC to 10°-15^OC in the fourth and fifth stages. In addition, the temperature range for compensation narrows from 10°-20^OC in the second and third stages to 15°-20^OC in the fourth and fifth stages. The megalops showed no compensation over the 10°-20^OC range.

    Cancer borealis larvae showed distinctly different M-T patterns. They are metabolically active over the entire temperature range of 5°-25^OC and there is no depression of the metabolic rate at warmer temperatures. The first and second stages showed compensation over a temperature range of 5°-20^OC and 5°-15^OC, respectively. Sensitivity was observed at warmer temperatures. The third stage showed no compensation over the entire temperature range of 5°-25^OC. The M-T patterns of the fourth, fifth, and megalops stage are

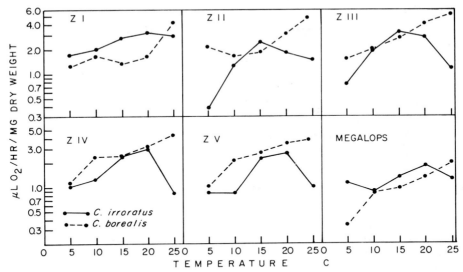

Fig. 5.   Metabolic-temperature response patterns
of larval stages of C. irroratus and
C. borealis (modified from Sastry and
McCarthy, 1973).

reversed from those of the first two with sensitivity to
colder temperatures and compensation to warmer climates.
     The larvae of these two species have different tolerance
ranges. Cancer irroratus larvae become progressively steno-
thermal with a narrowing of the temperature range for
compensation with development. Cancer borealis larvae
appear to be relatively eurythermal with the temperature
range for compensation gradually shifting from colder
temperatures in the early stages to warmer temperatures in
the later stages. Although the metabolic temperature
response patterns of C. borealis larvae are relatively eurythermal, complete larval development occurred only in 30 o/oo
at 20°C suggesting there may be other interacting factors
influencing larval development.
     The metabolic response patterns of P. pugio and C.
borealis larvae to temperature are shown in Figure 6 as Q_{10}
values. Palaemonetes pugio larvae are metabolically active
over a wide temperature range and at warmer temperatures than
C. borealis larvae. As mentioned before, the early larval
stages of C. borealis compensate at colder temperatures and
are sensitive to warmer temperatures. This pattern is
reversed in the fourth, fifth, and megalops stages.

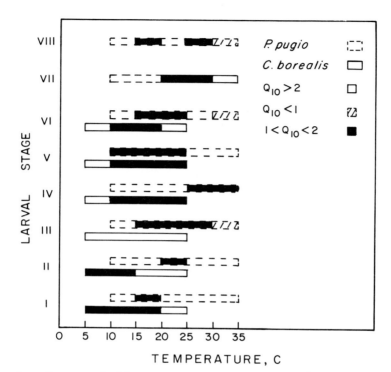

Fig. 6. Metabolic-temperature responses (as Q_{10} values) of larval stages of C. borealis and P. pugio cultured at constant 20°C and 30 o/oo salinity.

     The first and second larval stages of P. pugio compensate between 15°-20°C and 20°-25°C, respectively. The temperature range for metabolic compensation is much wider for the third through seventh stage larvae. In the fourth stage, compensation extends to warmer temperatures than in earlier stages. The fourth stage showed no compensation from 10°-25 C and compensates from 25°-35°C. The fifth stage is the reverse. In the sixth and seventh stages, metabolic compensation is observed at intermediate temperatures. In the eighth stage compensation occurs from 10°-15°C and 25°-30°C. At extremely warm temperatures (30°-35°C), the metabolic rate of the third, sixth, and eighth stages was depressed.
     The alteration in the metabolic responses to temperature of larvae cultured under daily temperature cycles compared to constant temperatures is shown for C. irroratus

(Fig. 7). The larvae cultured under cyclic temperatures showed an extension of the temperature range for metabolic compensation in at least some of the larval stages. For larvae cultured under a cyclic temperature regime the upper limits of temperature tolerance and depression of metabolic rate are increased compared to those reared under constant temperature regimes.

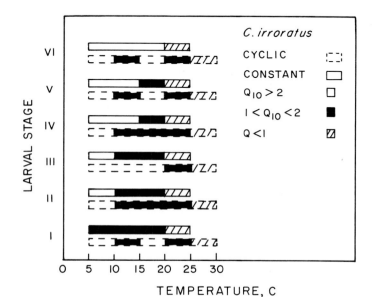

Fig. 7.   Metabolic-temperature responses (as Q_{10}
            values) of larval stages of C. irroratus
            cultured in 30 o/oo at constant 15°C and
            cyclic 10°-20°C.

Temperature Tolerance Limits

The larval stages of species may also exhibit varying capacities for resistance adaptation to environmental extremes (Vernberg and Vernberg, 1975). The upper temperature tolerance limits for 50% survival of C. irroratus, H. americanus, and P. pugio larvae cultured in their respective optimal temperature and salinity combinations are presented in Figure 8. The temperature tolerance limits for P. pugio larvae are much higher than those for the other species. Between the two sublittoral species, C. irroratus larvae have lower temperature tolerance limits than H. americanus larvae. The tolerance limits for

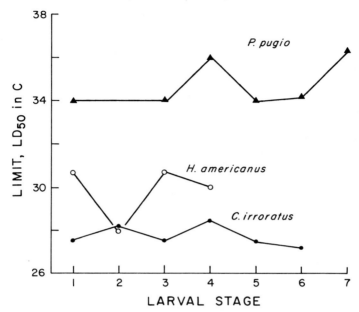

Fig. 8. Acute temperature tolerance limits for
three species of crustaceans cultured
under constant conditions optimal for
their survival. (C. irroratus, 15°C -
30 o/oo; H. americanus and P. pugio,
20°C - 30 o/oo).

successive larval stages remain fairly constant for each
species with the exception of certain stages. The limits for
the fourth and seventh stages of P. pugio are 2° and 2.3°C
higher respectively than for the other stages. Second stage
H. americanus larvae have a much lower tolerance limit than
any other stage. For C. irroratus larvae the second and
fourth stages tolerated slightly higher temperatures and the
megalops stage lower temperatures than the other stages.
Therefore, it would appear that although certain stages in
the larval development of a species may be slightly more
resistant or sensitive to temperature, the general
temperature tolerance limits for overall development of a
species remain about the same and are related to the
temperature range of their habitat.

CONCLUSIONS

The larval development of crustaceans, like all physiological processes, occurs within a range of environmental conditions which is characteristic for each species. Interacting environmental factors affect larval development which is reflected in varying survival and development rates in different combinations of temperature and salinity. A species may exhibit an optimal combination for survival and as environmental conditions deviate from this optimum survival is reduced.

Limits, as well as the optimal combination for development, vary interspecifically. Generally, sublittoral species completed development over a narrower range of temperature and salinity than estuarine species. These limits may vary from a single temperature and salinity combination as for C. borealis to a wide range of temperature and salinity as for P. longicarpus and P. pugio.

Limits for complete development also vary depending upon the prior experience of the eggs. Larvae resulting from eggs incubated during the period when they are normally released in the environment survive better than those at other seasons. Therefore, the stage of embryonic development at which eggs are incubated for hatching and their prior experience appears to have an effect on survival and the limits for complete development.

In addition to these variations, larvae cultured under constant and cyclic temperature regimes also differ in their survival and time for complete development. Those cultured under a suitable amplitude and rate of temperature change show an increase in survival compared to those at comparable constant temperatures. Beyond these limits as with the 15°-$25^{\circ}C$ cycle the development of C. irroratus larvae is delayed compared to that at $20^{\circ}C$.

Larvae from geographically separated populations also exhibit differences in their survival and time for complete development. Panopeus herbstii larvae from a southern geographical region developed slower than those from Narragansett Bay under comparable conditions. In contrast, R. harrisii larvae from northern latitudes appear to develop more slowly than those from more southern latitudes. For a valid assessment of these differences, there is a need for more rigorously controlled experiments.

Within the tolerance range of a species, the metabolic response to temperature may vary, but each species may have a characteristic overall response. For example, C. irroratus larvae showed a narrowing of the temperature range for metabolic compensation with development, but the temperature range for depression of metabolism remained the same.   In contrast, C. borealis larvae showed a shift in the temperature range of compensation from colder to warmer temperatures in the later stages.   These larvae also had a broader temperature range for metabolic compensation than its congeneric species, C. irroratus.   Larvae of P. pugio, an estuarine species, are metabolically active over a much wider temperature range than the two sublittoral species examined.   They are less tolerant of colder temperatures and more tolerant of warmer temperatures than those of C. irroratus, and showed metabolic compensation over a wider range.

The metabolic response patterns of the larvae also varied when they were cultured under cyclic temperatures. Although it is not clear how generalized these changes are for different species, cyclic temperatures increased the tolerance of C. irroratus larvae to warmer temperatures, and also extended their temperature range for metabolic compensation compared to those cultured under constant conditions.   A cyclic temperature regime also increased larval survival.

Acute temperature tolerance limits varied intra- and inter-specifically with estuarine species tolerating higher temperatures than sublittoral species.   These limits remained fairly constant for all C. irroratus larval stages, while some H. americanus and P. pugio stages showed differential sensitivity.

This variation in response with stage, season of hatching, species, culture conditions, and geographic origin should be considered in the application of laboratory bio-assay results to the assessment of the effects of heavy metal, organic, and thermal pollution on a population or community.

ACKNOWLEDGMENTS

This study was supported by a Grant (R-800981) from the Environmental Protection Agency.   The authors acknowledge the assistance of John Laczak, Sharon Pavignano, and Shirley Barton in conducting these studies.

LITERATURE CITED

Bookhout, C. G. and J. D. Costlow.   1975.   Effects of mirex on the larval development of blue crab.   Water, Air and Soil Pollut. 4:   113-126.

Broad, A. C. 1957. Larval development of Palaemonetes
   pugio Holthuis. Biol. Bull. 112: 144-161.
Costlow, J. D. and C. G. Bookhout. 1964. An approach to the
   ecology of marine invertebrate larvae. In: Symposium
   Expt. Ecol. Graduate School of Oceanography, Univ.
   Rhode Island, Kingston. Occasional publ. No. 2. pp. 69-
   75.
_____ and _____. 1965. The effects of environ-
   mental factors on larval development of crabs. In:
   Biological Problems in Water Pollution. pp. 77-86,
   ed. by C. M. Tarzwell. U. S. Dept. of Health,
   Education and Welfare, Cincinnati, Ohio (Publ. Hlth.
   Serv. Publs. Wash. 999-WP-25).
_____ and _____. 1971. The effect of cyclic
   temperatures on larval development in the mud crab,
   Rhithropanopeus harrisii. In: Fourth European
   Symposium on Marine Biology, pp. 211-220, ed. by D. J.
   Crisp. Cambridge: Cambridge Univ. Press.
_____, _____, and R. Monroe. 1962. Salinity and
   temperature effects on the larval development of the
   crab, Panopeus herbstii (Milne-Edwards), reared in the
   laboratory. Physiol. Zool. 35: 79-93.
_____, _____, and _____. 1966. Studies on
   the larval development of the crab, Rhithropanopeus
   harrissi (Gould). I. The effect of salinity and
   temperature on larval development. Physiol. Zool. 39:
   81-100.
DeCoursey, P. J. and W. B. Vernberg. 1972. Effect of
   mercury on survival, metabolism and behavior of larval
   Uca pugilator. Oikos 23: 241-247.
Forward, R. B. and J. D. Costlow. 1974. The ontogeny of
   phototaxis by larvae of the crab Rhithropanopeus
   harrisii. Mar. Biol. 26: 27-33.
Goldstein, A. 1964. Biostatistics: An Introductory Text.
   New York: MacMillan Co.
Grunbaum, B. W., B. V. Siegal, A. R. Schulz, and P. L. Kirk.
   1955. Determination of oxygen uptake by tissue growth
   in an all glass differential microrespirometer.
   Microchem. Acta 1955: 1069-1075.
Hicks, S. D. 1958. The physical oceanography of
   Narragansett Bay. Limnol. Oceanogr. 4: 316-327.
Hillman, N. S. 1964. Studies on the distribution and
   abundance of decapod larvae in Narragansett Bay, Rhode
   Island with a consideration of morphology and mortality.
   M. S. Thesis. University of Rhode Island, Kingston,
   R. I.

Jones, C. J. 1973. The ecology and metabolic adaptation of *Cancer irroratus* Say. M. S. Thesis. University of Rhode Island, Kingston, R. I.

Kalber, F. and J. D. Costlow. 1966. The ontogeny of osmoregulation and its neurosecretory control in decapod crustacean, *Rhithropanopeus harrisii* (Gould). *Amer. Zool.*, 6: 221-230.

Kinne, O. 1962. Irreversible non-genetic adaptation. *Comp. Biochem. Physiol.* 5: 265-282.

Mileikovsky, S. A. 1970. The influence of pollution on pelagic larvae of bottom invertebrates in marine near shore and estuarine waters. *Mar. Biol.* 6: 350-356.

Precht, H., H. Christophersen, L. Hensel, and W. Larcher. 1973. *Temperature and Life*. New York: Springer Verlag.

Roberts, M. H. 1971. Larval development of *Pagurus longicarpus* Say reared in the laboratory. II. Effects of reduced salinity on larval development. *Biol. Bull.* 140: 104-116.

Sandifer, P. A. 1975. The role of pelagic larvae in recruitment to populations of adult crustaceans in the York estuary and adjacent lower Chesapeake Bay, Virginia. *Est. Coastal Mar. Sci.* 3: 269-279.

Sastry, A. N. 1970. Culture of brachyuran crab larvae using a recirculating sea water system in the laboratory. *Helgol. wiss. Meeresunters* 20: 406-416.

_____. 1976a. Larval development of *C. irroratus* reared in the laboratory. *Crustaceana.* In press.

_____. 1976b. Larval development of *C. borealis* reared in the laboratory. *Crustaceana.* In press.

_____. 1976c. Effects of constant and cyclic temperature regimes on the larval development of a brachyuran crab. In: *Thermal Ecology Symposium*, ed. by Esch and McFarland. In press.

_____ and J. F. McCarthy. 1973. Diversity in metabolic adaptation in pelagic larval stages of two sympatric species of brachyuran crabs. *Neth. J. Sea Res.* 7: 434-446.

Scheltema, R. S. 1971. The dispersal of the larvae of shoal-water benthic invertebrate species over long distance by ocean currents. In: *Fourth European Symposium on Marine Biology*, pp. 7-28, ed. by D. J. Crisp. Cambridge: Cambridge Univ. Press.

Schneider, D. E. 1968. Temperature adaptation in latitudinally separated populations of the crab, *Rhithropanopeus harrisii*. *Amer. Zool.* 8: 772.

Vernberg, F. J. and J. D. Costlow. 1966. Studies on physiological variation between tropical and temperate zone fiddler crabs of the genus Uca. IV. Oxygen consumption of larvae and young crabs reared in the laboratory. Physiol. Zool. 39: 36-52.

Vernberg, F. J. and W. B. Vernberg. 1974. Synergistic effects of temperature and other environmental parameters on organisms. In: Thermal Ecology, pp. 94-99, ed. by J. W. Gibbons and R. Sharitz. U. S. Atomic Energy Commission, Conf. 730505.

_____ and _____. 1975. Adaptation to extreme environments. In: Physiological Ecology of Estuarine Organisms, pp. 165-180, ed. by F. J. Vernberg. The Belle W. Baruch Library in Mar. Sci., No. 3, Columbia: University of South Carolina Press.

Vernberg, W. B., P. J. DeCoursey, and J. O'Hara. 1974. Multiple environmental factor effects on physiology and behavior of the fiddler crab, Uca pugilator. In: Pollution and Physiology of Marine Organisms, pp. 381-425, ed. by F. J. Vernberg and W. B. Vernberg. New York: Academic Press.

Williams, A. B. 1965. Marine decapod crustaceans of the Carolinas. Fishery Bull. Fish. Wildl. Serv. 65: 1-298.

The Heterotrophic Potential Assay as an Indicator of Environmental Quality

PAUL L. ZUBKOFF and J. ERNEST WARINNER, III

Department of Environmental Physiology
Virginia Institute of Marine Science
Gloucester Point, Virginia   23062

Because estuarine and marine organisms are important natural and economic resources, the preservation and management of their populations and habitats are essential for maintaining the high productivity and balance of aquatic ecosystems. To do this requires a knowledge of the quality of water comprising their habitat. One approach is to employ a bioassay to measure the physiological response of the biological community. If such a measure could be made on readily obtainable samples and performed relatively quickly, large numbers of samples could be handled in space and time. Since primary producers and heterotrophic microorganisms have short generation times and are readily available, they are likely candidates for such an assay. These producers also constitute the base of complex food webs upon which the highly visible organisms of commercial importance ultimately depend for their food source.

Methods used for quantifying the functional capability of the aquatic environment are relatively few (Vollenweider, 1969). Those methods which employ standing crop analysis for assessing population diversities and species succession are most time consuming, often requiring highly specialized personnel, and do not permit replicate sampling or repeated samplings at close intervals in time. In contrast to such accepted procedures, we employed alternatives which are amenable to multiple analyses in space and time (Zubkoff and Warinner, in press). Such procedures, including the

425

heterotrophic potential assay (Vmax, glucose), in particular, may provide the means for a rapid determination of the state of the microbial ecosystem (Wright and Hobbie, 1965, 1966; Vaccaro and Jannasch, 1966).

The flowing waters of the estuary may be envisaged as having populations or living subunits with varying sizes and generation times, quite analogous to the fluid tissues of higher organisms. The viable subunits (organisms in the estuary and cells or macromolecules in the fluid tissues) are bathed by a fluid which is somewhat influenced by the physical environment. In both cases changes in the characteristics of the fluids or the subunits may be observed by appropriate methodology.

In a very general sense, a further analogy is possible based on the kinetics of enzyme action. All enzymatic systems, whether crude or highly purified, in their elementary analysis follow the Michaelis-Menten model of enzyme kinetics. In a similar manner, if the uptake of an organic substrate by the resident population of heterotrophic microorganisms is observed, the rate of substrate uptake may be interpreted as a reaction or a series of reactions which has a rate-limiting step. A feature common to these analogies is that specific measurements are made on appropriate fluids.

Although the direct application of the simple kinetic model has been questioned because natural microbial populations are heterogeneous assemblages of microorganisms with widely differing sizes, generation times, and nutritional requirements (Wright and Hobbie, 1965; Vaccaro and Jannasch, 1967; Munro and Brock, 1968; Wright, 1973; Williams, 1973), the observed phenomenon has been usefully employed for describing relative microbial activity in different water masses and within the same water mass (Hobbie et al., 1972; Paerl and Goldman, 1972; Crawford et al., 1973; Zeigler et al., 1974; Zubkoff and Warinner, in press; Sibert and Brown, 1975). In another application Albright et al. (1972) demonstrated that sublethal concentrations of heavy metals (Ag, AsO_3^-, Ba, Cr, Cu, Hg, Ni, Pb, Zn) produce a noncompetitive type inhibition of indigenous heterotrophic activity.

An assessment of existing and perturbed conditions of water quality may be reflected in the heterotrophic assay of the biological population. Because the physical and chemical measurements usually employed for characterizing a water mass are obtainable in a short period of time, an equally rapid measurement of the biological state of the waters is desired. Thus, the heterotrophic potential assay (Vmax, glucose) was employed as an indicator of environmental quality.

Heterotrophic Potential (Vmax)

The heterotrophic potential (Vmax) of the estuarine plankton community can be measured by the uptake of simple ^{14}C-labeled dissolved organic substrates added to natural water samples. When varying concentrations of substrate are employed, the response to an increase in substrate concentration resembles that of an enzyme-catalyzed reaction (Wright and Hobbie, 1965; Williams, 1973). As the concentration of a substrate, such as glucose, is increased, there is initially a linear increase in the rate of uptake by the heterotrophic organisms at low substrate concentration; at higher substrate concentrations, the rate of uptake reaches a maximum. If it is assumed that the attaining of the maximum rate of uptake at the increased substrate concentration is due to a saturation of the cell surface enzyme system responsible for uptake, then the maximum velocity (Vmax) should be proportional to the amount of available cell surface enzyme system, and, should serve as a relative physiological indicator of the biomass of the heterotrophic population (Hobbie, 1969). Under the controlled conditions of assay employed, this assumption relegates such phenomena as enzyme induction, functional efficiency of microbes in various life stages or of different communities of secondary or lesser importance with respect to the efficiency of the existing microbial population at the time of sampling.
The rate of uptake of the substrate at any given concentration is calculated using the pseudo-first order equation of enzyme kinetics (Parsons and Strickland, 1962):

$$v = \frac{c(Sn + Sa)}{C\mu t}$$

where

v = velocity of uptake of the substrate ($\mu g \ l^{-1}h^{-1}$)
c = counts taken up by heterotrophic population
Sn = natural substrate concentration
Sa = added substrate concentration
C = number of counts per μCi of substrate
μ = number of μCi added to incubation medium
t = time of incubation (hours)

When this equation is combined with a modified form of the Michaelis-Menten equation:

$$\frac{(Sn + Sa)}{v} = \frac{Kt}{Vmax} + \frac{(Sn + Sa)}{Vmax}, \text{ where}$$

K_t = transport constant
Vmax = maximum velocity of uptake at saturation of
substrate,

the resulting equation is:

$$\frac{C\mu t}{c} \text{ or } \frac{(Sn + Sa)}{v} = \frac{(K_t + Sn)}{Vmax} + \frac{1}{Vmax} (Sa)$$

When $\frac{C\mu t}{c}$ is plotted against Sa, the result is a
straight line with slope = $\frac{1}{Vmax}$. The value of Vmax,
previously defined as the heterotrophic potential and the
reciprocal of the slope, has units of μg glucose $l^{-1}h^{-1}$.
Although several kinetic parameters may be calculated, Vmax
is the most useful one for describing the relative functional
distributions of the microbial populations in this study.

The kinetic data obtained are undoubtedly the result of
heterogeneous assemblages of organisms which have active
transport systems. Vmax is the rate of uptake observed at a
substrate concentration high enough to completely saturate
the transport mechanisms of the natural microbial populations
under the experimental conditions. In these studies, Vmax,
glucose is an indicator which reflects the size of the viable
natural population of microbial organisms at the time of
sampling; it is an experimentally measured number which is a
resultant of the endemic community's cell size, number, and
state of viability as a function of temperature.

METHODS

The heterotrophic potential was determined by incubating
10 ml aliquots of an estuarine water sample with labeled ^{14}C-
glucose and carrier at final concentrations of 37.5, 75.0,
187.5, and 375 μg l^{-1} in the dark at ambient temperature for
two hours. Approximately 0.1 ml of 2% neutralized formalin
was added for the inactivated control sample and to terminate
the reaction. The ^{14}C-labeled particulate fraction was then
collected on cellulose-acetate filters (Millipore^R), treated
with NCS^R tissue solubilizer, and dissolved in a toluene-
based scintillation fluid containing 2,5-diphenyloxazole.
Counting was completed at 87-95% efficiency using a liquid
scintillation counter with external standardization. If
the calculation of Vmax, glucose using linear regression
analysis did not have an r value of 0.85 or greater for at
least 3 of the 4 concentrations of substrate used, the
results were termed erratic.

Since no provision was made to trap and measure the respired CO_2 from the assimilated ^{14}C-glucose, the calculated Vmax, glucose values represent only that portion of labeled substrate transformed into particulate form and is, therefore, a minimum estimate of the functional microbial community. Although the use of buffered formalin has been questioned in heterotrophic assays using ^{14}C-glutamate (Griffiths et al., 1974), this reagent was suitable for producing consistent results with ^{14}C-glucose under field conditions.

RESULTS AND DISCUSSION

In a series of observations on the plankton energetics at the mouth of the lower York River, Virginia from August 1971 to August 1974 (Fig. 1, Table 1), the seasonal change in

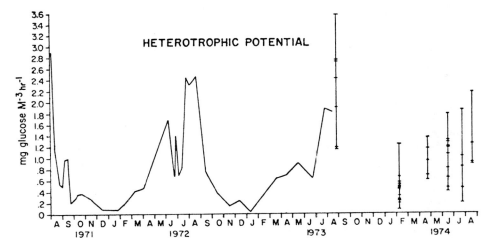

Fig. 1. Seasonal trend of heterotrophic potential (Vmax, glucose) of the waters at 1 meter below the surface of the mouth of the York River, Virginia. Individual measurements are connected by line (1971-1973). The ranges of values after August 1973 were obtained when the hydrographic station was continuously occupied for 2 to 3 tidal cycles (26-40 hours).

heterotrophic potential of the waters 1 meter below the surface was determined. The Vmax, glucose for these waters were clustered in five ranges: very low, <0.10; low 0.11-0.50; moderate, 0.51-1.00; high, 1.01-1.70; and very high,

Table 1
Seasonal pattern of heterotrophic potential (Vmax, glucose) of the waters at 1 meter depth at the mouth of the York River, Virginia (37°14.6'N, 76°23.4'W). 1971-1974.

Heterotrophic Potential (μg glucose $l^{-1}h^{-1}$)

	Very Low <0.10				Low 0.1–0.50				Moderate 0.51–1.00				High 1.01–1.70				Very High >1.70			
Observations	71	72	73	74	71	72	73	74	71	72	73	74	71	72	73	74	71	72	73	74
January		1																		
February			1	1	1			8*				3*				1*				
March					1	1			1			3				2				
April					1	1	1													
May					1											1				
June							1	4		1		2		2		4				
July							1		2	1						2			1	1
August												3		1				1		1
September					2					2	1						2	2		1
October					3	1				1										
November						1	1									3				
December						1		5												
Total Observations — Yearly	0	1	1	1	7	5	4	20	3	4	1	13	0	3	0	13	2	3	1	3
Total Observations — 3 Year		3				36				21				16				9		

*February 1974 was unseasonably warm and clear.

1.71-3.0 (Zubkoff, Grant, and Warinner, in preparation). Of
the 85 observations made, a distinct seasonal trend is
noted, with 57/85 determinations between 0.10 and 1.00 µg
glucose $l^{-1}h^{-1}$; the 3 low values occurring in the colder
months and the 9 very high values in the time periods of
highest temperatures. It is quite likely that some of
these very high values (i.e., August 1973) may also be
associated with patchiness of the plankton communities
observed over the 2-3 tidal cycles that the station was
occupied during the third year of the study. These values
are considerably lower than the range of 0.15-24.10 µg
$l^{-1}h^{-1}$ reported for the shallow Pamlico Sound estuary, one
of the most microbially active aquatic environments measured
(Crawford et al., 1973).

The observations on the heterotrophic potential of the
waters at 1 meter below surface of the York River mouth
(1971-1973) serve as a reference for the waters (1 meter
below surface) of the plankton stations of the lower
Chesapeake Bay assayed in the period immediately following
the deluge of fresh water into the Bay by Tropical Storm
Agnes in June 1972 (Zubkoff and Warinner, 1974). Because
the York River basin was less affected by Tropical Storm
Agnes than the rest of the Chesapeake Bay drainage system
(Davis, 1974), the York River mouth station serves as a
useful point of reference. These data (summarized in
Table 2) are interpreted as a shift in the heterotrophic
populations from low potentials (0.1-0.5 µg glucose $l^{-1}h^{-1}$)
to those of moderate potentials (0.51-1.00 µg glucose
$l^{-1}h^{-1}$). The higher heterotrophic potential values were
associated with the masses of lower salinity waters entering
the Chesapeake Bay from the tributary rivers. However,
masked in this overly simplified summary, are the significant
changes that occurred when pockets of lower salinity water
(<20 o/oo) became localized in areas which are usually of
higher salinity (>24 o/oo) (Zubkoff, Grant, and Warinner,
in preparation).

In another test of utility of the heterotrophic
potential assay as a responsive indicator of water quality,
the surface waters of the lower James River were surveyed
at 24 stations in March of 1974 (Table 3). The values of
12 stations were low (as expected for the season), nine were
moderate, and three were erratic with respect to the usual
data encountered in this type of assay. The three erratic
assays were from waters in the immediate proximity (within
20 meters) of sewage treatment plant outfalls. It was
concluded that the Vmax, glucose of these waters which was
highly influenced by the sewage treatment plant effluents
could either reflect a depletion of the microbial population

Table 2
Comparison of the heterotrophic potential (Vmax, glucose) of the waters at 1 meter depth of the York River Mouth (1971–1974) and lower Chesapeake Bay (1972).

Observations	Heterotrophic Potential (μg glucose $1^{-1}h^{-1}$)				
	Low 0.1–0.5	Moderate 0.51–1.00	High 1.01–1.70	Very High >1.70
York River Mouth (1971–1974)				
June	5	3	6	2
July	1	3	2	2
August	0	4	1	5
September	2	3	0	0
October	5	1	3	0
Total Observations (48)	13	14	12	9
Percentage	27.1	29.2	25.0	18.7
Lower Chesapeake Bay (1972)				
June	2	6	3	0
July	1	16	14	8
August	2	3	5	3
September	3	7	2	1
October	6	6	0	0
Total Observations (88)	14	38	24	12
Percentage	15.9	43.2	27.3	13.6

Table 3
Heterotrophic potential (Vmax, glucose) at selected stations
on the lower James River Basin (6 March 1974).

Station Identification	Total Carbon mg C l^{-1}	Vmax, glucose µg glucose $l^{-1}h^{-1}$
Open Waters		
1	25.3	0.36
2	24.2	0.60
3	25.0	0.61
4 S	24.0	0.50
M	24.2	0.32
B	26.2	0.37
5	24.0	0.22
6	23.2	0.36
7	22.2	0.15
Tributary Waters		
8	23.6	0.47
9	26.7	0.69
13	25.5	0.52
14	25.0	0.75
15	25.2	0.11
Waters Near Sewage Treatment Plant Outfalls		
LP	54.8	0.88
WB	47.5	0.76
CE	30.0	†
AB	35.2	†
P S	34.5	0.17
M	25.4	0.57
B	26.0	0.61
BH S	25.2	†
M	25.2	0.42
B	24.0	0.78

†Erratic data.

or drastically affect the uptake mechanisms of individual
organisms. Whether this heterotrophic response is caused by
chlorine and its transient by-products (chloramines) in
sewage treatment plant effluents or by other components in
the effluent (heavy metals, toxic organics) is unknown.
Further work should be undertaken using the heterotrophic
potential in ascertaining its usefulness as an indicator for
toxic components in receiving waters of the estuaries. It
would be particularly useful to know its reliability in
conjunction with bioassays carried out on the highly sensitive larval invertebrates (Roberts et al., 1975). In
addition, the heterotrophic assay may be helpful for
ascertaining the fate and, possibly, utilization of
industrial additions to the receiving waters after dilution
has occurred (Sibert and Brown, 1975).

In a single attempt to test the effects of an organic
addition (p-cresol at the 0.001 - 1.0 ppm level), the
heterotrophic potential was completely inhibited. Whether
one can work successfully at lower concentrations of organic
additions in order to ascertain the inhibition kinetics
(competitive or non-competitive) remains to be resolved.

Because the heterotrophic potential is a functional
measurement of the utilization of a substrate by an endemic
microbial community, it is subject to those meteorological,
hydrographic, and seasonal parameters which affect any
aquatic community. However, the relative microbial activity
within a water mass may be rapidly assessed by determining
the Vmax, glucose. It is quite possible that a greater
potential use of the heterotrophic assay lies in its
development for monitoring the response of the endemic
microbial community of receiving waters to industrial,
municipal, and non-point source additions. Under a given
set of conditions or as a function of time or distance from
a point-source discharge, is the growth of microbial
community of the receiving waters stimulated or inhibited?
Are there alternative treatments that may be applied to the
point-source discharge which will permit the microbial
communities in the receiving waters to respond in a
predictable manner? The use of the heterotrophic potential
assay to monitor discharges into the estuarine waters may
ultimately prove to be a valuable tool for assessing the
assimilative capacity of a receiving body of water. We
believe that with further testing the heterotrophic potential
measurement will be a useful tool in the environmental
manager's arsenal for assessing the functioning of the
microbial population in the estuarine environment.

SUMMARY

The heterotrophic potential assay has been employed for describing the ranges of Vmax, glucose throughout the year in a temperate zone mesohaline estuary. The range varied from 0.04 µg glucose $l^{-1}h^{-1}$ in the winter to greater than 2.9 µg glucose $l^{-1}h^{-1}$ in late summer.

In an assessment of the effects of a major perturbation (Tropical Storm Agnes) on the lower Chesapeake Bay waters, the heterotrophic potential fell to a level comparable to that of the York River mouth (0.10 - 2.9 µg glucose $l^{-1}h^{-1}$). Shifts to moderate density biomass populations occurred more frequently in the Bay waters, particularly during the summer months of 1972 following the storm.

With respect to human activity, as demonstrated by waters in the proximity of outfalls of sewage treatment plants, the heterotrophic assay is radically affected by producing unusually mixed uptake rates of various concentrations of substrate (Table 3, CE, AB, and BH 5). The heterotrophic activity of water samples further downstream are not as markedly affected and may possibly be higher than that of open waters because of stimulation through enrichment once dilution of the effluent has occurred.

ACKNOWLEDGMENTS

We thank Leonard W. Haas for helpful discussions and Margaret Bolus, Patricia A. Crewe, Richard A. Gleeson, Michael A. Gorey, and Linda Jenkins for technical assistance. This study was supported in part by the Chesapeake Research Consortium, Inc. under N.S.F. RANN Grant No. GI-29909 and GI-34869. The study of the James River was carried out in cooperation with Donald D. Adams, Old Dominion University.

LITERATURE CITED

Albright, L. J., J. W. Wentworth, and E. M. Wilson. 1972. Technique for measuring metallic salt effects upon indigenous heterotrophic microflora in a natural water. Water Res. 6: 1589-1596.

Crawford, C. C., J. E. Hobbie, and K. L. Webb. 1973. Utilization of dissolved organic compounds by microorganisms in an estuary. In: Estuarine Microbial Ecology, pp. 169-177, eds. L. H. Stevenson and R. R. Colwell. Columbia: Univ. S. Carolina Press.

Davis, J. 1974. The effects of Tropical Storm Agnes on the Chesapeake Bay Estuarine System. U. S. Army Corps of Engineers, Contract Report DACW 31-73-C-0189.

Griffiths, R. P., F. J. Hanus, and R. Y. Morita. 1974. The effects of various water-sample treatments on the apparent uptake of glutamic acid by natural marine microbial populations. Can. J. Microbiol. 20: 1261-1266.

Hobbie, J. E. 1969. A method for studying heterotrophic bacteria. In: I.B.P. Manual 12: Primary Production in Aquatic Environments, pp. 146-151, ed. R. A. Vollenweider. Oxford: Blackwell Scientific Publications.

_____, O. Holm-Hansen, T. T. Packard, L. R. Pomeroy, R. W. Sheldon, J. P. Thomas, and W. J. Wiebe. 1972. A study of the distribution and activity of microorganisms in ocean water. Limnol. Oceanogr. 17: 544-555.

Munro, A. L. S. and T. D. Brock. 1968. Distinction between bacterial and algal utilization of soluble substances in the sea. J. Gen. Microbiol. 51: 35-42.

Paerl, H. W. and C. R. Goldman. 1972. Heterotrophic assays in the detection of water masses at Lake Tahoe, California. Limnol. Oceanogr. 17: 145-148.

Parsons, T. R. and J. D. H. Strickland. 1962. On the production of particulate organic carbon by heterotrophic processes in sea water. Deep-Sea Res. 8: 211-222.

Roberts, M. H., R. Diaz, M. E. Bender. and R. Huggett. 1975. Acute toxicity of chlorine to selected estuarine species. J. Fish. Res. Bd. Can. 32: 2525-2528.

Sibert, J. and T. J. Brown. 1975. Characteristics and potential significance of heterotrophic activity in a polluted fjord estuary. J. Exp. Mar. Biol. Ecol. 19: 97-104.

Vaccaro, R. F. and H. W. Jannasch. 1966. Studies on the heterotrophic activity in sea water based on glucose assimilation. Limnol. Oceanogr. 11: 596-607.

_____ and _____. 1967. Variations in uptake kinetics for glucose by natural populations in sea water. Limnol. Oceanogr. 12: 540-542.

Vollenwieder, R. A., Ed. 1969. I.B.P. Manual 12: Primary Production in Aquatic Environments. Oxford: Blackwell Scientific Publications, Oxford.

Williams, P. J. LeB. 1973. The validity of the application of simple kinetic analysis to heterogeneous microbial populations. Limnol. Oceanogr. 18: 159-165.